Global Views on Climate Relocation and Social Justice

This edited volume advances our understanding of climate relocation (or planned retreat), an emerging topic in the fields of climate adaptation and hazard risk, and provides a platform for alternative voices and views on the subject.

As the effects of climate change become more severe and widespread, there is a growing conversation about when, where, and how people will move. Climate relocation is a controversial adaptation strategy, yet the process can also offer opportunity and hope. This collection grapples with the environmental and social justice dimensions of planned retreat from multiple perspectives, with cases drawn from Africa, Asia, Australia, Oceania, South America, and North America. The contributions throughout present unique perspectives, including community organizations, adaptation practitioners, geographers, lawyers, and landscape architects, reflecting on the potential harms and opportunities of climate-induced relocation. Works of art, photos, and quotes from flood survivors are also included, placed between sections to remind the reader of the human element in the adaptation debate. Blending art – photography, poetry, sculpture – with practical reflections and scholarly analyses, this volume provides new insights on a debate that touches us all: How we will live in the future and where?

Challenging readers' preconceptions about planned retreat by juxtaposing different disciplines, lenses and media, this book will be of great interest to students and scholars of climate change, environmental migration and displacement, and environmental justice and equity.

Idowu Jola Ajibade is an assistant professor in the Department of Geography at Portland State University.

A.R. Siders is an assistant professor in the Disaster Research Center, Biden School of Public Policy and Administration, and Department of Geography and Spatial Sciences at the University of Delaware.

Routledge Studies in Environmental Migration, Displacement and Resettlement

Repairing Domestic Climate Displacement
The Peninsula Principles
Edited by Scott Leckie and Chris Huggins

Climate Change Induced Migration and Human Rights
Law and Policy Perspectives
Edited by Andrew Baldwin, Dug Cubie, Dimitra Manou, Anja Mihr and Teresa Thorp

Migration and Environmental Change in the West African Sahel
Why Capabilities and Aspirations Matter
Victoria van der Land

Climate Refugees
Beyond the Legal Impasse?
Edited by Simon Behrman and Avidan Kent

Facilitating the Resettlement and Rights of Climate Refugees
An Argument for Developing Existing Principles and Practices
Avidan Kent and Simon Behrman

Communities Surviving Migration
Village Governance, Environment, and Cultural Survival in Indigenous Mexico
Edited by James P. Robson, Daniel Klooster and Jorge Hernández-Díaz

Climate Change Solutions and Environmental Migration
The Injustice of Maladaptation and the Gendered 'Silent Offset' Economy
Anna Ginty

Global Views on Climate Relocation and Social Justice
Navigating Retreat
Edited by Idowu Jola Ajibade and A.R. Siders

For more information about this series, please visit: https://www.routledge.com/Routledge-Studies-in-Environmental-Migration-Displacement-and-Resettlement/book-series/RSEMDR

Global Views on Climate Relocation and Social Justice

Idowu Jola Ajibade and
A.R. Siders

First published 2022
by Routledge
2 Park Square, Milton Park, Abingdon, Oxon OX14 4RN

and by Routledge
605 Third Avenue, New York, NY 10158

Routledge is an imprint of the Taylor & Francis Group, an informa business

© 2022 selection and editorial matter, Idowu Jola Ajibade and A.R. Siders individual chapters, the contributors

The right of Idowu Jola Ajibade and A.R. Siders to be identified as the authors of the editorial material, and of the authors for their individual chapters, has been asserted in accordance with sections 77 and 78 of the Copyright, Designs and Patents Act 1988.

With the exception of chapter 1, no part of this book may be reprinted or reproduced or utilised in any form or by any electronic, mechanical, or other means, now known or hereafter invented, including photocopying and recording, or in any information storage or retrieval system, without permission in writing from the publishers.

Chapter 1 of this book is available for free in PDF format as Open Access from the individual product page at www.routledge.com. It has been made available under a Creative Commons Attribution-Non Commercial-No Derivatives 4.0 license.

Trademark notice: Product or corporate names may be trademarks or registered trademarks, and are used only for identification and explanation without intent to infringe.

British Library Cataloguing-in-Publication Data
A catalogue record for this book is available from the British Library

Library of Congress Cataloguing-in-Publication Data
Names: Ajibade, Idowu Jola, editor. | Siders, A. R., editor.
Title: Global views on climate relocation and social justice : navigating retreat / Idowu Jola Ajibade and A.R. Siders.
Description: Milton Park, Abingdon, Oxon ; New York, NY : Routledge, 2022. | Includes bibliographical references and index.
Identifiers: LCCN 2021025117 (print) | LCCN 2021025118 (ebook) |
Subjects: LCSH: Environmental refugees. | Relocation (Housing--Environmental aspects. | Climatic changes--Social aspects.
Classification: LCC HV640 .G557 2022 (print) | LCC HV640 (ebook) | DDC 305.9/06914--dc23
LC record available at https://lccn.loc.gov/2021025117
LC ebook record available at https://lccn.loc.gov/2021025118

ISBN: 978-0-367-69344-2 (hbk)
ISBN: 978-0-367-69348-0 (pbk)
ISBN: 978-1-003-14145-7 (ebk)

DOI: 10.4324/9781003141457

Typeset in Goudy
by MPS Limited, Dehradun

Dedicated to the future generations whose world we are shaping.

Contents

List of figures xi
Notes on editors xiii
List of contributors xiv

1 Introduction: Climate change and planned retreat 1
IDOWU JOLA AJIBADE AND A.R. SIDERS

PART I
Definitions and legal landscapes 17

2 Rethinking process and reframing language for climate-induced relocation 19
KRISTIN BAJA

3 The role of international governance to reduce maladaptive climate relocation 34
THEA DICKINSON AND IAN BURTON

4 Charting a justice-based approach to planned climate relocation for the world's refugees 48
LAURA E.R. PETERS AND JAMON VAN DEN HOEK

Interlude 1   *Origins of Limestone* 62
MARTHA LERSKI

viii Contents

PART II
Shifting lands, resistance, and acceptance 65

5 Breaking the borderscape: Migration, resettlement, and
   citizenship on the Anthropocene Brahmaputra 67
   KEVIN INKS

6 Losing ground: Rethinking land loss in the context of
   managed retreat 78
   MAGGIE TSANG AND ISAAC STEIN

7 Resistance, acceptance, and misalignment of goals in
   climate-related resettlement in Malawi 88
   HEBE NICHOLSON

8 Land is Life: A poem of the Philippines Lumad 99
   NIKKI C.S. DELA ROSA

   Interlude 2   *Flood experiences in the United States* 104
   Interlude 3   *Rock that Fell to Earth: Storm Series* 108
   MARTHA LERSKI

PART III
Navigating transitions 111

9 Moving to higher ground: Planning for relocation as an
   adaptation strategy to climate change in the Fiji Islands 113
   BEATRICE RUGGIERI

10 Voices of Arraigo: Redefining relocation for landslide-
   affected communities in the informal settlements of
   Bogota, Colombia 127
   DUVAN H. LÓPEZ MENESES, ARABELLA FRASER, AND SONIA HITA
   CAÑADAS

11 The climate crisis is a housing crisis: Without growth we
   cannot retreat 142
   DEBORAH HELAINE MORRISMCP

12 Voices of Ghoramara Island, India: The case for planned
   relocation 152
   OANA STEFANCU

13 The need for a resettlement pathway for Guyana's vulnerable coastal communities 163
DINA KHADIJA BENN

14 Mobile livelihoods and adaptive social protection: Can migrant workers foster resilience to climate change? 180
HAORUI WU AND CATHERINE BRYAN

15 Identity and power: How cultural values inform decision-making in climate-based relocation 194
RACHEL ISACOFF

Interlude 4 *Tool Sharpening* 209
MARTHA LERSKI

Interlude 5 *Untitled, 2013: Storm Series* 212
MARTHA LERSKI

PART IV
Finding hope 215

16 Voices of Enseada da Baleia: Emotions and feelings in a preventive and self-managed relocation 217
GIOVANNA GINI, ERIKA PIRES RAMOS, AND COMUNIDADE ENSEADA DA BALEIA

17 Hope, community, and creating a future in the face of disaster 232
CLAIRE-LOUISE VERMANDÉ

Interlude 6 *Gratitude* 242
MARTHA LERSKI

PART V
Future directions 245

18 Retreating from the waves 247
ORRIN H. PILKEY, SARAH LIPUMA, AND NORMA LONGO

19 Climate-induced relocation as a third wave of response to climate change 263
PATRICK MARCHMAN

20 Waves of grief and anger: Communicating through the
 "end of the world" as we knew it 273
 SUSANNE C. MOSER

Interlude 7 Dialogue of the Shattered 289
MARTHA LERSKI

Index 291

Figures

4.1	Maps of (a) UNHCR-managed refugee settlements and host countries as of mid-2019; (b) mean annual surface temperature for 2040–2059; and (c) change in annual precipitation for 2040–2059 relative to mean annual precipitation from 1986 to 2005	50
Int 1.1 and 1.2	*Origins of Limestone*, marble	63-64
6.1	Property on barrier islands is frequently literally underwater	79
6.2	Barrier islands natural migrate landward through overwash	80
6.3	Development and coastal fortification prevent overwash and eventually barrier islands disappear	81
6.4	Map of projected "repetitive loss properties" in the Arch Creek Basin, Miami, FL, USA	83
6.5	Good Neighbor Stormwater Park, Miami, FL, USA	85
7.1	Map of Malawi showing political districts and study site	91
Int 2.1	"Swampy" the South Carolina, USA, flood activist	105
Int 2.2	United Flooded States of America	105
Int 2.3	Repetitive floods in a home	106
Int 2.4	Flood activist for managed retreat	106
Int 2.5	Flood loss means more than loss of property	107
Int 3.1	*Rock That Fell to Earth: Storm Series*, oak	109
9.1	Mural in Natalei Eco Lodge, Fiji with memorial to TC Winston	114
9.2	Aerial view of Fiji	118
9.3	Village of Tabuya, Fiji, rebuilt uphill	120
9.4	Homes in the Village of Tabuya, Fiji	120
10.1	Putting down roots in the Codillera neighborhood	129
10.2	A landscape of informality in the south of Bogota	130
10.3	Housing project by Rosa Buenaventura	133

10.4	Housing gradients in Bogota, Colombia	133
10.5	Ruins and residences in Altos de la Estancia	135
10.6	Mariela's orchard	136
12.1	A tree sinks off the coast of Ghoramara Island, India	153
12.2	Traditional gender roles in house and workforce	155
12.3	House on Ghoramora Island, India, soon to be lost to erosion	157
12.4	Temporary embankment to slow erosion	158
12.5	Life on Ghoramora Island, India	159
12.6	Community of Ghoramora Island, India	160
13.1	Map of the Guyana coast	164
13.2	Felled coconut palms on the Almond Beach	165
13.3	Schematic of irrigation and drainage on the Guyana coast	169
13.4	Spring tide waves crash over a seawall defending Anna Catherina, Guyana	172
Int 4.1 and 4.2	*Tool Sharpening*, acacia wood	210 and 211
Int 5.1	*Untitled, 2013: Storm Series*, marble	213
16.1	Coast of Comunidade de Enseada, Brazil	219
16.2	Between the beach and mountains in Comunidade de Enseada, Brazil	220
16.3	Trees as connection to family and land	221
16.4	Nova Enseada, Brazil	222
Int 6.1 and 6.2	*Gratitude*, marble	243 and 244
18.1	Flooded home in Centerport, New York, USA	248
18.2	Erosion scarp on a beach in Nags Head, North Carolina, USA	249
18.3	Urban development on the shore in Honolulu, Hawai'i, USA	252
18.4	Relocation of Cape Hatteras Lighthouse	257
20.1	Diagram of the emotional landscape	276
Int 7.1 and 7.2	*Dialogue of the Shattered*, marble	290

Notes on editors

Idowu Jola Ajibade is an assistant professor of Geography at Portland State University. She applies an environmental justice and political ecology lens to study how individuals, communities and cities respond to climate change and their different capacities for adaptation and transformation. Specifically, she examines the politics of adaptation, demonstrating how historical injustices, state practices, relocation programs, and Utopian solutions to climate change intertwined with exclusionary planning policies and development patterns to undermine disadvantaged and marginalized communities. Dr. Ajibade emphasizes the articulation of multiple solutions and alternative pathways to climate adaptation that includes partnering with grassroots coalitions, frontline communities, Indigenous groups, cooperatives, social entrepreneurs, and small businesses. Her research has been published in the *Journal of Global Environmental Change*, *Climatic Change*, *Climate and Development*, among other outlets. Dr. Ajibade has also been featured in Science Friday, NPR, Yale Environment 360, New Internationalist, and Vice.

A.R. Siders is an assistant professor at the University of Delaware in the Disaster Research Center, Biden School of Public Policy and Administration, and Department of Geography and Spatial Sciences in the College of Earth, Ocean, and Environment. Her research focuses on climate change adaptation decision-making and evaluation: How and why communities decide when, where, and how to adapt and how these decisions and processes affect risk reduction and equity. Current projects focus on managed retreat and adaptation equity. Her work has appeared in journals such as *Science*, *Climatic Change*, and *One Earth* and has been featured in global news outlets.

Contributors

Anthropocene Alliance – The Anthropocene Alliance is a national nonprofit that educates and organizes individuals and communities harmed by environmental abuse and climate change. They are the largest coalition of frontline communities in the United States fighting for climate and environmental justice.

Kristin Baja – Kristin Baja ("Baja") is a national leader in climate resilience, sustainability, and racial equity. She supports proactive equity-centered climate action through innovative solutions, shifting power, and collaboration. She holds two Masters degrees from the University of Michigan and was named an Obama *Champion of Change* in 2016.

Dina Khadija Benn – Dina Khadija Benn is a member of the Department of Geography, Faculty of Earth and Environmental Sciences, University of Guyana. She holds an MSc in Geoinformatics and a BA in Geography. Her research interests include community-based mapping and geospatial analysis for land management.

Catherine Bryan – Catherine Bryan is an anthropologist and faculty member at Dalhousie University's School of Social Work. Her work focuses on migration, transnational livelihoods, and social reproduction through the lens of feminist political economy.

Ian Burton – Ian has been a pioneer in research on human adjustment to natural hazards and adaptation to climate change in numerous countries. He served as the first Director of the Climate Adaptation Group in Environment Canada and as Director of the Institute for Environmental Studies at the University of Toronto.

Sonia Hita Cañadas – Sonia Hita is an early-career audio–visual communicator and social researcher, strongly committed to social activism. Sonia has provided wide support to make visible the injustices and struggles of Arraigo, the community platform of people affected by risk and resettlement in Bogotá, Colombia.

Nikki C.S. Dela Rosa – Nikki is a community organizer and freelance writer specializing in cultural work to uplift unheard voices. She worked with LA Food Bank and Heal the Bay through Kiwanis International. Her experiences grew to serve grassroot organizations in land rehabilitation and human rights concerns.

Thea Dickinson – Thea Dickinson (PhD) is a climate change adaptation specialist. Her research focuses on the effectiveness of national and international adaptation policy, and the disaster-climate policy interface. She was contributing author for the IPCC Working Group II and the Special Report on Extremes.

Comunidade Enseada de Baleia – The community of Enseada da Baleia is a traditional Caiçara community located in Ilha do Cardoso, in Cananéia, São Paulo State, Brazil. Born in 1845 with the main activity of drying fish and artisanal fishing, its working process was modified in 2010 to a solidarity economy managed by women which included community-based tourism and handicrafts. In 2017, an intense erosion forced the entire community to relocate to a new area on the same island, which has been a State Park since 1962.

Arabella Fraser – Arabella Fraser is a social and political scientist whose research specializes in the politics of urban resilience in the global South. She is a Research Fellow in the School of Geography at the University of Nottingham, UK. Her doctoral dissertation focused on the politics of risk assessment in informal settlements of Bogota, Colombia.

Giovanna Gini – Giovanna Gini is a PhD student working on mobilities in the Anthropocene with a focus on South America. She holds a BA in Political Science from the University of Trento, Italy; where also in 2018 she completed her MA in European and International Studies. Since 2018 she has been part of the Mobile People PhD program, a collaboration between Queen Mary University of London and Leverhulme Trust Doctoral. She also collaborates with the South American Network for Environmental Migrations (RESAMA).

Kevin Inks – Kevin Inks is a PhD student in the Department of Geography at the University of Wisconsin-Madison. His work deals with human migration and the cartography of shifting landscapes.

Rachel Isacoff – Rachel currently focuses on equity in the United States at The Rockefeller Foundation. Previously, she worked on climate justice at the White House Council on Environmental Quality, energy policy at HUD, and climate adaptation strategies at HR&A Advisors. She is part of the Climigration Network and a Visiting Assistant Professor at the Pratt Institute.

Martha Lerski – Sculptor and Lehman College librarian, Lerski explores cultural heritage as it relates to climate change. San Francisco-born and

xvi *Contributors*

NYC-based, Martha Lerski studied at the University of Pennsylvania, LSE, GSLIS CUNY, Syracuse iSchool, CUNY Graduate Center, and the Art Students League.

Sarah Lipuma – Sarah Lipuma studies climate change adaptation and community resilience to natural hazards at Duke University. She aims to help communities to work through tough decisions as climate change aggravates challenging land use issues. She hails from New Jersey and has always treasured living between the Pine Barrens and the Jersey Shore.

Norma Longo – Norma Longo is a long-time technical research associate of Orrin Pilkey at the Nicholas School of the Environment. She has co-authored two other books and numerous articles with him over the years. Also, many of her photographs of beaches have appeared in books and articles by Pilkey and others.

Duvan H. López Meneses – Duván López is doctoral student in sustainability, dedicated to research on risk environmental conflicts and people relocation, initially, as geologist and urban practitioner in Bogotá, Colombia; and afterwards from myriad perspectives nurtured by social activism, environmental studies, contemporaneous philosophy and even, theological reflections concerning risk taking attitude and hope.

Patrick Marchman – Patrick Marchman founded and led the Climate Migration and Managed Retreat group with the American Society of Adaptation Professionals from 2019 to 2021. He is currently the Climate Resilience Practice Lead for Kleinfelder, and lives with his wife and daughters in Kansas City, Missouri, USA.

Deborah Helaine Morris – Deborah Helaine Morris is an urban planner and designer focused at the nexus of climate change and social equity. She received a Master's in City Planning from the Massachusetts Institute of Technology and a Bachelor of the Arts from the University of Michigan. Deborah was a Loeb Fellow at the Harvard Graduate School of Design.

Susanne C. Moser – Susanne Moser is an independent scholar and consultant whose work focuses on adaptation to climate change, science–policy interactions, climate change communication, and psychosocial resilience in the face of the traumatic and transformative challenges associated with climate change.

Hebe Nicholson – Hebe is a researcher and activist working on issues related to the environment, migration, homelessness, and social justice. Hebe is currently mixing a part-time postdoc at the University of St. Andrews with frontline coaching work at Mayday Trust.

Laura E.R. Peters – Laura E.R. Peters is a Postdoctoral Research Fellow at University College London. She studies how divided societies act upon contemporary social-environmental changes and challenges to support health, sustainability, justice, and peace. Laura completed her PhD in Geography at Oregon State University.

Orrin H. Pilkey – Orrin Pilkey began his career as an oceanographer studying deep sea features, especially abyssal plains. After his parents' retirement home in Waveland, Mississippi, was destroyed by Hurricane Camille, he began to emphasize coastal studies and has written extensively concerning the impact of impending intensification of storms and sea level rise.

Erika Piers Ramos – Erika Pires Ramos is a Brazilian Public Lawyer and Researcher. She is co-founder of the South American Network for Environmental Migrations – RESAMA. She holds a PhD in International Law from the University of São Paulo (USP), Brazil. Currently she is involved with the Latin American Observatory on Human Mobility, Climate Change and Disasters (MOVE-LAM), an initiative developed in partnership between RESAMA and University for Peace – UPEACE – in Costa Rica.

Beatrice Ruggieri – Beatrice Ruggieri holds a Master Degree in Geography and is currently a PhD Candidate in Global Histories, Cultures and Politics (University of Bologna). In her thesis, she explores climate-induced (im)mobilities focusing on planned relocation as a contested form of adaptation.

Oana Stefancu – Oana Stefancu is PhD student at University of Exeter, UK. Her research focuses on well-being and justice aspects of planned relocations in the context of climate change.

Isaac Stein – Isaac Stein is a landscape architect and co-founder of Dept, a landscape architecture and urban design studio working extensively at the crossroads of urbanism and infrastructure. He holds a Bachelor of Architecture from University of Miami and Master in Landscape Architecture and Master of Design Studies in Risk and Resilience from Harvard Graduate School of Design.

Maggie Tsang – Maggie Tsang is an architect and urbanist. She is co-founder of Dept, a landscape architecture and urban design studio working extensively at the crossroads of urbanism and infrastructure. Maggie received her Master of Architecture from Yale University and Master of Design Studies in Urbanism, Landscape, and Ecology from Harvard Graduate School of Design.

Jamon Van Den Hoek – Jamon Van Den Hoek leads the Conflict Ecology lab at Oregon State University. He studies refugee–environment relationships using satellite imagery and open geospatial data. Jamon was a NASA Postdoctoral Fellow and completed his PhD in Geography at the University of Wisconsin-Madison.

Claire-Louise Vermandé – Claire-Louise Vermandé is a commercial real estate professional with a background in journalism. She is the editor and publicist for the memoir *Rising from the Flood: Moving the Town of Grantham*. She was raised in a country town, Bowral, in regional Australia and now lives in Brisbane, Queensland.

Haorui Wu – With an interdisciplinary background (social work, architecture, urban planning), Dr. Wu's community participatory research and practice have comprehensively examined various socioecological vulnerabilities of residents in disaster-stricken regions, advancing community resilience and sustainability in the global context of climate change, disaster, and willful acts of violence.

1 Introduction: Climate change and planned retreat

Idowu Jola Ajibade and A.R. Siders

Climate change is already redefining the landscapes of risk across the globe: from rising seas and shoreline erosion in small island states to heat waves and massive flooding in Europe, Asia, and Africa, and expanding wildfires and heatdome in the American West. These events are intensifying patterns of displacement, migration, and relocation within and between countries. In the last two decades, over 480 million people were displaced globally by climate-related disasters (IDMC, 2018; UNDRR, 2020). From 2000 to 2019, over 7,000 climate-related disasters killed an estimated 1.23 million people and caused 2.97 trillion (USD) in economic losses (UNDRR, 2020). During this time, an average of 24 million people were displaced per year globally (IDMC, 2018). These displacements are not experienced in isolation but as part of the complex intersecting economic, social, political, and environmental crises that puts severe strain on individual and community well-being across the world. By 2050, as many as one billion people could be displaced by a combination of climate change impacts, extreme events, and environmental degradation (IEP, n.d.), and thus raising critical concerns about finding appropriate climate adaptation and disaster risk reduction strategies.

To date, climate adaptation efforts have primarily focused on enabling people to remain in their homes – to adapt *in situ* (Jamero et al., 2019). However, in light of relatively unambitious climate change mitigation by cutting greenhouse gas emissions, and with increasing but widespread disasters, some adaptation practitioners, policy makers, and communities have begun to consider planned retreat – that is, proactive and coordinated efforts to relocate people, infrastructure, and assets from hazardous areas and resettling them in relatively safer locations (Greiving et al., 2018; Hino et al., 2017; King et al., 2014). Around the world, governments and communities have retreated, are in the process of doing so, or are planning for a future when retreat may be inevitable. While some planned retreat programs empower and benefit individuals and communities, others ignore people's rights, entrench inequities, and perpetuate risk, vulnerability, and harm on already marginalized communities and groups. This lack of attention to equity and justice can undermine the potential of planned retreat as a viable adaptation strategy.

This volume contributes to an emerging body of literature on planned retreat and socioenvironmental justice. It aims to help researchers, policy makers, practitioners, students, affected communities, and the public to explore climate-induced relocations from a multidimensional justice perspective. Using justice-based approaches as a framework and an analytical lens has a potential to advance

DOI: 10.4324/9781003141457-1

a deeper understanding of how retreat might support the rights, self-determination, livelihoods, physical health, and sociocultural needs of individuals and communities facing the most severe impacts of climate change. We argue that such approaches must be rooted in an understanding of communities' past experiences, current challenges and needs, and their visions for the future.

1.1 Planned retreat: Why is it important?

Planned retreat (also called planned relocations, managed retreat, planned resettlement, or assisted migration) is not new. Communities across the globe have relocated throughout history in response to climatic drivers (McLeman & Smit, 2006; Warner et al., 2013). If we consider just the 20th century, there are numerous examples from every corner of the globe. The Banaban community relocated from present-day Kiribati to Fiji and the Vaitupuans moved from Tuvalu to Fiji (McAdam, 2014). In the 21st century, towns in Australia and the United States relocated to avoid repetitive and/or coastal flooding (Forsyth & Peiser, 2021; Pinter & Rees, 2021; Sipe & Vella, 2014). Communities in Canada, China, Ethiopia, Germany, India, Italy, Japan, Mozambique, New Zealand, Peru, the Philippines, Tanzania, Thailand, Uganda, the United Kingdom, and Vietnam also relocated due to floods, storms, erosion, and other climatic hazards (Arnall, 2019; Greiving et al., 2018; Marter-Kenyon, 2020; Reisinger et al., 2014). These forms of relocations differ from climate migration in the degree of planning, government intervention, funding, legal protection, and claims to property rights (Ajibade et al., 2020; Miller, 2020).

There is no single pattern for how planned retreat or climate relocation occurs. It may be voluntary or forced (Farbotko et al., 2020; King et al., 2014), community or state-led (Albert et al., 2018; Cronin & Guthrie, 2011), and in-country or cross border (McAdam, 2014; McMichael & Katonivualiku, 2020). It is usually implemented through building restrictions (Reisinger et al., 2014), property acquisitions or buyouts (Mach et al., 2019; Siders, 2019a, 2019b; Thaler & Fuchs, 2020), social housing provision (Ajibade, 2019; See & Wilmsen, 2020), farmland swaps (Arnall, 2019; Gebauer, & Doevenspeck, 2015), and construction of new residential areas or towns (Bower & Weerasinghe, 2021; Forsyth & Peiser, 2021). Although, retreat is a universal strategy in response to environmental change, it is most prevalent in the Global North (Bower & Weerasinghe, 2021; Niven & Bardsley, 2013) and expanding in the Global South (Arnall, 2019; Marter-Kenyon, 2020; Piggott-Mckellar et al., 2020).

Depending on how planned retreat occurs, it can have a variety of positive and negative outcomes for the same individuals or for different groups. At its best, relocation can protect lives, avoid costly efforts to remain in place, reduce mental stress, and allow land to be used for community activities and/or nature-based ecosystem restoration (Ferris & Weerasinghe, 2020; Kochnower et al., 2015; Koslov et al., 2021; Zavar et al., 2016). At its worst, it can disconnect people from their livelihoods, exacerbate poverty and food insecurity, disrupt place attachment and identity, and splinter communities (Ajibade, 2019; Connell & Lutkehaus, 2017; Hammond, 2008), and thus perpetuating social inequality and vulnerability (Afifi et al., 2012). For instance, the relocation of a self-sufficient community from a frequently flooded but fertile ecosystem in eastern Uganda to a drier location in the western part of the country, transformed the social reproduction of farmers such that they became wage laborer and experienced livelihood fragility and economic vulnerability (Mafaranga, 2021). Relocations can also affect people emotionally and culturally. Place attachment, for example, can be profound in the case of Indigenous peoples whose identity is tied to the land (Albert et al., 2018; Huang, 2018). Relocation may also contribute to marginalization and disempowerment. For example, when informal settlers are moved from visible places (i.e., riverbanks and popular urban centers) to uninhabited land or rural areas, where the problem of poverty becomes more difficult to see and residents are less likely to receive support (Alvarez & Cardenas, 2019; Hammond, 2008). Wealthy elites may also take over spaces formerly occupied by the poor, thus contributing to wealth disparities and unequal access to social services (Ajibade, 2019). Put differently, some individuals and communities may gain and feel empowered as a result of relocation, but others may lose and feel disempowered. These feelings of loss and gain may also occur simultaneously for some people as they grieve the loss of their former home and embrace the opportunities in a new location (McNamara et al., 2018). Planned retreat therefore presents a number of complex logistical, social, political, ethical, and cultural challenges (Bower & Weerasinghe, 2021; McNamara et al., 2018; Siders and Ajibade, 2021; Thaler & Fuchs, 2020).

Decisions about retreat can be very complex as it typically involves multiple households, government agencies, civic organizations, and the private sector. Group decision-making requires balancing power dynamics among unequal actors and addressing trade-offs among different needs such as economic efficiency, human security, ecological preservation, and cultural heritage. For some communities the decision is whether to move or stay (Seebauer & Winkler, 2020); for some residents, it is when to move, where, and who or what should move (Ajibade, 2019; Linnenluecke et al., 2011); and for others, it is about acquiring the financial resources, technical assistance, and political support needed for relocation (Marino, 2018; McNamara et al., 2018). For example, in this volume Giovanni, Ramos, and the Enseada community in Brazil describe how their village's historical lack of political power made identifying a relocation site and obtaining relocation support more difficult. Finally, when populations wish to relocate but are unable to access resources, they may become trapped-in-place, leading to feelings of abandonment and continued exposure to multiple risks (Das & Hazra, 2020; Marino, 2018).

1.2 Planned retreat and the justice challenge

The notion of justice exists in different cultures and has developed through the ages as a basis for social institutions, economic relations, religion, politics, environmental protection, and climate stewardship. Justice is a fundamental political element about how people are treated and what claims they can make with respect to freedom, opportunities, resources, and social goods (Barry, 1989; Rawls, 1971; Schlosberg & Collins, 2014). In the context of planned retreat, one might ask: Who is most at risk from which climatic hazards and why? Who has access to resources to adapt in place or to relocate? And who has the political or economic power to determine whether they stay or leave? These questions intersect with different concepts of justice.

Retreat intersects with *environmental justice* (EJ). EJ goes beyond the equitable distribution of environmental goods (i.e., green amenities and infrastructure) and environmental bads (i.e., pollution, toxic chemicals, and urban heat) to include *procedural justice*, which involves formal participation of affected communities in decision-making about retreat; and *distributive justice*, which argues against the uneven distribution of the benefits or harms caused by relocation (Ajibade, 2019; Bullard, 1996; Bullard & Wright, 2009). *Social justice* is similarly implicated in retreat by focusing on the allocation of resources and a broader set of goods such as affordable housing, access to livelihoods, preservation of culture and heritage, wealth distribution, and power dynamics in the political economy.

Ecological justice is also crucial. It urges consideration of the rights and needs of ecosystems and nonhuman species in decisions and implementation of retreat (Davis et al., 2018; Parks & Roberts, 2006). Without reviving degraded ecosystems through planting trees, cleaning riverbanks, or giving nature space to recover, it may be difficult to achieve other justice goals such as equitable distribution of environmental amenities including clean air and water. *Recognition justice* in retreat requires the acknowledgment of historic wrongs such as slavery, settler's colonialism, redlining, segregationist policies, and disinvestments patterns that shape current conditions and people' experience of marginality in different facets of life (Pulido, 2000; Schlosberg, 2003). Blacks, Indigenous, and other communities of color have a legacy of disinvestment that has increased their exposure to risk and decreased their access to healthy environments. *Restorative justice* in retreat seeks to tackle these problems by ensuring that relocation programs ameliorate not perpetuate historical wrongs (McCauley & Heffron, 2018).

The different aspects of justice discussed in this section often agglomerate for historically marginalized communities facing climate threats and relocation decisions. In the United States, for example, concerns about community safety goes beyond matters of land use and hazard mitigation to include questions about systemic racism, housing inequities, police brutality, unequal burdens of pollution, gentrification, exclusionary development patterns, neoliberal policies, and extractive practices that contribute to climate change (Tessum et al., 2019). These problems have consequences that linger for centuries (Davis et al., 2018;

Schell et al., 2020). In other words, the multiple injustices of climate change and relocation is fundamentally and intricately linked to questions about the social production of humans and ecosystem evolution.

1.3 Gender, planned retreat, and adaptation labor

One aspect of justice that has been insufficiently explored in the academic literature and public discourse on planned retreat is gender. Yet the multiscalar, micropolitical, and differentiated effects of climate change and climate-induced disasters are often gendered (Ajibade et al., 2013; Butterbaugh, 2005; Lama et al., 2020; Vaz-Jones, 2018). In many parts of the world, women's lives are inextricably tied to climate and weather conditions. Women in rural Africa and Latin America, for example, are involved in the agricultural sector as the main producers of stable foods, making their livelihoods vulnerable to climate variability and change (Koubi et al., 2016; Yila & Resurreccion, 2013). The ecological and health burdens of hazards such as flooding or water scarcity are also disproportionately borne by women because of their domestic duties and gendered roles in the household (Ajibade et al., 2013; Sultana, 2011). In times of socioeconomic instability and destroyed harvest following natural disasters, women are mostly responsible for finding in-place solutions as men migrate to urban areas in search of opportunities (Abel, 2018; Koubi et al., 2016). Men's migration, in turn, increases the burden of responsibilities on women such as their share of agricultural work, water management, and household chores (Nizami et al., 2019; Yila & Resurreccion, 2013). These intersecting problems of climate change, disasters, migration, and gender have been well discussed in the migration literature (Lama et al., 2020; Vincent et al., 2021), but there are limited data on the gendered experiences and impacts of planned retreat.

The available literature on planned retreat suggests men determine relocation decisions by virtue of their position as head of the household (Neef et al., 2018), community leaders, and landowners (see chapter by Ruggeri in this volume). By contrast, women are typically renters or land-users – they perform much of the household, agricultural, and low-paying commercial work but do not own land for themselves (Vaz-Jones, 2018). Relocation programs based on property or *dejure* land ownership, therefore, may ignore women's needs and customary rights. Women may also be overlooked in relocation negotiations due to domestic constraints on their time and their limited experience engaging with high level government agencies or emergency managers, many of whom are men. Furthermore, female-headed households and single-mothers may receive less support during relocation because women have less bargaining power than men due to flawed perceptions of their contribution to household well-being (Smyth & Sweetman, 2015). In other cases, women are leading the charge for adaptation and shouldering the actual labour required to ensure the safety of families and communities and their access to livelihoods (Dube et al.,2017). Ignorance of these dynamics in retreat programs and of women's adaptation labour may create new forms of gendered invincibility by reinforcing existing patriarchal structures that prioritizes men's voices and needs. Questions about how gender shapes relocation and how relocation in turn shapes the lived experiences of men and women and their access to resources remains crucial as retreat programs gain new grounds.

We argue that it is important not to essentialize women or men, or overgeneralize their experiences, as other axes of differentiation and overlapping identities such as race, class, ethnicity, socioeconomic status, age, marital status, religion, health rights, caregiving and parental status shape an individual, family, or community's experience of relocation (Ajibade et al., 2013; Crenshaw, 1991; Lama et al., 2020; Vaz-Jones, 2018). For example, in Ithemba, South Africa, a government-led relocation and expropriation of land revealed how race, gender, and class inequalities intersect to exacerbate the experience of relocation for women (Vaz-Jones, 2018). Also, in Haiti, resettlement and land appropriation for banana plantations after the 2010 earthquake intensified poverty and food insecurity for women who were largely absent from the relocation decision (Steckley & Steckley, 2019). The out-migration of men as a result of the land appropriation led to a loss of solidarity within communities and an increased divorce rate; consequently, a higher number of rural women became heads-of-households where they were forced to assume the responsibility of family subsistence needs in an increasing context of risk and uncertainty. A justice-oriented planned retreat perhaps may be a panacea to such problems, especially when people relocate as a family or community as opposed to when they migrate independently. For example, in New Zealand, government-relocation programs for families in response to earthquakes, improved the quality of life for women and their families (Hoang & Noy, 2020). Also, in coastal Vunidogoloa, Fiji, women reported high benefits from climate-related relocation because they were involved in planning processes, and the resettlement allowed villagers to maintain physical, sociocultural, ancestral, and spiritual attachment to place as well as access to land and livelihood resources (McMichael et al., 2019).

We encourage researchers in the planned retreat field to consider a feminist decolonial approach (Wijsman & Feagan, 2019) that can open up novel lines of exploration, inquiries, methods, and a deeper understanding of the gendered and intersectional implications of climate relocation. Such an approach does not only challenge dominant knowledge production, typically connected to neoliberal hegemonic masculinity, but also rejects narrow solutions that perpetuate all kinds of injustices. Furthermore, this approach calls for transforming global and local systems as well as institutions and structures that foster uneven class, gender, and racialized experiences of climate disasters and relocation in response to those disasters.

1.4 A diversity of perspectives

This edited volume draws attention to historical and contemporary structures, policies, and practices that create differentiated social, gender, and racialized landscapes of risk and how these landscapes intersect with the complex experiences of communities and individuals confronted with planned retreat as a climate adaptation strategy. Our book includes global examples (from Australia, Brazil, Canada, Colombia, Fiji, Guyana, India, Myanmar, Malawi, the Philippines, and the United States) of communities who have relocated, are in the process of relocating, remain partially in place while members relocate, or have been unable to relocate. Contributors include academics, community members, social activists, lawyers, adaptation practitioners, landscape architects, poets, sculptors, and communication specialists. Each author provides different lenses through which to consider the justice implications of planned retreat. Sculptures and commentaries from individuals facing climate hazards and relocation are spaced between sections to remind readers that relocation is a deeply personal and emotional process affecting the daily lives of people.

We begin with an exploration of the *legal and historical landscapes* in the United States. Kristin Baja offers a "thick analysis" of recognition justice by showing how power, race, class, and language shape who leads, manages, and experiences relocation. This chapter centers the importance of recognizing how historical injustices have contributed to why Black, Indigenous, and People of Color have persistently been in harm's way and how the current structure and implementation of planned retreat through federal property acquisition programs continues this racialized system. Baja argues for planned relocations to include reparative actions that account for historical and contemporary injustices by ensuring improved access to livelihoods, cultural connections to land, and robust support for community health and well-being.

Moving from national to international scale, Thea Dickinson and Ian Burton note how the lack of international agreements, policy incoherence, increasing nationalism and closed borders, and prolonged adjudication of legal cases involving climate-induced relocations has not only put resettling communities at risk, but also deny them the protection and safety nets necessary for climate adaptation. Dickinson and Burton draw on planned relocations in the Maldives to demonstrate how climate relocations can become a form of maladaptation. Specifically, they question whether intranational relocations in the Maldives create a false sense of security and permanence that may ill-prepare citizens for international relocation and the legal challenges it will entail.

Laura Peters and Jamon Van Den Hoek continue this critique of international law by focusing on the injustices of climate-induced risk in refugee camps. They demonstrate how international policies, such as the practice of "warehousing," trap people in precarious conditions by preventing settlement in new areas located out of harm's way or restrict refugees from moving to safer, neighboring communities. The authors also note the translocation of refugees may create its own set of vulnerabilities. They describe, for example, the plight of a million Rohingya refugees in Bangladesh who were displaced due to human rights violations and extreme violence in Myanmar, but who now live in overcrowded refugee camps exposed to flooding and landslides. While their conditions in the refugee camps are untenable, their planned relocation to the floating Island of

Bhasan Char in the Bay of Bengal raises several justice concerns. This is because the island is vulnerable to tidal waves and tropical cyclones and could be inundated in a few years. Peters and Van Den Hoek offer a framework for designing planned relocations that incorporates multiple dimensions of justice (distributive, procedural, and restorative) and promotes the agency, dignity, and security of refugees.

Legal frameworks offer opportunities to support relocation, but they can also be a constraint when rigid formality fails to adapt to physically *shifting landscapes*. Along the Brahmaputra River in India, Kevin Inks demonstrates how coastal and river landscapes do not conform to formal cadastral surveying and legal interpretations of land ownership or property rights. The Brahmaputra is one of the most geophysically dynamic braided rivers on the planet, and its land is frequently swallowed and recreated through an interplay of erosion and deposition. Faced with seasonal flooding and temporary displacement, riverine communities in the area seek to relocate permanently, but the government resettlement program based on formal practices of land surveyance and documentation preclude many residents from eligibility for resettlement support. Inks offers an insightful critique of the cartographic technologies employed by the state and proposes a fluid understanding of coastal landscapes and land ownership in the context of climate-induced resettlement programs.

Maggie Tsang and Isaac Stein build on this concept of land-in-motion by providing a fresh perspective on how we might rethink retreat through decoupling concepts of land, loss, and property value. Based on case studies in Hatteras Island, North Carolina, and Miami-Dade County, Florida, the authors demonstrate the mutability of land resulting from natural undulation and urbanization processes. The authors argue that retreating coastlines may be a natural coastal defense that protects cities from flooding; meaning land losses should be viewed from a geological and ecological perspective rather than an economic one focused on real estate market growth and municipal budgets.

Flexible legal tools and frameworks may enable communities to relocate, but they may also inspire resistance when community needs are not addressed. Focusing on three communities in the Lower Shire Valley of Malawi, Hebe Nicholson illustrates how a government-labeled "no-go-zone" and disinvestment invigorated people to fight against relocation rather than persuade them to leave. Others used resistance strategies to reappropriate autonomy in retreat plans and re-center their needs. Through poetry, Nikki Dela Rosa conveys the efforts of the Lumad, an Austronesian Indigenous people in the Philippines, to prevent forcible relocation from their mineral-rich coastal island. Mining activities have decimated the community, and temporary relocation destabilizes their daily lives and exposes them to assault, discrimination, and violence. Despite these challenges, the community continues to fight for their rights as the original custodians of the land.

Relocation is a process that requires *navigating transition* and experiences. Beatrice Ruggieri draws on a social justice framework and gender lens to examine relocation decisions in the Tabuya coastal community in Fiji. Wanting to maintain autonomy over the process, and thereby protect their culture and livelihoods, villagers initially rejected external assistance for relocation. However, considering the enormous cost and logistics required in community relocation, the village eventually made the controversial decision to request government support. Decision-making in rural Fijian communities, Ruggieri notes, have frequently been subject to hierarchical and seniority systems that privilege men and older people (see also Neef et al., 2018). In Tabuya, the process allowed for performative inclusion and participation by women but decision-making remained dominated by men who are traditional landowners.

Intracommunity relations and power dynamics play an important role in relocation. Duvan López Meneses, Arabella Fraser, and Sonia Hita Cañadas present testimonies of Arraigo members in Bogota, Colombia – a network of neighborhood organizations, social leaders, scholars, volunteers and activists, and informal settlers – to demonstrate how the right to stay or resettle is mediated by power and political discourses. In particular, how risk is defined and by whom has significant consequences for people. For example, children may be taken away from parents living in areas defined as "nonmitigable risk." Building improvements may also be prohibited in such areas, thus exposing residents to future risk. The decision to relocate is not a simple one: Some residents had to wait more than a decade for support and some never received support or received too little. Such actions can entrench existing inequalities while trapping residents in a state of liminality and destitution.

Lack of support during and after relocation is central to Deborah Morris's description of how planned retreat programs in New York after Hurricane Sandy exacerbated social vulnerability. Renters displaced by property acquisitions struggled to find safe and affordable housing in a city plagued by a housing crisis. Relocation programs are legally required to provide resettlement assistance and temporary rent subsidies, but administrators applied these provisions inconsistently, and support was often insufficient to enable tenants to find permanent housing. Morris draws from her experience administering one of these programs to argue that planned retreat efforts must consider a wide range of social vulnerabilities beyond risk exposure and must relate to larger debates about affordable housing, poverty reduction, and access to social services.

Inability to access resettlement resources can result in *displacement without relocation*, as Oana Stefancu's study on Ghoramara Island, India, shows. Although residents have identified wholesale relocation as their preferred strategy to address persistent flooding, disappearing lands, and depleted livelihoods, the community has received no external support. The marginality and obscurity of these impoverished communities have led to state abandonment, which the community sees as a form of disposability and violence. People on the island describe themselves as *"trapped"* - as the environmental threat increases, and their ability to escape decreases. In the case of Guyana, Dina Khadija Benn documents how government-led resettlement plans failed to address resident concerns about increased impoverishment, food insecurity, and splintering of community ties. The community designed their own relocation plans but, just as in Fiji (Ruggieri) and India (Stefancu), these plans relied on funding and logistical support from government that was not forthcoming, thus leaving residents in a state of limbo.

In some cases, relocation of part of the community provided adaptation resources for the remaining residents, as in the case of migrant remittances. Haorui Wu and Catherine Bryan offer a narrative of the lived experience of Filipino migrant workers in Canada and how their remittances shaped disaster recovery and reconstruction in Leyte, Philippines, following Typhoon Haiyan. Wu and Bryan blur the lines between migration and planned retreat by demonstrating how formal policies and structures shape not only the migrant experience but also their relations with distant families. Indeed, treating migration as an individual-centered event may complicate consideration of important group dynamics such as extended family, community ties, and culture. Rachel Isacoff explores how competing cultural values, identities, and worldviews shape relocation decisions and implementation. Using case studies of Kivalina, Alaska; Isle de Jean Charles, Louisiana; and Staten Island, New York, she juxtaposes the techno-managerial, individualistic, and econometric values that inform relocation decisions of experts and policy makers with local knowledge, group identity, and livelihood-based approaches prioritized by communities.

Although relocation poses numerous challenges to culture, community, livelihoods, and place attachment, relocation may also offer a space to *find hope*. Giovanna Gini and Erika Pires Ramos assemble the voices of members from the Comunidade Enseada da Baleia to explore the emotions that arose during their proactive and self-planned relocation on the Island of Cardoso, Brazil. Led by a coalition of women, a traditional artisanal fishing community faced with severe erosion, environmental degradation, and disappearing livelihoods. This group of women rechanneled their sense of loss into a fight for justice and for relocation on their terms. Their successful creation of Nova Enseada still involves nostalgia for their old home but also hope for the future. Such examples can provide hope not only for residents in communities facing relocation but also for former community members. Claire-Louise Vermandé reflects on how learning about the relocation of Grantham, Australia, affected her personal reflections on the potential loss and relocation of her hometown following major brushfires. Through sculpture and reflective essays, interspersed between sections of this volume, Martha Lerski explores the emotions of loss, nostalgia, family, identity, and the possibility of hope through change and the passage of time. Photos and comments from Anthropocene Alliance members facing severe and repetitive flood loss similarly illustrate a mix of loss and hope.

Finally, looking to the future, three papers explore how policy reforms, reconceptualization, and communication strategies can inform planned retreat *moving forward*. Orrin H. Pilkey, Sarah Lipuma, and Norma Longo discuss the need for coherent relocation policies to account for historical injustices, cultural context, and heritage sites. While acknowledging the cultural and structural challenges ahead, they draw hope for reform in the United States from examples where retreat has preserved heritage. They argue for future plans to take a robust and long-term perspective to relocation. Situating retreat in a history of adaptation and techno-optimism, Patrick Marchman explores the emerging discourse of climate relocation as a third wave of response compared to earlier discussions on mitigation and adaptation. He questions the idea that climate change can be planned in ways that do not threaten current development patterns and economic growth pathways. Unlike adaptation in place, relocation forces people to reckon with the physical, economic, cultural, and social reality of climate change, and this, the author argues, may be a force for good, leading people to embrace a simpler, just, and more sustainable existence.

In the final chapter, Susanne Moser offers a variety of strategies for communicating relocation. Rather than focusing on the "right words" (Chapter One by Baja), Moser calls for an investigation of human needs throughout the relocation process, noting these needs are complex and complicated by underlying histories, racist legacies, current socioeconomic realities, attachments to place, and personal emotions. This chapter takes us on a journey that sketches out the deeply human, psychological, and relational needs entangled in communicating relocation.

The chapters of this volume present global examples of the complicated processes and contexts in which planned retreat will occur. Individually, each chapter introduces a new case, story, or lens on the relocation discussion. As a collection, the volume aims to challenge readers' pre-conceptions about planned retreat by juxtaposing different disciplines, lenses, and media – and by consistently grounding the conversation in the human experience. Leaders will increasingly be called upon – at local, national, and international levels – to support, prevent, direct, or facilitate movement within and across borders in all of these complicated contexts. Each of our authors provide recommendations through their contributions, based on extensive research or experience, for how leaders (academics, practitioners, and policy makers) can improve future climate relocation programs. Although each case of climate-induced movement is unique, forged by the history of the land and people, common recommendations include empowering local communities and marginalized groups, adopting long-term planning horizons that make space for impermanence, designing relocations to address well-being beyond physical security, and explicitly centering values and justice in decision-making processes. These recommendations are by no means exhaustive; rather than purport to provide a blueprint for planned retreat, this volume seeks to complicate the discussion and highlight how the past and present will shape our future.

References

Abel, G. J. (2018). Estimates of global bilateral migration flows by gender between 1960 and 2015. International Migration Review, 52, 809–852. 10.1111/imre.12327.

Afifi, T., Govil, R., Sakdapolrak, P., & Warner, K. (2012). *Climate change, vulnerability and human mobility: Perspectives of refugees from the East and Horn of Africa*. UNHCR UN Refugee Agency.

Ajibade, I. (2019). Planned retreat in Global South megacities: Disentangling policy, practice, and environmental justice. *Climatic Change*, 157(2), 299–317.

Ajibade, I., McBean, G., & Bezner-Kerr, R. (2013). Urban flooding in Lagos, Nigeria: Patterns of vulnerability and resilience among women. *Global Environmental Change*, 23(6), 1714–1725.

Ajibade, I., Sullivan, M., & Haeffner, M. (2020). Why climate migration is not planned retreat: Six justifications. *Global Environmental Change*, 65, 102187.

Albert, S., Bronen, R., Tooler, N., Leon, J., Yee, D., Ash, J., Boseto, D., & Grinham, A. (2018). Heading for the hills: Climate-driven community relocations in the Solomon Islands and Alaska provide insight for a 1.5 C future. *Regional Environmental Change*, 18(8), 2261–2272.

Arnall, A. (2019). Resettlement as climate change adaptation: What can be learned from state-led relocation in rural Africa and Asia? *Climate and Development*, 11(3), 253–263.

Alvarez, M. K., & Cardenas, K. (2019). Evicting slums, 'building back better': Resiliency revanchism and disaster risk management in Manila. *International Journal of Urban and Regional Research*, 43(2), 227–249.

Barry, B. (1989). *Theories of justice: A treatise on social justice*. Harvester-Wheatsheaf.

Bower, E., & Weerasinghe, S. (2021). *Leaving place, restoring home: Enhancing the evidence base on planned relocation cases in the context of hazards, disasters, and climate change*. Platform on Disaster Displacement and Andrew & Renata Kaldor Centre for International Refugee Law.

Butterbaugh, L. (2005). Why did Hurricane Katrina hit women so hard? *Off Our Backs*, 35(9/10), 17–19.

Bullard, R. D. (1996). Environmental justice: It's more than waste facility siting. *Social Science Quarterly*, 77(3), 493–499.

Bullard, R. D., & Wright, B. (Eds.). (2009). *Race, place, and environmental justice after Hurricane Katrina: Struggles to reclaim, rebuild, and revitalize New Orleans and the Gulf Coast*. Perseus Books.

Connell, J., & Lutkehaus, N. (2017). Environmental refugees? A tale of two resettlement projects in coastal Papua New Guinea. *Australian Geographer*, 48, 1–17.

Crenshaw, K. (1991). Mapping the margins: Intersectionality, identity politics, and violence against women of color. *Stanford Law Review*, 43(6), 1241. 10.2307/1229039

Cronin, V., & Guthrie, P. (2011). Community-led resettlement: From a flood-affected slum to a new society in Pune, India. *Environmental Hazards*, 10(3-4), 310–326.

Das, S., & Hazra, S. (2020). Trapped or resettled: Coastal communities in the Sundarbans Delta, India. *Forced Migration Review*, 64, 15–17.

Davis, J., Moulton, A., Van Sant, L., & Williams, B. (2018). Anthropocene, capitalocene, ... platationocene? A manifesto for ecological justice in an age of global crises. *Geography Compass*, 13, e12438.

Dube, T., Intauno, S., Moyo, P., & Phiri, K. (2017). The gender-differentiated impacts of climate change on rural livelihoods labour requirements in Southern Zimbabwe. Journal of Human Ecology, 58, 48–56.

Farbotko, C., Dun, O., Thornton, F., McNamara, K. E., & McMichael, C. (2020). Relocation planning must address voluntary immobility. *Nature Climate Change*, 10(8), 702–704.

Ferris, E., & Weerasinghe, S. (2020). Promoting human security: Planned relocation as a protection tool in a time of climate change. *Journal on Migration and Human Security*, 8(2), 134–149.

Forsyth, A., & Peiser, R. (2021). Lessons from planned resettlement and new town experiences for avoiding climate sprawl. *Landscape and Urban Planning*, 205, 103957.

Gebauer, C., & Doevenspeck, M. (2015). Adaptation to climate change and resettlement in Rwanda. *Area*, 47, 12168. 10.1111/area.12168

Greiving, S., Du, J., & Puntub, W. (2018). Planned retreat—A strategy for the mitigation of disaster risks with international and comparative perspectives. *Journal of Extreme Events*, 5, 1850011.

Hammond, L. (2008). Strategies of invisibilization: How Ethiopia's resettlement programme hides the poorest of the poor. *Journal of Refugee Studies*, 21(4), 517–536.

Hino, M., Field, C. B., & Mach, K. J. (2017). Planned retreat as a response to natural hazard risk. *Nature Climate Change*, 7(5), 364–370.

Hoang, T., & Noy, I. (2020). Wellbeing after a managed retreat: Observations from a large New Zealand program. *International Journal of Disaster Risk Reduction*, 48, 101589.

Huang, S. M. (2018). Heritage and post-disaster recovery: Indigenous community resilience. *Natural Hazards Review*, 19(4), 05018008.

International Displacement and Monitoring Center. (2018). Global Report on Internal Displacement. https://www.internal-displacement.org/global-report/grid2018/

Institute for Economics & Peace (IEP). (n.d.). *IEP: Over one billion people at threat of being displaced by 2050 due to environmental change, conflict and civil unrest.* Retrieved April 30, 2021, from https://www.prnewswire.com/news-releases/iep-over-one-billion-people-at-threat-of-being-displaced-by-2050-due-to-environmental-change-conflict-and-civil-unrest-301125350.html

Jamero, Ma. L., Onuki, M., Esteban, M., Chadwick, C., Tan, N., Valenzuela, V. P., Crichton, R., & Avelino, J. E. (2019). In-situ adaptation against climate change can enable relocation of impoverished small islands. *Marine Policy*, 108, 103614.

King, D., Bird, D., Haynes, K., Boon, H., Cottrell, A., Millar, J., Okada, T., Box, P., Keogh, D., & Thomas, M. (2014). Voluntary relocation as an adaptation strategy to extreme weather events. *International Journal of Disaster Risk Reduction*, 8, 83–90.

Kochnower, D., Reddy, S. M. W. W., & Flick, R. E. (2015). Factors influencing local decisions to use habitats to protect coastal communities from hazards. *Ocean and Coastal Management*, 116, 277–290.

Koslov, L., Merdjanoff, A., Sulakshana, E., & Klinenberg, E. (2021). When rebuilding no longer means recovery: The stress of staying put after Hurricane Sandy. *Climatic Change*, 165(3), 1–21. 10.1007/s10584-021-03069-1.

Koubi, V., Spilker, G., Schaffer, L., & Böhmelt, T. (2016). The role of environmental perceptions in migration decision-making: Evidence from both migrants and non-migrants in five developing countries. *Population and Environment*, 38(2), 134–163.

Lama, P., Hamza, M., & Wester, M. (2020). Gendered dimensions of migration in relation to climate change. *Climate and Development*, 13(4), 326–336.

Linnenluecke, M. K., Stathakis, A., & Griffiths, A. (2011). Firm relocation as adaptive response to climate change and weather extremes. *Global Environmental Change*, 21(1), 123–133.

Mach, K. J., Kraan, C. M., Hino, M., Siders, A. R., Johnston, E. M., & Field, C. B. (2019). Planned retreat through voluntary buyouts of flood-prone properties. *Science Advances*, *5*(10), eaax8995.

Mafaranga, H. (2021). Landslides mar the "Pearl of Africa". *EOS*, *102*. 10.1029/2021 EO157124. Published on 19 April 2021.

Marino, E. (2018). Adaptation privilege and voluntary buyouts: Perspectives on ethnocentrism in sea level rise relocation and retreat policies in the US. *Global Environmental Change*, *49*, 10–13.

Marter-Kenyon, J. (2020). Origins and functions of climate-related relocation: An analytical review. *The Anthropocene Review*, *7*(2), 159–188.

McAdam, J. (2014). Historical cross-border relocations in the Pacific: Lessons for planned relocations in the context of climate change. *The Journal of Pacific History*, *49*(3), 301–327.

McCauley, D., & Heffron, R. (2018). Just transition: Integrating climate, energy and environmental justice. *Energy Policy*, *119*, 1–7.

McLeman, R., & Smit, B. (2006). Migration as an adaptation to climate change. *Climatic Change*, *76*(1), 31–53.

McMichael, C., & Katonivualiku, M. (2020). Thick temporalities of planned relocation in Fiji. *Geoforum*, *108*, 286–294.

McMichael, C., Katonivualiku, M., & Powell, T. (2019). Planned relocation and everyday agency in low-lying coastal villages in Fiji. *The Geographical Journal*, *185*(3), 325–337.

McNamara, K. E., Bronen, R., Fernando, N., & Klepp, S. (2018). The complex decision-making of climate-induced relocation: Adaptation and loss and damage. *Climate Policy*, *18*(1), 111–117.

Miller, F. (2020). Exploring the consequences of climate-related displacement for just resilience in Vietnam. *Urban Studies*, *57*(7), 1570–1587.

Neef, A., Benge, L., Boruff, B., Pauli, N., Weber, E., & Varea, R. (2018). Climate adaptation strategies in Fiji: The role of social norms and cultural values. *World Development*, *107*, 125–137.

Niven, R. J., & Bardsley, D. K. (2013). Planned retreat as a management response to coastal risk: A case study from the Fleurieu Peninsula, South Australia. *Regional Environmental Change*, *13*(1), 193–209.

Nizami, A., Ali, J., & Zulfiqar, M. (2019). Climate change, hydro-meteorological hazards and adaptation for sustainable livelihood in Chitral Pakistan. *Sarhad Journal of Agriculture*, *35*(2), 432–441.

Parks, B. C., & Roberts, J. T. (2006). Environmental and ecological justice. In M. M. Betsill, K. Hochstetler, & S. Demitris (Eds.), *Palgrave advances in international environmental politics* (pp. 329–360). Palgrave Macmillan.

Piggott-McKellar, A. E., Pearson, J., McNamara, K. E., & Nunn, P. D. (2020). A livelihood analysis of resettlement outcomes: Lessons for climate-induced relocations. *Ambio*, *49*(9), 1474–1489.

Pinter, N. & Rees, J. C. (2021). Assessing planned flood retreat and community relocation in the Midwest USA. *Natural Hazards*, *107*, 497–518.

Pulido, L. (2000). Rethinking environmental racism: White privilege and urban development in Southern California. *Annals of the Association of American Geographers*, *90*(1), 12–40.

Rawls, J. (1971). *A theory of justice*. Harvard University Press.

Reisinger, A., Lawrence, J., Hart, G., & Chapman, R. (2014). From coping to resilience: The role of planned retreat in highly developed coastal regions. In B. Glavovic, R. Kaye, M. Kelly, & A. Travers (Eds.), *Climate change and the coast: Building resilient communities* (pp. 285–310). CRC Press.

Schell, C. J., Dyson, K., Fuentes, T. L., Des Roches, S., Harris, N. C., Miller, D. S., Woelfle-Erskine, C. A., & Lambert, M. R. (2020). The ecological and evolutionary consequences of systemic racism in urban environments. *Science*, 369(6510), eaay4497.

Schlosberg, D. (2003). The justice of environmental justice: Reconciling equity, recognition, and participation in a political movement. *Environmental Politics*, 13(3), 517–540.

Schlosberg, D., & Collins, L. B. (2014). From environmental to climate justice: Climate change and the discourse of environmental justice. *Wiley Interdisciplinary Reviews: Climate Change*, 5(3), 359–374.

See, J., & Wilmsen, B. (2020). Just adaptation? Generating new vulnerabilities and shaping adaptive capacities through the politics of climate-related resettlement in a Philippine coastal city. *Global Environmental Change*, 65, 102188.

Seebauer, S., & Winkler, C. (2020). Should I stay or should I go? Factors in household decisions for or against relocation from a flood risk area. *Global Environmental Change*, 60, 102018.

Siders, A. R. (2019a). Planned retreat in the United States. *One Earth*, 1(2), 216–225.

Siders, A. R. (2019b). Social justice implications of US planned retreat buyout programs. *Climatic Change*, 152(2), 239–257.

Siders, A. R., & Ajibade, I. (2021). Introduction: Social and environmental justice challenges and practices in global planned retreat. *Journal of Environmental Studies and Sciences*, 1–7. 10.1007/s13412-021-00700-6.

Sipe, N., & Vella, K. (2014). Relocating a flood-affected community: Good planning or good politics? *Journal of the American Planning Association*, 80(4), 400–412.

Smyth, I., & Sweetman, C. (2015). Introduction: Gender and resilience. *Gender & Development*, 23(3), 405–414.

Steckley, M., & Steckley, J. (2019). Post-earthquake land appropriations and the dispossession of rural women in Haiti. *Feminist Economics*, 25(4), 45–67.

Sultana, F. (2011). Suffering for water, suffering from water: Emotional geographies of resource access, control and conflict. *Geoforum*, 42(2), 163–172.

Tessum, C. W., Apte, J. S., Goodkind, A. L., Muller, N. Z., Mullins, K. A., Paolella, D. A., Polasky, S., Springer, N. P., Thakrar, S. K., Marshall, J. D., & Hill, J. D. (2019). Inequity in consumption of goods and services adds to racial–ethnic disparities in air pollution exposure. *Proceedings of the National Academy of Sciences*, 116(13), 6001–6006.

Thaler, T., & Fuchs, S. (2020). Financial recovery schemes in Austria: How planned relocation is used as an answer to future flood events. *Environmental Hazards*, 19(3), 268–284.

United Nations Disaster Risk Reduction. (2020). Annual report capturing output and impact for 2020. https://www.undrr.org/publication/undrr-annual-report-2020

Vaz-Jones, L. (2018). Struggles over land, livelihood, and future possibilities: Reframing displacement through feminist political ecology. *Signs: Journal of Women in Culture and Society*, 43(3), 711–735.

Vincent, K., de Campos, R. S., Lázár, A. N., & Begum, A. (2021). Gender, migration and environmental change in the Ganges-Brahmaputra-Meghna delta in Bangladesh.

In A. Hans, N. Rao, A. Prakash, & A. Patel (Eds.), *Engendering climate change* (pp. 152–171). Routledge.

Warner, K., Afifi, T., Kälin, W., Leckie, S., Ferris, B., Martin, S. F., & Wrathall, D. (2013). *Changing climate, moving people: Framing migration, displacement and planned relocation.* UNU-EHS.

Wijsman, K., & Feagan, M. (2019). Rethinking knowledge systems for urban resilience: Feminist and decolonial contributions to just transformations. *Environmental Science & Policy*, 98, 70–76.

Yila, J. O., & Resurreccion, B. P. (2013). Determinants of smallholder farmers' adaptation strategies to climate change in the semi-arid Nguru Local Government Area, Northeastern Nigeria. *Management of Environmental Quality: An International Journal*, 24(3), 341–364.

Zavar, E., Hagelman, R. R., & Hagelman III, R. R. (2016). Land use change on U.S. floodplain buyout sites, 1990–2000. *Disaster Prevention and Management*, 25(3), 360–374.

Part I

Definitions and legal landscapes

2 Rethinking process and reframing language for climate-induced relocation

Kristin Baja

2.1 Introduction

I am a white woman and my white body has provided me with unearned privileges my entire life. My white skin has impacted the way the world interacts with me and what people believe they know of me at first glance. I also acknowledge that I write this from land that was stolen from the Piscataway Tribe by European colonizers. My privilege has also allowed me to make choices about where I live and if I would like to move somewhere else; a privilege thousands of Americans do not have.

The climate change field, just as the environmental field, is deeply rooted in racism and injustices (Sierra Club, 2020). For too long, the top–down approach in this field has led to an overabundance of climate plans, academic papers, and sets of recommendations developed in ivory towers while frontline communities continue to live the harsh realities of a rapidly changing climate. This chapter focuses on the massive shift that is needed in this field. It looks at two key elements that can help the field begin down a path of greater humility, respect, and awareness: (1) How climate relocation as a form of adaptation needs process change and (2) modified and co-developed language.

Traditional climate work has ignored the reality that the United States was designed for one user group – white male landowners (Williams-Rajee, 2020). Centuries of discriminatory policies and practices have created inequitable structures that put wealth and decision-making power into the hands of a few at the expense of many. The environmental and climate fields of practice are no exception. Each is grounded in decades of conventional methodologies which have resulted in increased inequities, amplified distrust by frontline communities, a cycle of plan development and paper writing disconnected from action, and a system centered on reaction and response. Candidly, the climate field, which includes climate-induced relocation and migration, continues to embody a culture of white supremacy and dominance over people (Hardy et al., 2017) and nature (Milfont et al., 2013); the same components that underpin the founding of this country. If progress is to be made, the field must reckon with this fact and seek to reframe the process and shift to a just and human-centered approach.

DOI: 10.4324/9781003141457-2

Being explicit about racism and white supremacy is one step toward process change. Another key element is the reassessment of language and the role it plays in this field. Language is power. Although it does not solve the problems associated with "voluntary" or forced climate relocation, academics, local governments, and others stakeholders working in the field have a responsibility to recognize how important language and the naming of experiences matters to the relocation of those on the frontlines.

In this chapter, I argue for a change in this field; an uncomfortable and dramatic shift from the norm to an approach that centers the experiences of Black, Indigenous, and people of color (BIPOC) and other frontline communities. This chapter provides thoughts and reflections from a practitioner's perspective. It is grounded in over 12 years of experience working with local government and direct interactions with community members across the nation who are on the forefront of the climate crisis. This chapter does not attempt to provide all the answers or address all the issues. It is intended to challenge the current approach to climate-related relocations and to help identify first steps in how academics and practitioners alike can shift to a more respectful and just approach and develop culturally respectful shared language and process.

2.2 Climate relocations and process

Impacts from our rapidly warming planet have become increasingly difficult to ignore (Hoegh-Guldberg et al., 2018). The world is collectively experiencing the consequences of increased destruction of nature in the form of a global pandemic, wildfires, storms, flooding, drought, and other extreme events that continue to escalate in frequency and intensity (IFRC, 2020). Although historically, climate has not been documented as the primary cause for relocation, studies find it has been an underlying theme for both migration and relocation of people across the world and that it is likely to play a bigger role as the planet rapidly warms (Lustgarten, 2020). For many communities, choosing to relocate from high-risk areas is economically unmanageable. For others, these critical conversations and need for action are becoming unavoidable. However, even with the increased frequency and intensity of natural disruptions, climate-induced relocations in America has remained a relatively white academic discussion centered on data, a few well-worn case studies, and traditional planning models. Very little has been done to acknowledge and integrate the nation's racist history and humanize the topic.

The traditional approach to climate action has been to gather and assess data, create a draft plan, "engage the community," and to finalize the plan. In some cases, plans lead to a few infrastructure-centered pilot projects, however, more often than not, plans are developed and very few actions are realized. The field of climate-induced relocation has followed in the footsteps of traditional climate planning. Much of the field is focused on recommendations for large-scale infrastructure projects, building retrofits, and land-use policy that perpetuate inequity and inaction by centering engineering and "one-size-fits-all" frameworks over human experience and need.

Although data and scientific research are critical elements that aid in understanding risk and informing policy making, the decision to proactively move away from high-risk areas requires a different approach; one grounded in empathy, healing, human-centered design, and trauma-informed reparative action. The field must reckon with its roots of systemic racism before developing more toolkits and recommendations.

Government and academic partners have an opportunity to restructure the approach and collectively support this transition by recognizing that this country is founded on a legacy of extraction, from both people and nature and then acknowledging the systems and structures that attempt to solve this issue perpetuate inequities and white supremacy (Williams-Rajee, 2020). Studies show that the protection of property, especially expensive corporate property, is prioritized over protection of people by US disaster policy (Allen, 2020; Siders & Keenan, 2020). In acknowledging that systems and structures have intentionally harmed many for the benefit of a few, that imbalance of power can begin to be addressed and transferred as part of the process.

Current approaches around what is presently referred to as "managed retreat" is reminiscent of archaic, short-sighted and discriminatory resettlement patterns in the United States (Kahrl, 2012). Existing retreat programs in the United States are mostly in form of buyouts. Typically, buyouts are a combination of federal, state and local dollars prioritized in areas that have experienced substantial or repetitive damage and are focused on economic elements over human health and well-being (Cartier, 2019). Most buyout programs are activated after a major disaster and take advantage of residents and businesses who lack resources for proactive action and are reeling from devastating impacts. Federal buyout programs are reactive and built on a white racial hierarchy that ultimately deepen racial and class inequities (Powell, 2012). How buyouts are administered and by whom plays a major role in who is prioritized and who benefits. This often translates to wealthy, white and more densely populated areas receiving and implementing buyout programs (Mach et al., 2019) and, although undocumented, personal experience confirms that political connections play a role in who wins and who loses.

Rather than centering empathy and holistic value, buyout programs focus on the value of land and structures. The reimbursement provided is for the market value without considering redlining and how racist practices created systematically depressed communities with lower market values (Solomon et al., 2019). This approach is often dehumanizing; diminishing people's incredibly emotional and traumatic experience that has financial, quality of life, and life expectancy consequences and further exacerbating trauma and harm. Research on post-disaster floodplain buyout programs has shown that BIPOC communities and lower-income residents are misinformed about flood risk (Muñoz, 2016), pressured into buyouts (Vries, 2012), and at times, relocated into areas that are equally at risk (Jimenez-Magdaleno, 2017). From a combined social justice and systems perspective, buyout programs mimic forced relocation. Forced relocation exacerbates trauma causing depression, grief, and illness – all of which create cascading impacts on healthcare and other systems, especially given that those most impacted by climate disaster are already likely to experience comorbidities and limited access to essential life resources. Reparative action, therefore, would require a change in the system of valuation starting with reimbursing for the costs of a similar piece of land and similar structure in a non-redlined community while considering social elements such as connection of space to livelihood, cultural connection to the land, or the benefits of being part of a community. These elements that are difficult to quantify but it is critical to integrate them into financial considerations.

Additionally, distrust of government is intensified when there is a disconnection between different levels of government and when loopholes are provided for developers to build new luxury housing in place of what was there (Nonko, 2020). After a disruption, buyout programs offer homeowners with financial assistance to rebuild or repair their homes. However, buyouts are not prioritized based on a property's vulnerability to future impacts and program timelines (which take an average of five years) are too long to benefit most homeowners who need to make decisions quickly. Often homeowners who sign up for buyout programs end up selling to investors for under market value which leads to continued personal loss and transfer of risk to people who are unaware of the property's risk profile (Zavar, 2019). This was demonstrated in Houston after Hurricane Harvey when investors purchased flooded properties and converted them into rental housing putting people back in harm's way without disclosing the property's flooding history (Zavar, 2019).

Climate disruption is a multiplier of racial inequities. It compounds historical trauma and consolidates wealth and essential life resources among the wealthy and most privileged. As a 2019 NPR investigation found, "after a disaster, rich people get richer and poor people get poorer. Federal disaster spending appears to exacerbate these wealth inequality" (NPR, 2019). The current response-based process perpetuates distrust and rips open old wounds related to when government played an active role in locating people in high-risk areas through policies such as redlining and/or indirectly played a role such as allowing wealthy people to build and often rebuild second homes in high-risk areas. This has been documented in low-lying barrier island areas such as North Carolina's outer banks and Dauphin Island, Alabama, where millions of dollars in federal aid have allowed wealthy second homeowners to rebuild in the exact same location post-disaster (Gaul, 2019). When these actions are allowed to continue and federal aid is prioritized for second homeowners, BIPOC and low-income community members continue to lose wealth, power, access to resources, and faith in the system.

Beyond regulatory loopholes, equitable climate relocation faces an additional challenge, capital. Typically, projects that center human quality of life, respect historical injustices, and attempt to rectify those injustices are not upfront moneymakers. Investors seek the highest rate of return and cash-strapped local governments look for the lowest cost projects. This combination exacerbates inequities and perpetuates ineffective cycles that place short-term "band-aid" solutions over long-term effective change. As the field shifts toward reframing the process in a culturally respectful manner, it should work with investors to integrate non-monetized benefits and utilize a comprehensive cost-benefit analysis that integrates consideration of avoided damage, stormwater management benefits, public green space, natural flood reduction, and enhanced human health. Project prioritization through traditional cost-benefit analysis is not objective but rather deeply political. Many of the most important costs, such as the loss of community or sites of cultural significance, are not easily monetized. These costs are often felt locally whereas benefits such as avoided disaster recovery costs or lower insurance premiums are accrued across larger scales (Shreve & Kelman, 2014). Furthermore, federally funding often drives these projects, but localized elements of utmost value to those directly affected may be overlooked, furthering the experience of loss of control.

Loss of control is directly related to another challenge, humans fear of change. Change challenges our sense of autonomy, increases uncertainty, creates a sense of loss and often requires more work. Change creates disruption, something humans tend to dislike at the personal level. Often people need to experience something personally before they are willing to act, even if they are aware of all the facts. Cognitive dissonance – the uneasiness humans feel when trying to make sense out of contradictory ideas or between an idea and behavior – can lead to resistance to change because humans do not like being challenged on preexisting convictions and often would rather take increased risk rather than accept information that would lead to undesirable change (Aronson & Tavris, 2020). Thus, being personally impacted by a disruption is often what it takes for people to act, perpetuating the response-based cycle. This is not just with individuals; management and legal frameworks are also structured for preserving the status quo.

Politicians and decision-makers, often those who hold most tightly to control, must be approached differently. Similar to their constituents, political leaders are part of a system that is not easy (or in many cases, willing) to change. From experience, decision-makers are more likely to support proactive action if they are given the opportunity to come to terms with the facts emotionally. This often has to happen behind closed doors in a safe space where they can take time to process realities. This type of processing does not come from frameworks or guidance materials, it comes from interactions and relationships grounded in approachable data and stories that include social, economic, and environmental consequences of inaction.

As processes evolve, it is essential to utilize a culturally respectful targeted universalism approach (Powell, 2019) that recognizes that all people matter but the needs of those who have been disenfranchised and intentionally put in harm's way due to systemic racism must be prioritized for both funding and proactive support. Instead of investing in high income areas and locations with majority second homeowners, resources should be prioritized in BIPOC and low-income communities that do not have the ability to relocate without support. This cannot be in the form of an equity statement. All levels of government need to reckon with systemic racism and ensure that those with "friends in high places" do not continue to benefit first because they support a politician or golf with the "right people."

Additionally, leaders in the field should bring in trauma specialists, grief counselors, respected elders, and neighborhood-based community partners as members of the core project team which will help to bridge the divide between government and frontline communities and re-center this difficult and often traumatic experience in reparative action rooted in community needs and priorities. These core team members can help with recognition of, and empathy for, historic and current injustices and humans inherent fear of change while also understanding the climate-induced relocation is deeply personal and emotional. As part of a process shift, they can help co-develop common language that reflects the culture of the community and resonates with the diverse set of stakeholders on the frontlines.

2.3 Language

Adjusting the language about climate relocation is not going to fix the complex issues around trauma and fear of the unknown. However, current language (such as managed retreat) utilized to discuss movement away from high-risk areas is a barrier to engaging those most impacted. Language reflects culture and connects customs, beliefs and values. It is critical to understand how important terminology is when discussing deeply personal and difficult change. All humans have subconscious prejudice or implicit bias which impacts how situations are interpreted and words utilized (Kirwan Institute, 2015). For example, calling a group of people "vulnerable" is an exclusionary language that applies a deficit-based lens, implies a reduced quality of life, and suggests assumptions regarding their class, racial, ethnic, and other identities. Inclusive language acknowledges past harms, existing strengths and assets, expresses dignity, respect and empathy, and rejects stereotyping (Linguistic Society of America, 2016).

The Latin word *vulnus* means "wound" and is the origin of the word vulnerable. Although all humans are vulnerable in one way or another, assigning this word uniformly to many different groups of people does not take into consideration their unique situational differences including the root causes for those differences. Those who have been intentionally put in higher risk areas, also known as sacrifice zones, have been made more vulnerable due to centuries of discrimination that increased poor health conditions and trauma and intentionally placed people in harm's way. Sacrifice zones (Lerner, 2010) is a term developed by government officials to designate areas that were massively contaminated by hazardous materials. There are also areas where BIPOC were intentionally located leading to disparate health and economic impacts (Bullard, 2011). The term also applies to land use and zoning injustices that placed BIPOC into areas at higher risk from disasters and climate change impacts. This leads to questions: Who America is willing to sacrifice and who does it want to protect? Language should reflect the fact that most "vulnerabilities" exist from the systems and structures people are forced to operate in (Preto, 2018).

There is risk of creating more harm by adopting one blanket term for people facing greater exposure to climate change. The climate relocation field has an opportunity to incorporate historic injustices and current inequities as well as local knowledge and abilities into language choices. Different communities may prefer terms that better reflect the sociopolitical and sociohistorical factors for their increased exposure such as "frontline" or "impacted" rather than vulnerable. When language is adjusted to reflect past experience and culture, people are more likely to engage in a redesigned process where they feel heard, respected, valued, as well as having a sense of belonging and ownership. As Nelson Mandela said, "If you talk to a man in a language he understands, that goes to his head. If you talk to him in his language, that goes to his heart." Ultimately, respectful co-developed language provides an opening for co-development of holistic solutions with higher levels of commitment to implementation.

2.3.1 Is the term "managed retreat" a problem?

The term *managed retreat*, while frequently used in academic literature, does not resonate well with community members, government practitioners or politicians for a range of reasons (Gibbs, 2016). First, the word managed. Typically, people do not want to be managed. Especially in the United States where BIPOC communities have been "managed" through slavery [via transatlantic slave trade (Bertocchi, 2016)]; through inhumane and extractive working conditions [e.g., exploitation of Chinese immigrants to build the Transcontinental Railroad (Kennedy, 2020)]; and other oppressive practices for centuries. For example, slaveholders utilized modern management techniques to objectify and dominate Black people (Johnson, 2018). Moreover, colonial settlers managed Indigenous peoples through theft of land and resources, horrific massacres, as well as enslaving them. Crimes against humanity were therefore "justified" under the guise of modernity and civilization through various management strategies (Jalata, 2011). These same methods are used today to perpetuate oppression such as prioritizing projects through traditional cost-benefit analysis and land-use laws that allow takings and lead, once again, to people being removed against their will from a place they call home (Melo, 2020).

In addition to extraction and exploitation of people and land for the benefit of wealthy white people, BIPOC currently living in high-risk areas were intentionally "placed" in those areas because they were categorized as inherently "less than" those who colonized the United States and were forced into areas with lower property values and less accessibility. An example is Biloxi-Chitimacha-Choctaw tribe of Isle de Jean Charles (IDJC), Louisiana. The 1830 Indian Removal Act forced them to migrate west of the Mississippi settling in Louisiana's mainland. Later, the tribe was forced to relocate to the coast in IDJC. Today, IDJC is inhabitable due to erosion caused by climate change and land subsidence caused by oil extraction (Boyd, 2019; Herrmann, 2017). Multiple injustices, such as those suffered by the Biloxi-Chitimacha-Choctaw tribe, along with redlining and other forms of structural racism, systematically reduced the economic power of BIPOC communities while amplifying the economic power of white communities. This also forced many BIPOC communities into high-risk areas out of necessity.

Furthermore, the United States has a deeply ingrained individualistic culture in which the term managed evokes feelings of being controlled and manipulated which threatens people's sense of independence and autonomy. One of the foundational components of this country (right or wrong) is that *certain* individuals have rights (to own land, bear arms, etc.) and freedom to do what they want (within a prescribed set of limits). It is necessary to clarify that these "rights" were intended for white people living in the United States, not all people (Albert, 2020). Additionally, there is a much lower emphasis on the common good and welfare than we see in many other parts of the world. For example, in the current global pandemic, the combative and individualistic response Americans had to being asked to wear a mask demonstrates a disconnection with collective care and lack of understanding of the interdependence we have on each other as humans and on our socio-ecological systems (Vargas, 2020). The perception of being managed, especially by the government, is contentious and often, a difficult barrier to overcome. This touches on the critical need to utilize co-developed language that does not exacerbate trauma and integrates justice and human rights.

Similarly, *retreat* is another loaded and often "forbidden" term. Simplistically, *retreat* is defined as pulling back or withdrawing. Often, the concept of retreat is associated with weakness. For example, retreating in battle meant losing ground to the enemy or admitting defeat. There is demonstrated aversion to retreat with politicians in response to a major disaster – where the romanticized promise to rebuild back better and stronger rather than stepping back and strategically assessing whether there are lessons to be learned and different approaches to take. Former NYC Mayor's response to Hurricane Sandy was to say that "As New Yorkers, we cannot and will not abandon our waterfront, it's one of our greatest assets. We must protect it, not retreat from it" (Nonko, 2020). Until recently, there were many executive branch politicians who have been clear that they do not view retreat as an option worth considering, at least not publicly. As Roman Gastesi, Monroe County Administrator said in a 2019 interview, "It's that word nobody likes to use – retreat" (Harris, 2019). This language stops the conversation with the community before it begins, especially in states like California and Florida where practitioners avoid using language that suggests moving away from the coast due to the impacts on property values and tax base (USDN, 2020).

Moreover, when retreat is used in land-use discussions it has heightened negative associations related to being uprooted involuntarily. Those facing relocation do not have the comfort of returning to the familiarity of home turf. Rather, they may be moved to an equally or even more vulnerable space (Ajibade et al., 2020), but without the community ties and support system they had before (Ajibade, 2019). Similarly, retreating from impacts related to the climate, including rising sea levels, riverine flooding from heavy precipitation and wildfire-prone areas taps into the psyche of giving up and/or losing control over one's destiny.

Humans, especially those in the United States, have been conditioned to believe that we can control nature and our natural systems and thus, do not need to consider retreat. This belief can be traced back to the Ancient Greeks and is deeply tied to western forms of Christianity which is anthropocentric and states "it is God's will that man exploit nature" (White, 1967). This conviction is evident in major infrastructure projects such as dams, seawalls, and channelization of rivers – all amazing feats of engineering that allowed for exponential industrial growth, and over time have proven to degrade the environment and serve those with more access and privilege over disenfranchised communities. For example, the Grand Coulee Dam project in Washington State disrupted the natural systems and livelihood of local tribes who were forced to relocate due to flooding-related to the dam development (Bureau of Reclamation, 2020). In 1935, the Bureau of Reclamation committed to compensation for the land and promised hydropower revenues and money for loss of fishery; however, in 1940 the Spokane Tribe was paid $4700 and disregarded (UCUT, 2016) and the dam continued to damage the ecosystem. In 1994, Congress signed a bill to make payments to the local tribes. While beneficial, the bill does not undo generations of trauma, disruption, loss of culture and way of life, early death, and comorbidities nor repair the natural ecosystem. This country's attempts to control natural systems are costly, short-sighted and have fiscal, environmental, and social cascading costs over time.

2.3.2 Is climate relocation, a preferred language?

Managed Retreat is not the only term to reconsider in this work. *Relocation*, referenced several times in this chapter for lack of more agreeable language, is another term often used without consideration of the significant role it has played in the white supremacist history of the United States. Relocation has been purposefully used to destroy non-White community identities and cultures to maintain a racial hierarchy with whiteness dominating over all others. In 1830, the Indian Removal Act was passed and was masked as a "voluntary relocation program." The Removal Act forcibly relocated Native Americans out of land east of the Mississippi while simultaneously imprisoning, raping and murdering those that resisted. Those that survived were provided with little to no provisions and forced to walk over 1000 miles to establish new Territories. Thousands died both on the "Trail of Tears" march from east to west and also in the Black Hawk war that opened millions of acres of land in the Midwest that had belonged to other native nations. Over 46,000 Native Americans were removed from the land their ancestors had occupied and cultivated for generations, opening 25 million acres to White settlement, cotton growth, and slavery (National Geographic Society, 2020).

In addition to atrocities associated with The Removal Act, during World War II the Roosevelt Administration adopted Executive Order 9066 that forced relocation of Americans of Japanese ancestry and established internment camps. This brutal violation of American civil rights affected more than 117,000 people, the majority of whom were American citizens (History.com Editors, 2020). Americans of Japanese ancestry were arrested without evidence, had their assets seized, and were forcefully relocated without the ability to inform their families. Although the dictionary definition of relocation is "to move to a new location," the traumatic experiences and atrocious violation of basic human rights associated with the word and its use to oppress entire groups of people must be carefully considered. In truth, the "relocation" of communities due to climate change is a domino effect of the white supremacist system that relegated them to vulnerable lands in the first place. This system is engineered to perpetuate the status quo. While it is necessary to plan for moving communities, the act itself will be traumatic, especially if done with a traditional planning and policy approach.

Current US relocation policies lack cultural sensitivity and have created intergenerational trauma that require thoughtful reparative action. Forced relocation continues today in association with hydropower and other major infrastructure projects such as proposed oil pipeline projects being cited in tribal communities, permanently altering their landscape which would lead to either unmanaged relocation or lifestyle changes. Many of these relocations are being completed without sensitivity to culture, compassion for the people being moved, or long-term planning for compensation of shifting livelihoods. For instance, villages that were traditionally subsistence or fishing oriented are moved to locations without fishstocks.

2.3.3 Strategic transition or community-led and support relocation

Strategic transition is another word utilized in these discussions. Strategic has more constructive associations with it. Americans like the idea of being thoughtful, deliberate, and intelligent. In working in-depth with the four county governments in Hawaii over the past two years, I have found that substituting the word "strategic transition" for "managed retreat" while also altering the approach opened doors for dialogs with community members and partners. However, the ability to strategically move to a lower-risk area tends to be an option mostly available to wealthy people whereas lower-income people are more likely to be forced into relocation. With BIPOC communities making up the majority of those impacted by coastal and riverine flooding, the term strategic is also not ideal, even if it may be more accepted by decision-makers. No matter what, relocation will lead to further disruption, trauma, and disenfranchisement. Centering community priorities and equity in both process and language can create an opening to heal, emphasize humans right to self-determination, and redistribute the costs humankind must pay to respond to climate disruption.

There is a wide range of proposed language around climate-forced migration. Managed retreat, strategic relocation, planned realignment, transformational adaptation, or resilient relocation to name a few. It is worth exploring new terminology coupled with reparative and respectful collaborative efforts to shift away from high-risk areas. A one-size-fits-all approach is inadvisable. Communities, and the people who live in them, are unique. Just as the term resilience has benefitted from having different definitions that support bipartisan action on preparedness, this concept will also benefit from multiple characterizations determined by those most impacted. Communities have different sets of circumstances, historical context, challenges and needs; a native tribe inhabiting a small island in Alaska is likely to have different desires and concerns than Latinx agriculture workers who are faced with high drought and wildfire risk.

Language that resonates and connects culturally should be determined by the people most impacted and being asked to relocate. Language can help to center racial healing and autonomy while supporting empathetic approaches and transformation. For example, in Isle de Jean Charles, Louisiana, the term "re-settlement" is utilized and explicitly notes that "Resettlement cannot be driven solely by economic and operational objectives, but must incorporate a comprehensive, holistic and open-ended approach" (Louisiana Office of Community Development, 2020). The traditional structures of academia and government prefer a list of case studies citing "best practices" and providing a foundation for others to replicate. However, that approach neglects consideration of unique community characteristics and community-determined language that resonates based on those characteristics. Additionally, consider that the majority of mainstream documented case studies and examples are rooted in colonialist frameworks. Therefore, relying on those would be futile toward transforming to more just approaches to climate action.

While it is recommended to co-develop appropriate language with frontline communities, shared language will be necessary when developing national policy and improving government programs. As communities, government, academia, and other stakeholders attempt to take proactive action, a potential language option for national and regional policy is difficult to pinpoint while respecting the unique history and diverse set of reasons for certain communities being on the frontlines. There is no silver bullet in this space; however, framing around "*community-supported relocation*" is one option that integrates self-determination and yet centers empathy and the reality that BIPOC community members must be respected and prioritized.

2.4 Conclusion

This chapter intentionally stresses the need to work with frontline communities, in particular BIPOC communities, to alter language and process with thoughtful, empathic recognition of past and current injustices. It stresses the need to shift away from the traditional climate planning approaches and to actively shift power to frontline communities while remedying unjust and archaic federal programming. Determination of process is power. Currently, the field of climate relocation mimics inequitable top–down structures and approaches that have proven to be ineffective. Researchers and practitioners alike are deeply disconnected from the ongoing trauma and lived experience of people on the frontlines of the climate crisis. Climate relocation should be anchored in restorative justice and repairing past harm in partnership with those harmed and to the natural environment. Acknowledging and shifting power as core elements of the process helps to break down systems of oppression and dismantle structures that perpetuate trauma and inequality. Additionally, language is power. A key step in the altering process is effective language. Top–down elitist language that ignores the history of extraction of land and people in America will exacerbate distrust and continue to support historic power dynamics. I challenge all readers of this book to assess their individual and institutional power and to shift away from outdated language and approaches and move to a more just transition.

2.5 Recommendations

In adjusting the process and language for climate relocation, this chapter identifies six key recommendations:

- **Acknowledge Historic and Current Power.** In North America, systems and structures have intentionally harmed many for the benefit of a few. This imbalance of power needs to be addressed and shifting as part of relocation processes.
- **Utilize Comprehensive Cost-Benefit Analyses.** Work with investors to integrate social, environmental, and economic benefits and non-monetized benefits into cost-benefit assessments. This should include consideration of avoided damage, stormwater management benefits, public green space, natural flood reduction, and enhanced human health.
- **Restructure the System of Valuation.** Reparative action requires a change in the system of valuation and consideration of social elements such as connection of space to livelihood, cultural connection to the land, or the benefits of being part of a community; all elements that are difficult to quantify but critical to integrate into reimbursement considerations.
- **Utilize a Proactive Targeted Approach.** Resources should be prioritized and timelines shortened in BIPOC and low-income communities that do not have the ability to relocate without support. Shift to a proactive approach rather than reactive.
- **Co-develop Appropriate Language with Frontline Communities.** Different communities may prefer terms that better reflect the sociocultural, sociopolitical, and sociohistorical factors for their increased exposure. Collaborate to co-develop language that resonates and is culturally appropriate.
- **Consider New Language for National and Regional Policy.** Frame national language around community needs and integrate empathy, respect, humility and reparations. Consider *"community-supported relocation"* or options that do not have historic negative connotations.

References

Ajibade, I., Sullivan, M., & Haeffner, M. (2020). Why climate migration is not managed retreat: Six justifications. *Global Environmental Change, 65*, 102187.

Ajibade, I. (2019). Planned retreat in Global South megacities: Disentangling policy, practice, and environmental justice. *Climatic Change, 157*(2), 299–317.

Albert, R. (2020, June 30). Time to update the language of the Constitution. *The Hill.*

Allen, M. (2020, October 14). Protect for the rich, retreat for the poor. *Hakai Magazine.*

Aronson, E., & Tavris, C. (2020, July 12). *The role of cognitive dissonance in the pandemic.* Retrieved from The Atlantic: https://www.theatlantic.com/ideas/archive/2020/07/role-cognitive-dissonance-pandemic/614074/

Bertocchi, G. (2016). The legacies of slavery in and out of Africa. *IZA Journal of Development and Migration, 5*, 24.

Boyd, R. (2019, September 23). *Jean Charles Are Louisiana's First Climate Refugees-but They Won't Be the Last.* National Resources Defense Council. https://www.nrdc.org/stories/people-isle-jean-charles-are-louisianas-first-climate-refugees-they-wont-be-last

Bullard, R. D. (2011). Sacrifice zones: The front lines of toxic chemical exposure in the United States. *Environmental Health Perspectives, 119*(6).

Bureau of Reclamation. (2020, May 11). *Pacific Northwest region, Grand Coulee Dam cultural history.* Retrieved from Bureau of Reclamation: https://www.usbr.gov/pn/grandcoulee/history/cultural/index.html

Cartier, K. M. (2019, October 9). *Equity concerns raised in Federal flood property buyouts.* Retrieved from Eos Science News by AGU: https://eos.org/articles/equity-concerns-raised-in-federal-flood-property-buyouts

Gaul, G. (2019, September 5). *On the Alabama Coast, the unluckiest island in America.* Retrieved from YaleEnvironment360: https://e360.yale.edu/features/on-the-alabama-coast-the-unluckiest-island-in-america

Gibbs, M. (2016). Why is coastal retreat so hard to implement? Understanding the political risk of coastal adaptation pathways. *Ocean & Coastal Management, 130*, 107–114.

Hardy, R. D., Milligan, R., & Heynen, N. (2017). Racial coastal formation: The environmental injustice of colorblind adaptation planning for sea-level rise. *Geoforum, 87*, 62–72.

Harris, A. (2019, December 6). *Florida Keys may abandon some roads to sea rise rather than raise them.* Retrieved from Tampa Bay Times: https://www.tampabay.com/news/environment/2019/12/06/florida-keys-may-abandon-some-roads-to-sea-rise-rather-than-raise-them/

Herrmann, V. (2017). Americaas first climate change refugees: Victimization, distancing, and disempowerment in journalistic storytelling. *Energy Research & Social Science, 31*, 205–214.

History.com Editors. (2020, February 21). *Japanese internment camps.* Retrieved from History.com: https://www.history.com/topics/world-war-ii/japanese-american-relocation

Hoegh-Guldberg, O., Jacob, D., Taylor, M., Bindi, M., Brown, S., Camilloni, I., Diedhiou, A., Djalante, R., Ebi, K. L., Engelbrecht, F., Guiot, J., Hijioka, Y., Mehrotra, S., Payne, A., Seneviratne, S. I., Thomas, A., Warren, R., & Zhou, G. (2018). Impacts of 1.5°C global warming on natural and human systems. In V. Masson-Delmotte, P. Zhai, H.-O. Pörtner, D. Roberts, J. Skea, P. R. Shukla, A. Pirani, W. Moufouma-Okia, C. Péan, R. Pidcock, S. Connors, J. B. R. Matthews, Y. Chen, X. Zhou, M. I. Gomis, E. Lonnoy, T. Maycock, M. Tignor, & T. Waterfield (Eds.), *Global warming of 1.5°C. An*

IPCC Special Report on the impacts of global warming of 1.5°C above pre-industrial levels and related global greenhouse gas emission pathways, in the context of strengthening the global response to the threat of climate change, sustainable development, and efforts to eradicate poverty, In Press.

IFRC. (2020, November 17). *World Disasters Report 2020: Come heat or high water.* International Federation of Red Cross and Red Crescent Societies, Geneva.

Jalata, A. (2011). Indigenous peoples in the capitalist world system: Researching, knowing, and promoting social justice. *Sociology Publications and Other Works*, 82, 9–33.

Jimenez-Magdaleno, K. (2017, November 30). *FEMA-funded property buyouts: The impacts on land and eeople.* Retrieved from North Carolina Buyouts Impacts Literature Review: https://ncgrowth.unc.edu/wp-content/uploads/2018/01/Buyouts_Impact_LIteratureReview_Final.pdf

Johnson, W. (2018, February 20). To remake the world: Slavery, racial capitalism, and justice. *Boston Review*, pp. 1–5.

Kahrl, A. W. (2012). *The land was ours: African American beaches from Jim Crow to the Sunbelt South.* Harvard University Press.

Kennedy, L. (2020, April 30). Building the transcontinental railroad: How 20,000 Chinese immigrants made it happen. *History.*

Kirwan Institute. (2015). *Implicit bias.* Retrieved from Understanding Implicit Bias: http://kirwaninstitute.osu.edu/wp-content/uploads/2015/05/2015-kirwan-implicit-bias.pdf

Lerner, S. (2010). *Sacrifice zones.* The MIT Press.

Linguistic Society of America. (2016, November). Guidelines for Inclusive Language.

Louisiana Office of Community Development. (2020, June 9). *Resettlement of Isle De Jean Charles background and overview.* Retrieved from Resettlement of Isle De Jean Charles: http://isledejeancharles.la.gov/sites/default/files/public/IDJC-Background-and-Overview-6-20_web.pdf

Lustgarten, A. (2020, July 23). The great climate migration. *The New York Times Magazine.*

Mach, K.J., Kraan, C.M., Hino, M., Siders, A.R., Johnston, E.M., & Field, C.B. (2019). Managed retreat through voluntary buyouts of flood-prone properties. *Science Advances*, 5(10). 10.1126/sciadv.aax8995.

Melo, L. (2020). Land use regulations: Racial oppression by another name. *The Morningside Review*, 16, 1–2.

Milfont, T., Richter, I., Sibley, C., Wilson, M., & Fischer, R. (2013). Environmental consequences of the desire to dominate and be superior. *Personality & Social Psychology Bulletin*, 39, 1127–1138. 10.1177/0146167213490805.

Muñoz, C. E. (2016). Unequal recovery? Federal resource distribution after a midwest flood disaster. *International Journal of Environmental Research and Public Health*, 13(5), 507.

National Geographic Society. (2020, April 6). *May 28, 1830 CE: Indian Removal Act.* Retrieved from Resource Library, This Day in History: https://www.nationalgeographic.org/thisday/may28/indian-removal-act/#:~:text=More%20than%2046%2C000%20Native%20Americans,and%20exposure%20to%20extreme%20weather.

Nonko, E. (2020, January 2). *NYC's coastline could be underwater by 2100. Why are we still building there?* Retrieved from Curbed New York: https://ny.curbed.com/2020/1/2/21046581/new-york-city-climate-change-managed-retreat-development

NPR. (2019, March 5). How federal disaster money favors the rich.

Powell. (2019, May 08). *Targeted universalism: Policy & practice.* Retrieved from Othering & Belonging Institute: https://belonging.berkeley.edu/targeteduniversalism

Powell, J. A. (2012). Interrogating privilege, transforming Whiteness. In J. A. Powell (Ed.), *Racing to justice: Transforming our conceptions of self and other to build an inclusive society* (pp. 75–101). JSTOR.

Preto, B. C. (2018, March 19). *Exploring the concept of vulnerability in health care.* Retrieved from CMAJ: https://www.cmaj.ca/content/190/11/E308

Shreve, C. M., & Kelman, I. (2014). Does mitigation save? Reviewing cost-benefit analyses of disaster risk reduction. *International Journal of Disaster Risk Reduction, 10*(Part A), 213–235.

Siders, A. R., & Keenan, J. M. (2020). Variables shaping coastal adaptation decisions to armor, nourish, and retreat in North Carolina. *Ocean & Coastal Management, 183,* 105023.

Sierra Club. (2020, July 22). *Statement on systemic and pervasive racism within the environmental field.* Retrieved from: https://www.sierraclub.org/sites/www.sierraclub.org/files/uploads-wysiwig/Statement%20on%20systemic%20and%20pervasive%20racism%20within%20the%20environmental%20field.pdf

Solomon, D., Maxwell, C., & Castro August, A. (2019, August 7). Systemic inequality: Displacement, exclusion, and segregation. How America's housing system undermines wealth building in communities of color. *Center for American Progress.* Retrieved from: https://www.americanprogress.org/issues/race/reports/2019/08/07/472617/systemic-inequality-displacement-exclusion-segregation/

UCUT, U.C. (Director). (2016). *Grand Coulee and the Forgotten Tribe* [Motion Picture].

USDN. (2020, February). Urban Sustainability Directors Network (USDN) strategic relocation & transition interviews (K. Baja, Interviewer).

Vargas, E. (2020, August 31). American individualism is an obstacle to wider mask wearing in the US. *Brookings Institution, Up Front.*

Vries, D. D. (2012). Citizenship rights and voluntary decision making in post-disaster U.S. floodplain buyout mitigation programs. *International Journal of Mass Emergencies and Disasters, 30,* 1–33.

White, L. (1967). The historical roots of our ecological crisis. *Science, 155* (3767) (1203–1207), 5–6.

Williams-Rajee, D. (2020). Equity and climate theory of change. Kapwa Consulting (K. Baja, Interviewer).

Zavar, E. (2019, May 28). After disaster: Why home buyout programs fizzle out. *Gov1,* 1–3.

3 The role of international governance to reduce maladaptive climate relocation

Thea Dickinson and Ian Burton

3.1 Planned relocation

"We don't want to leave this place. We don't want to leave, it's our land, our God given land, it is our culture, we can't leave. People won't leave until the very last minute."

– Paani Laupepa[1]

Paani Laupepa of Tuvalu captured the heartbreaking complexity of relocation when he said, "people won't leave this place until the very last minute." The past 50 years have demonstrated that there is substantial difficulty in planned relocation and significant potential for maladaptation. The interconnected and cascading effects of climate change do not end at changes in temperature or rising sea levels. Nor do they end with reclaiming islands and the migration of entire populations since these effects have yet to explore potential maladaptations such as loss of cultures and Indigenous languages that may suddenly vanish or slowly die out in new lands. Moreover, the current reluctance in the European Union and the United States to accept refugees from conflict zones paints a concerning picture for environmentally displaced populations.

Migration is at the core of most contemporary and historical societies, and many of the vast transformations of human geography over the past 60,000 years were prompted by changes in climate. Much of this historic migration occurred without formal understanding of climate change, without dire headlines, without legal architecture, without experts scouting for areas of "safe" relocation, or government procurements of land in nearby countries. The concept of "planned" migration is relatively new in terms of the collective history of 60,000+ years of unplanned migration. Yet, even planned migration can create untenable challenges for host communities, migrating populations, governments, and institutional systems attempting to cultivate these dramatic transformations.

In 2018, over 70.8 million people were forcibly displaced worldwide as the result of persecution, conflict, violence, and human rights violations (UNHCR, 2019). A rapidly changing climate adds to these drivers and can promote or lead to an expansion of human mobility and displacement throughout the globe

DOI: 10.4324/9781003141457-3

(Adger et al., 2014). The effect of climate change on migration is unprecedented in modern history.

Planned relocation requires informed policy and supportive governance structures backed by multilateral institutions. None of these measures are presently in place. Current policies are place-based and piecemeal, and the only legal frameworks are based on dated legalization (such as the *1951 Refugee Convention*). Recently developed institutional agreements (e.g., *Global Compact for Migration*) fail to include climate as a driver of displacement. Merone and Tait (2018) exclaim, "there is no long-term plan to support those who face environmental displacement." With increasing nationalism and closed borders, the international community seems unwilling to add climate as a "new" driver to migration. Across the globe, localities and even entire countries are facing unprecedented choices without systematic international guidance. By leaving nations to act in isolation to relocate populations, we are paving the way for increased climate change maladaptation in migration. In this chapter, we refer to the definition in Juhola and colleagues (2016) who define maladaptation as the, "result of an intentional adaptation policy or measure directly increasing vulnerability for the targeted and/or external actor(s), and/or eroding preconditions for sustainable development by indirectly increasing society's vulnerability."

3.1.1 Migration and planned relocation

To understand planned relocation, one must first understand the complex concept of migration. Migration is an overarching term that refers to movement of persons to a new place of residence, commonly involving the crossing of political or administrative borders (Shryock & Siegel, 1980). Migration can be internal (within national boundaries) or external (across international borders). It may be voluntary or involuntary (forced); assisted or unassisted; and temporary or permanent. Migration is often the result of a multitude of *push–pull* drivers (Lee, 1966). Pull factors – ones that draw migrants to other geographic locations – include increased standard of living, greater resource access, occupational and educational opportunities, family reunification, and greater rights and freedoms (Bernzen et al., 2019; Stojanov et al., 2017). Push-factors – which drive persons from their usual place of residence – include persecution, violence and war, famine, political instability, rapid or slow-onset environmental change, or disasters (that may be technological, climatological, biological, geophysical) (Black et al., 2013; Van Hear et al., 2018).

Planned relocation is a form of migration whereby persons or communities are assisted in moving away from their homes and resettled in a new geographic location. Planned relocation is often carried out under government authority and may be internal or external. Planned relocation is commonly undertaken to protect communities from risks associated with disasters and rapid or slow-onset environmental change, including the effects of climate change. Currently, there is no formal international guidance or governance on how planned relocation should be carried out.

3.2 Evolution of the concept of migration and adaptation

3.2.1 Migration viewed as a failure of climate change adaptation

When the concept of migration as a form of adaptation was initially proposed, like the concept of adaptation itself, it was seen as a failure (Campbell, 2008; Kelman, 2015; King et al., 2014; Stojanov et al., 2014; Zoomers, 2012). There was tremendous opposition from the adaptation community about acknowledging migration as an acceptable method of adapting to climate change. Miller and colleagues note that early on, "migration was not discussed in the context of adaptation to climate change, because many developing countries considered it a mechanism of last resort, reflecting the failure of developed countries adequately to address the mitigation challenge" (2013). In 2009, Barbier and colleagues (2009) examined adaptation strategies of farmers in northern Burkina Faso and discovered that, "when asked about their adaptation to future droughts, few farmers consider migration" as an option. Sward and Codjoe (2012) analyzed the National Adaptation Program of Action (NAPAs) plans submitted by 51 least developed countries (LDCs) and found that 13 did not discuss migration. When migration was mentioned it was often in the context of preventing migration. Birk (2012) noted that many believed the mere suggestion of migration was a failure [of the imagination] by planners and policy makers to identify alternative adaptation options.

If migration occurs due to force, it could be considered a form of climate change dispossession. Being forcibly displaced by a changing climate from one's ancestral land is akin to a level of loss and damage that could be viewed as an international human rights violation. Hence, the mere suggestion of migration as a form of adaptation may be on a par with a sense of collective trauma that some may not want equated with adaptation.

As time passed, it was realized that even with vast amounts of mitigation, climate change impacts were unavoidable. Warner (2010) discusses a spectrum of responses to environmental change where, "some forms of environmentally induced migration may be adaptive, while other forms of forced migration and displacement may indicate a failure of the social–ecological system to adapt."

3.2.2 Migration viewed as climate change adaptation

In 2006, McLeman and Smit (2006) wrote a paper entitled, "Migration as an adaptation to climate change" that started the exploration of migration as a viable adaptation option. Probing questions were examined about "what conditions lead to migration instead of alternative adaptation options" and how "the factors that influence whether members of a given population may adapt through migration…[can] be identified." In the past decade, many of these questions have begun to be answered, mostly as a result of climate shocks requiring migration as a necessary adaptation option. Now "a transition has taken place where migration is seen as more than a coping strategy: It is also adaptation" (Ober, 2014). There are many reasons why migration as a form of adaptation can and should be supported. Miller et al. (2013) write, "in the face of increased and repetitive climate-induced disasters, it is becoming clear that some climate change impacts are unavoidable and that it will not be possible to protect certain communities from damage in their current locations." Many drivers of migration can be positive; as Mortreux and Barnett (2009) note, migration as an adaptation strategy can lead to increases in quality of life and economic opportunity.

3.3 Maladaptation in the context of planned relocation

Many definitions of maladaptation exist (Magnan et al., 2016). The IPCC defines maladaptation as "an adaptation that does not succeed in reducing vulnerability but increases it instead" (McCarthy et al., 2001). The nuances of maladaptation are far more complex than the definition suggests. For instance, palliative adaptation occurs when adaptations to make vulnerable populations of this generation safe increase damages for future generations (Dickinson & Burton, 2014). The potentially negative consequences of adaptation should therefore always be assessed (Davis & Ali, 2014). Maladaptation in the context of planned relocation can be obscured under layers of other socioeconomic factors, but maladaptation may occur, for instance, if the area for relocation[2]:

1. is temporary or palliative;[3]
2. decreases quality of life and standard of living;
3. exposes the community to new environmental hazards (climatic changes);
4. is not economically viable for resources or livelihood;
5. increases vulnerability for host community (e.g., introduction of novel diseases, economic competition, reduce livelihood opportunities, and resource strain);
6. causes xenophobic conflicts;
7. results in silent genocides (loss of culture or language); and
8. encourages economic activity that increases emissions.

Planned relocation involves costs and benefits for relocated populations. Tamir (2000) studied the 1974 forced relocation of Navajos to Pinon, Arizona and concluded, "relocation is ... successful only when relocatees restore or expand their economic production activities" and "forced relocation hurts people and no method can weigh the suffering of relocatees." The World Bank Operational Directive on Involuntary Resettlement from 1990 begins with a similar tone, stating that displacing persons from their homes can give rise to "severe economic and social" problems. However, the document provides a caveat: "*unless appropriate measures are carefully planned and carried out*" (World Bank, 1990).

While not all climate-related migration is forced migration, there are cases where even voluntary relocation contains some level of trauma and loss. Kenneth Knudson in several publications (1964, 1977) analyzed the loss of cultural identity in Solomon Islanders, through what could now be described as a series of maladaptive planned migrations. In the 1930s, the government of the Gilbert and Ellice Islands developed a program aimed at reducing overpopulation. Families volunteered to be relocated in the unoccupied Phoenix Islands. The relocation began in 1938, creating "new and relatively self-sufficient communities" (Knudson, 1977). By the 1950s, the male elders of the community requested to be relocated again, as a series of extended droughts had created severe hardships. However, the new location in Titiana eroded "community structures" and resulted in the "loss of cultural identities" (Birk, 2012; Hastrup & Olwig, 2012).

This raises the question of how long does a community need to be relocated for it to be considered a success? Campbell (2008) discussed relocation in terms of "permanent or long term" but does not define a length of time. Does the planned relocation of a population to a new location where severe drought or flooding are predictable in the foreseeable future indicate maladaptation? How quickly would climate disaster need to recur in a relocated community for the relocation to be judged maladaptive? Unlike climate refugees who are in a waystation, should communities participating in planned relocation have a realistic expectation of permanency? And if a relocation does not have a reasonable expectation of permanency, is it palliative, failed, or maladaptive?

The aim from a local and international policy perspective is to understand the potential for maladaptation and attempt to mitigate maladaptation. Safeguards should be proactively in place in the form of international and institutional governance, and preferably also as laws, to prevent harms from maladaptation or palliative adaptation.

3.3.1 Case study: Maldives safe islands?

To further explore these concepts, consider the development of artificial islands in the Maldives. There are two adaptation measures in this example: The construction of the artificial islands themselves and the planned relocation of residents to the new. The question arises whether both measures are adaptive, maladaptive, or palliative.

The Maldives is said, through oral tradition, to have been populated for the past 2500 years. In 2004, nearly every one of its approximately 1200 islands were struck by the Indian Ocean Tsunami. Aid for reconstruction flowed in almost immediately, one of the strongest outpourings for a disaster to date. The International Federation of the Red Cross and Red Crescent Societies constructed hundreds of homes for displaced residents (Fuller, 2009). The government began developing artificial islands by draining seafloors, relocating coral, and infilling with dredged sediment, rocks, and sand. These artificial islands included the lavishly envisioned "designer islands" of Hulhumalé on the Malé Atoll. Over 100,000 Maldivians have moved to these "safe islands," with Phase 2 of reclamation intending to draw thousands more. At the time of the 2004 disaster, the population in the Maldives was estimated to be around

300,000, with roughly 200 of the almost 1200 islands inhabited. As the economy of the Maldives rapidly grew with expanding tourism, the population growth rate increased to almost five times the global average. In 2019, there was estimated to be over half a million residents (World Bank, 2019a). The increase in reclaimed land in the Maldives has further allowed for an influx of migrants. In 2007, net migration was 17,994; by 2012 it was 58,358 (World Bank, 2019b). The densely populated and extremely vulnerable island nation situated precariously in the middle of the Indian Ocean is *attracting* migrants, not resettling them in other nations.

While the massive reconstruction projects of the Maldives include better sea defenses, elevated lands and buildings, and disaster education, the question remains: Will this be enough? To boost economic activity, the government has hinted at plans to geo-engineer more artificial islands, relocate populations and attract tourists by creating 50 more resorts. Yet, these "safe islands" will be exposed to severe impacts from climate change – the extent of which depends on whether the Paris Agreement goals are met (Brown et al., 2020). In addition, the Maldives have other vulnerabilities, and the dispersed geography of the Maldives makes disaster response extremely difficult, as the almost 200 inhabited islands are hard to reach. In 2014, a fire destroyed the only desalination facility in Malé, leading to a water crisis. The President stated, "We did not have any fall-back plan for any disaster of this magnitude." The Maldives declared a state of emergency and had to rely on neighboring countries flying in bottled water. The 10-day disaster cost USD 20 million in relief operations (Schafer, 2019). Thus, the question is raised: Are these relocations truly adaptive; are they maladaptive, as suggested by Magnan and colleagues (2016); or are they a palliative adaptation (defined in Dickinson & Burton, 2014) that gives a false sense of security to populations who will eventually require migration to new countries?

At what point do we determine whether the Maldives relocation is a success? Scudder's resettlement model (Scudder, 1985) infers that communities are relocated to a stable host environment. The model does not make room for rapid or even slow-onset environmental change, such as that which could be experienced in the Maldives, including: Unmanageable flooding, beach erosion, coastal inundation and land loss, rising temperatures, and further degradation of precious coral reef and fishing ecosystems, all of which can lead to a decline in local food security and negatively affect tourism (the hub of their economy). Is the relocation a success once all these hazards have reached the shores of the Maldives and they have weathered all potential crises?

The creation of the artificial islands was built on the premise that, "the need for development outweighs the risks" (Darby, 2017). The island nation placed a bet on short-term economic gain, potentially investing in *disaster risk creation*. If the new islands are maladaptive, will the situation evolve into an ongoing humanitarian crisis or a "refugee crisis" where displaced persons seek refuge outside of their sovereign borders?

3.4 International agreements on migrants and refugees

Currently, "there is no international agreement on a term to be used to describe persons or groups of persons that move for environment related reasons" (IOM, 2019). This is the case even though the term *environmental refugees* first appeared 35-years ago (El-Hinnawi, 1985), and publications on environmentally displaced persons are littered with ponderings on "climate refugees" (Berchin et al., 2017; Biermann & Boas, 2010; Farbotko & Lazrus, 2012). Climate-induced migrants are trapped in an unrecognized and unprotected valley between international climate policy and international refugee law. The identification of *refugee status* is based on a very narrow definition from *The 1951 Refugee Convention*, that a refugee is a person with a "well founded fear of being persecuted for reasons of race, religion, nationality, membership of a particular social group or political opinion" and that person is "outside the country of [their] nationality." Internal displacement is not covered, nor is any climate or environmental variable. Thus, *climate refugees* as a defined legal term do not yet exist.

Such was the case when Ioane Teitiota from Kiribati went before the New Zealand Immigration and Protection Tribunal (IPT) to argue for refugee status, "on the basis of changes to his environment in Kiribati caused by sea-level-rise associated with climate change." After losing in New Zealand courts, Teitiota appealed to the United Nations Human Rights Committee, who upheld New Zealand's decision but declared that states should consider whether climate change is causing human rights violations when making refugee status determinations. This case took over a decade to resolve (UNHRC, 2020). While the committee did not declare his deportation unlawful (because his life was not in immediate danger), if the risk is foreseeable, as with climate change, does this not warrant immediate protection? Or do the communities need to wait until their lives are fully engulfed in catastrophe before they can seek refuge? Can foreseeable climate risk be sufficient? If artificial islands turn maladaptive, what becomes of those who are displaced for environmental reasons?

In December 2018, two international agreements on migration were opened for signature: The *Global Compact on Refugees* (GCR) and the *Global Compact for Safe, Orderly and Regular Migration* (GCM). However, both agreements failed to address climate migrants. Alice Thomas, Climate Displacement Program manager at Refugees International, stated, "There's a long way to go before any of this language translates to tangible benefits for climate migrants" and reiterated "the GCM has yet to yield substantive regional or national policy changes aimed at climate migrants" (McDonnell, 2019). Further, neither Compact aims to prevent palliative nor maladaptive measures from being implemented. In fact, the GCR voids or rejects responsibility by the international community:

> 8. *Large-scale refugee movements and protracted refugee situations persist around the world. Protecting and caring for refugees ... needs to be accompanied by dedicated efforts to address root causes. While not in themselves causes of refugee*

movements, climate, environmental degradation and natural disasters increasingly interact with the drivers of refugee movements. **In the first instance, addressing root causes is the responsibility of countries at the origin of refugee movements.** (Section iv, 8, emphasis added)

The line, "addressing root causes is the responsibility of countries at the origin of refugee movements" conflicts with *common but differentiated responsibility* in the United Nations Framework Convention on Climate Change (UNFCCC). While not all climate relocation is dispossession; climate change dispossession should be viewed as a human rights violation of the right to adequate housing free from state evictions (OHCHR, 2008). The envisioned aim of these Compacts was to cover, "all dimensions of international migration in a holistic and comprehensive manner" (UN General Assembly, 2018). Instead, the Compact provides the language to allow developed and neighboring nations, who may hold partial responsibility for the root causes of migration to avoid responsibility for addressing those causes or dealing with the resulting migration.

The UNFCCC does not reference the concept of migration directly. Rather the concept began to emerge within the negotiation texts of the annual Conference of the Parties (COPs) beginning most noticeably with the Cancun Agreement produced during COP 16 held in Mexico in 2010. By 2011, the Nansen Initiative on Disaster-Induced Cross-Border Displacement was catalyzed by section 14(f) of the Cancun Agreement. The desire was to, "initiate a process for developing a similar convention with respect to people who might be designated refugees due to climate change and natural disasters" (Cook, 2020). Despite being held on the 60th anniversary of the 1951 the *Convention Relating to the Status of Refugees*, the proposal was rejected by almost all governments except Costa Rica, Germany, Mexico, Norway, and Switzerland. Arguments against the initiative included a belief that there was no place in international law for "climate refugees" and "it was too early to talk about developing soft-law frameworks for climate change displacement" (Hall, 2016). By 2015, the initiative had gained endorsements and the *Agenda for the Protection of Cross-Border Displaced Persons in the Context of Disasters and Climate Change* was drafted. In 2016, the *Platform on Disaster Displacement* (PDD) replaced the Nansen Initiative. Motivated by, "no one will be left behind" it acknowledged the challenges of disaster displacement including, "planned relocation and migration as adaptation, [requiring] integrated approaches that address the drivers as well as the consequences and protection needs of such human mobility" (PDD, 2019).

The Nansen Initiative, the PDD, the GCM, the Office of the United Nations High Commissioner for Human Rights (OHCHR) and other international agreements repetitively use the language, "planned relocation as a measure of last resort." This is a misnomer for several reasons:

1. By relegating planned relocation as a "last resort," *the language may instead promote parties to attempt a series of palliative and maladaptive measures.* These

adaptation measures may become *disaster risk creation* for localities already exposed to environmental hazards. When planned relocation is considered during the adaptation strategy development process it can be weighed against potential maladaptive options such as rebuilding on a vulnerable floodplain. Such careful assessment at the outset may prevent further catastrophic loss of life or livelihood.

2. *The language does not incentivise a greater understanding of the maladaptive elements of planned relocation.* As described earlier, relocations have significant potential for maladaptation for the people who relocate and for the origin and destination communities. However, with appropriate long-term planning, an inclusive and welcoming community is possible.

3. *The language fails to offer expanded support or promote proactive planning for relocation.* The international agreements currently provide minimal guidance, governance, or lessons learned from early adapters for planned relocation. The language overlooks the potential capacity that relocation can have to transform lives *when innovative measures that aim to promote successful relocation are proactively implemented.* Migration with Dignity, presented by President Anote Tong of Kiribati, intends to preserve cultural identify and provide educational access and skills to potential displaced Kiribati nationals (Tong, 2014). The plan highlights the need for pragmatic conceptualization of these transitions well before the potential date of relocation. Pilot Programs such as the AUD20.8 million Kiribati-Australia Nursing Initiative (KANI) at Griffith University, take considerable forethought and years to implement and execute. They cannot be implemented as a "last resort" idea on the reefs of tumultuous seas. Similarly, the Jordan Compact for Syrian refugees has the hallmarks of potentially transforming the refugee landscape. Special economic zones, such as the King Hussein Bin Talal Development Area (KHBTDA), would provide employment for a percentage of displaced Syrian nationals, while benefiting host communities. Last resort policies steal the freedom and opportunities of displaced persons.

4. Given the length of time required for safe and dignified relocation, *the language "last resort" falsely implies that there is ample time to consider several adaptation alternatives.* The perception that relocation is quickly feasible is inherently maladaptive. Ferris (2017) asserts,

> "planned relocation can often span multiple generations... involve.... successive governments, divergent political priorities, and, potentially, multiple changes in policy... [thus] a clear, coherent, and comprehensive legal framework, incorporating human rights principles, ... in accordance with national laws and policies... and complying with an appropriate legal framework ... is critical."

If not considered during conceptualization of adaptation measures for a given locality, it will be impossible to move migration as an adaptation forward. It has taken upwards of years if not decades to relocate communities. Proactive measures require anticipatory foresight; *last resort* relocation is a roadmap for maladaptation.

3.5 Moving forward on climate-induced migration

Reducing maladaptation in climate-change-induced migration and planned relocation requires a collective effort. The developed economies responsible for the dispossession of land are avoiding responsibility and failing to create inclusive policies and supportive migration laws. There are more displaced persons now than at any time since World War II (UNHCR, 2019). The drivers of displaced persons need to expand to legally include climate as a variable. The current fragmentation and disconnect between communities of practice, intergovernmental organizations, legal frameworks, and international agreements leads to and will continue to lead to increases in maladaptation in climate-induced migration. Unless and until substantial change is made to support environmentally displaced persons, countries will continue to go unguided, making attempts at maladaptive measures to keep their citizens safe. The proverbial head-in-the sand will only serve humanity for so long: Until the sea rises and societies are struggling for air.

3.6 Recommendations

- *Encourage Bilateral Partnerships*: Developed nations should seek bilateral (or trilateral) partnerships with vulnerable nations and develop training programs, educational scholarships, and other forms of assistance for displaced persons to aid in entering market economies (e.g., New Zealand and Australia partnering with Pacific Island nations).
- *Modify International Agreements*: The language of international agreements on migrants should clearly include climate and environmental change as a driver of migration and remove language that describes planned relocation as a measure of last resort.
- *Amend Migration Policies*: Countries with higher levels of greenhouse gas emissions should modify migration policies to be more inclusive and accepting of environmentally displaced persons as an extension of the Common but Differentiated Responsibilities under the UNFCCC.
- *Supportive Multilateral Institutions*: In countries where they have ongoing operations, multilateral development banks and bilateral aid agencies should provide financial support and assistance for the development of lessons learned guidance documents on planned relocation (see IOM, 2017 as an example).
- *Assessing International Development Assistance*: International Development Assistance should investigate and assess how development funding is contributing to disaster risk creation, take steps to control it, and avoid funding short-term unsustainable development that increases vulnerability and displacement in the longer run.

Notes

1 Former assistant secretary at Tuvalu's Ministry of Natural Resources, Energy and Environment.
2 List is based on 12 key elements of climate change adaptation identified in Dickinson (2019).
3 Palliative adaptation is the use of adaptation to make vulnerable populations of this generation safe but at the costs of creating major problems and catastrophic circumstances for future generations. This short-term risk reduction is a form of unsustainable development. In some cases, permanent migration will need to take place. See Dickinson and Burton (2014).

References

Adger, W. N., Pulhin, J. M., Barnett, J., Dabelko, G. D., Hovelsrud, G. K., Levy, M., Oswald Spring, Ú., & Vogel, C. H. (2014). Human security. In C. B. Field, V. R. Barros, D. J. Dokken, K. J. Mach, M. D. Mastrandrea, T. E. Bilir, M. Chatterjee, K. L. Ebi, Y. O. Estrada, R. C. Genova, B. Girma, E. S. Kissel, A. N. Levy, S. MacCracken, P. R. Mastrandrea, & L. L. White (Eds.), *Climate change 2014: Impacts, adaptation, and vulnerability. Part A: Global and sectoral aspects* (pp. 755–791). Contribution of Working Group II to the Fifth Assessment Report of the Intergovernmental Panel on Climate Change. Cambridge University Press.

Barbier, B., Yacouba, H., Karambiri, H., Zoromé, M., & Somé, B. (2009). Human vulnerability to climate variability in the Sahel: Farmers' adaptation strategies in northern Burkina Faso. *Environmental Management, 43*(5), 790–803.

Berchin, I. I., Valduga, I. B., Garcia, J., & de Andrade, J. B. S. O. (2017). Climate change and forced migrations: An effort towards recognizing climate refugees. *Geoforum, 84*, 147–150.

Bernzen, A., Jenkins, J. C., & Braun, B. (2019). Climate change-induced migration in coastal Bangladesh? A critical assessment of migration drivers in rural households under economic and environmental stress. *Geosciences, 9*(1), 51.

Biermann, F., & Boas, I. (2010). Preparing for a warmer world: Towards a global governance system to protect climate refugees. *Global Environmental Politics, 10*(1), 60–88.

Birk, T. (2012). Relocation of reef and atoll island communities as an adaptation to climate change: Learning from experience in Solomon Islands. In K. Hastrup & K. Fog Olwig (Eds.), *Climate change and human mobility: Challenges to the social sciences* (pp. 81–109). Cambridge University Press.

Black, R., Arnell, N. W., Adger, W. N., Thomas, D., & Geddes, A. (2013). Migration, immobility and displacement outcomes following extreme events. *Environmental Science & Policy, 27*, S32–S43.

Brown, S., Wadey, M. P., Nicholls, R. J., Shareef, A., Khaleel, Z., Hinkel, J.,... & McCabe, M. V. (2020). Land raising as a solution to sea-level rise: An analysis of coastal flooding on an artificial island in the Maldives. *Journal of Flood Risk Management, 13*, e12567.

Campbell, J. (2008). *International relocation from Pacific Island countries: Adaptation failure?* Bonn, Germany: International Conference on Environment, Forced Migration and Social Vulnerability.

Cook, I. (2020). *The politics of the final hundred years of humanity (2030-2130)*. Springer Nature.

Darby, M. (2017). *Maldives regime imperils coral reefs in dash for cash.* Climate Home News. March 20, 2017. https://www.climatechangenews.com/2017/03/20/maldives-regime-imperils-coral-reefs-dash-cash/ [Accessed October 2020].

Davis, P., & Ali, S. (2014). Exploring local perceptions of climate change impact and adaptation in rural Bangladesh (Vol. 1322). Intl Food Policy Res Inst.

Dickinson, T. & Burton, I. (2014). Palliative climate change planning and its consequences for youth. In J. Elder (Ed.), *The challenges of climate change: Children in the front line, Innocenti Insight.* UNICEF Office of Research, Florence.

Dickinson, T. (2019). *Advancing climate change adaptation (Doctoral dissertation).* University of Toronto.

El-Hinnawi, E. (1985) *Environmental refugees.* United Nations Environment Programme, UNEP Office, Nairobi, Kenya. 40 pp.

Farbotko, C., & Lazrus, H. (2012). The first climate refugees? Contesting global narratives of climate change in Tuvalu. *Global Environmental Change*, 22(2), 382–390.

Ferris, E. (2017). *A toolbox: Planning relocations to protect people from disasters and environmental change.* Institute for the Study of International Migration, UNHCR, The UN Migration Agency: Georgetown University, Washington DC.

Fuller, P. (2009). Red Cross Red Crescent completes 44,000 new homes for tsunami survivors. August 5, 2009. https://www.ifrc.org/en/news-and-media/news-stories/asia-pacific/maldives/red-cross-red-crescent-complete-44000-new-homes-for-tsunami-survivors/ [Accessed October 2020]

Hall, N. (2016). *Displacement, development, and climate change: International organizations moving beyond their mandates.* Routledge.

Hastrup, K., & Olwig, K. F. (2012). *Climate change and human mobility: Challenges to the social sciences.* Cambridge University Press.

IOM. (2017). *Planned relocation for communities in the context of environmental change and climate change: A training manual for provincial and local authorities.* IOM Mission in Vietnam.

IOM. (2019). *International migration law: Glossary on migration* (No. 34). International Organization for Migration.

Juhola, S., Glaas, E., Linnér, B. O., & Neset, T. S. (2016). Redefining maladaptation. *Environmental Science & Policy*, 55, 135–140.

Kelman, I. (2015). Difficult decisions: migration from small island developing states under climate change. *Earth's Future*, 3(4), 133–142.

King, D., Bird, D., Haynes, K., Boon, H., Cottrell, A., Millar, J., ... & Thomas, M. (2014). Voluntary relocation as an adaptation strategy to extreme weather events. *International Journal of Disaster Risk Reduction*, 8, 83–90.

Knudson, K. (1964). *Titiana: a Gilbertese community in the Solomon Islands.* Mimeo, Eugene: Department of Anthropology, University of Oregon.

Knudson, K. (1977). *Sydney Island, Titiana, and Kamaleai: Southern Gilbertese in the Phoenix and Solomon Islands* (pp. 195–241). Exiles and Migrants in Oceania.

Lee, E. S. (1966). A theory of migration. *Demography*, 3(1), 47–57.

Magnan, A. K., Schipper, E. L. F., Burkett, M., Bharwani, S., Burton, I., Eriksen, S., ... & Ziervogel, G. (2016). Addressing the risk of maladaptation to climate change. *Climate Change*, 7(5), 646–665.

McCarthy, J. J., Canziani, O. F., Leary, N. A., Dokken, D. J., & White, K. S. (Eds.). (2001). *Climate change 2001: Impacts, adaptation, and vulnerability: Contribution of*

Working Group II to the third assessment report of the Intergovernmental Panel on Climate Change (Vol. 2). Cambridge University Press.

McDonnell, T. (2019). *Climate migrants face a gap in international law*. Centre for International Governance. February 12, 2019. https://www.cigionline.org/articles/climate-migrants-face-gap-international-law [Accessed October 2020]

McLeman, R., & Smit, B. (2006). Migration as an adaptation to climate change. *Climatic Change, 76*(1-2), 31–53.

Merone, L., & Tait, P. (2018). 'Climate refugees': Is it time to legally acknowledge those displaced by climate disruption? *Australian and New Zealand Journal of Public Health*.

Millar, I., Gascoigne, C., & Caldwell, E. (2013). Making good the loss: An assessment of the loss and damage mechanism under the UNFCCC process. Threatened island nations: An assessment of the loss and damage, 433–472.

Mortreux, C., & Barnett, J. (2009). Climate change, migration and adaptation in Funafuti, Tuvalu. *Global Environmental Change, 19*(1), 105–112.

Ober, K. (2014). Migration as adaptation: Exploring mobility as a coping strategy for climate change. Climate and Migration Coalition Briefing. https://studydirect.sussex.ac.uk/course/view.php [Accessed October 2020]

OHCHR. (2008). *Frequently asked questions on economic, social and cultural rights*. Office of the United Nations High Commissioner for Human Rights. https://www.ohchr.org/Documents/Issues/ESCR/FAQ%20on%20ESCR-en.pdf [Accessed October 2020]

PDD. (2019). *Platform on disaster displacement (PDD) strategy 2019-2022*. https://disasterdisplacement.org/wp-content/uploads/2019/06/26062019-PDD-Strategy-2019-2022-FINAL_to_post_on_website.pdf [Accessed October 2020]

Schafer, H. (2019). *Bracing for climate change is a matter of survival for the Maldives*. World Bank: End Poverty in South Asia. January 20, 2019. https://blogs.worldbank.org/endpovertyinsouthasia/bracing-climate-change-matter-survival-maldives

Scudder, T. (1985). A Sociological framework for the analysis of new land settlements. Putting people first: Sociological variables in rural development, 121–153.

Shryock, H. S., & Siegel, J. S. (1980). The methods and materials of demography (Vol. 2). *Department of Commerce, Bureau of the Census*.

Stojanov, R., Duží, B., Kelman, I., Němec, D., & Procházka, D. (2017). Local perceptions of climate change impacts and migration patterns in Malé, Maldives. *The Geographical Journal, 183*(4), 370–385.

Stojanov, R., Kelman, I., Martin, M., Vikhrov, D., Kniveton, D., & Duží, B. (2014). *Migration as adaptation. Population dynamics in the age of climate variability*. Global Change Research Centre, The Academy of Sciences of the Czech Republic, Brno.

Sward, J., & Codjoe, S. (2012). *Human mobility and climate change adaptation policy: A review of migration in National Adaptation Programmes of Action (NAPAs)*. Migrating out of Poverty RPC Working Paper 6. Migrating out of Poverty Consortium, University of Sussex, Brighton, UK. 44 pp.

Tamir, O. (2000). Assessing the success and failure of Navajo relocation. *Human Organization, 59*(2), 267–273.

Tong, A. (2014). *Permanent Mission of Kiribati to the United Nations, Statement by H.E. President Anote Tong*, 69th UNGA, 26th September 2014, New York. https://www.un.org/en/ga/69/meetings/gadebate/pdf/KI_en.pdf [Accessed October 2020]

UN General Assembly. (2018). *Final draft of the global compact on migration*. https://refugeesmigrants.un.org/sites/default/fles/180711_fnal_draft_0.pdf [Accessed October 2020]

UNHCR. (2019). *Forced displacement in 2018*. UNHCR: Geneva, Switzerland.

UNHRC. (2020). *Ioane Teitiota v. New Zealand*. CCPR/C/127/D/2728/2016, UN Human Rights Committee (HRC), January 7, 2020. https://www.refworld.org/cases,HRC,5e26f7134.html [Accessed October 2020]

Van Hear, N., Bakewell, O., & Long, K. (2018). Push-pull plus: Reconsidering the drivers of migration. *Journal of Ethnic and Migration Studies*, 44(6), 927–944.

Warner, K. (2010). Global environmental change and migration: Governance challenges. *Global Environmental Change*, 20(3), 402–413.

World Bank. (2019a). *World development indicators: Population, total – Maldives* [Data File]. https://data.worldbank.org/indicator/SP.POP.TOTL?locations=MV [Accessed October 2020]

World Bank. (2019b). *World development indicators: Net migration – Maldives* [Data File]. https://data.worldbank.org/indicator/SM.POP.NETM?locations=MV [Accessed October 2020]

World Bank. (1990). *Operational Directive 4.30: Involuntary resettlement*. World Bank.

Zoomers, A. (2012). Migration as a failure to adapt? How Andean people cope with environmental restrictions and climate variability. *Global Environment*, 5(9), 104–129.

4 Charting a justice-based approach to planned climate relocation for the world's refugees

Laura E. R. Peters and Jamon Van Den Hoek

4.1 Introduction

The geographer Yi-Fu Tuan famously wrote that "place is security, space is freedom" (Tuan, 1977). For the world's refugees, this axiom does not hold true. Violent conflict and political persecution often result in insecurity and social instability that compel individuals, families, and whole communities to leave their home country and seek asylum abroad in search of new spaces of safety. Yet, even after settling in a refugee camp in a new country, refugees may find neither freedom nor refuge. State policies often strip refugees of their agency and decision-making power over livelihoods and employment and confine refugees within the camp. As current and expected future climate change impacts raise concerns over the viability of long-term human habitation in many of the world's refugee camps, the looming need for the relocation of refugee camps must be considered. This chapter addresses the pending collision of these two options of last resort: settlement in refugee camps and relocation due to climate change effects. We provide an overview of refugee displacement; anticipated climate change effects in refugee-hosting regions; and the ways that climate change is already affecting refugee camps. We also discuss three dimensions of social justice relevant for refugee relocation and conclude by detailing a justice-based approach to camp relocation that supports refugee agency, dignity, and security.

4.2 Protracted temporariness

The world has never had more refugees than in the 21st century. As of mid-2019, the United Nations High Commissioner for Refugees (UNHCR) documented 26 million refugees globally (UNHCR, 2020). Turkey, Pakistan, and Uganda collectively host 6.4 million refugees, populations primarily made up of those displaced by long-term armed conflict in Syria, Afghanistan, and South Sudan, respectively. The persistence of civil wars, persecution, and statelessness leaves refugees with few options to voluntarily repatriate, and once a person becomes a refugee, they usually remain in asylum for the rest of their life. In 2019, only 317,200 refugees returned to their countries of origin, while an additional 107,800 refugees were resettled in a third country. Countries that host refugees, meanwhile, are often impoverished, and their requests for financial support from UNHCR to provide refugees refugees with basic services – water, food, and shelter – are typically only partially met.

After being forcibly displaced from their home communities due to persecution and violence, refugees typically secure international protection and asylum across national borders (i.e., in countries where they do not have citizenship). With the attendant rights and protections afforded to refugees, some integrate into host communities (usually just over the national border), and others transit through to settle in a third country. Approximately 2.6 million (10%) refugees settle in camps managed by UNHCR, even while camps are considered by UNHCR as a last resort (UNHCR, n.d.). Refugee camps may be thought of as a temporary solution to forced displacement, yet the average stay in a refugee camp is more than 10 years (Devictor, 2019). In 2019, nearly 16 million refugees (77%) were living in a "protracted refugee scenario" (PRS) in which more than 25,000 refugees of the same nationality live in a camp for more than 5 years (UNHCR, 2020). Common in PRS is the practice of "warehousing" that restricts refugee migration from camps, effectively isolating refugees from neighboring communities, preventing settlement in urban areas or third countries, and broadly trapping refugees in a camp for an indefinite duration at the whim of the state (USCRI, 2019). Together, PRS and warehousing turn what should be a temporary solution into a lasting reality for millions of people.

4.3 Edged out of the climate niche

In an ignominious conclusion to the hottest decade on record (2010–2019), the National Aeronautics and Space Administration (NASA) and the National Oceanic and Atmospheric Administration (NOAA) measured 2019 as the second hottest year since modern record keeping began in 1880 (NOAA, 2020). Rising temperatures are shifting the geography of the "climate niche" where mean annual temperatures lie between 11 and 15°C (52 and 59°F) and most of humanity has lived for the past 6,000 years (Xu et al., 2020). With the next 50 years of business-as-usual carbon emissions, one to three billion people are expected to be left out of the climate niche as extreme temperatures push human physiology beyond the range of effective and safe functioning (Xu et al., 2020). Refugee-hosting countries across sub-Saharan Africa through the Middle East to South Asia are broadly poised to see above average temperature increases and extreme variation in precipitation in the coming decades (Figure 4.1).

While the consequences of climate-change-induced shifts in temperature and precipitation on refugee lives and livelihoods will vary region to region based on ecological sensitivity to climate perturbations (Conway et al., 2019; Nadeau et al., 2017) as well as by socioeconomic inequality (Füssel, 2010; Islam & Winkel, 2017), increased temperatures are expected to strain food production (Connolly-Boutin & Smit, 2016; Schmidhuber & Tubiello, 2007), and more erratic rainfall may alternately contribute to flash floods, landslides, and drought

(a)

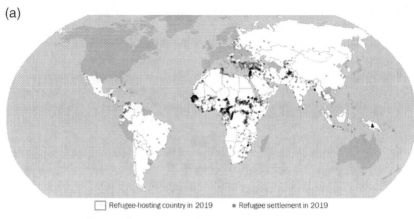

Refugee-hosting country in 2019　　　* Refugee settlement in 2019

(b)

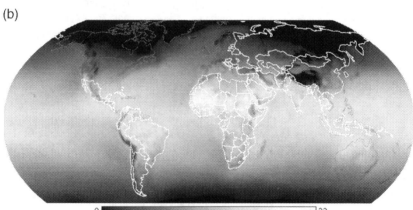

0　　　　　　　　　　　　　　33
Projected mean annual surface temperature (°C) in 2040-2059

(c)

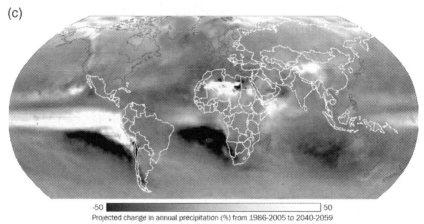

-50　　　　　　　　　　　　　　50
Projected change in annual precipitation (%) from 1986-2005 to 2040-2059

(*caption on next page*)

Figure 4.1 Maps of (a) UNHCR-managed refugee settlements (black dots) and host countries (in white) as of mid-2019; (b) projected mean annual surface temperature for 2040–2059 (with host countries outlined in white); and (c) projected change in annual precipitation for 2040–2059 relative to mean annual precipitation from 1986 to 2005 (with host countries outlined in white). Both (b) and (c) are based on CMIP5 (Coupled Model Intercomparison Project 5) multi-model ensemble projections with the "business-as usual" emissions scenario (RCP8.5). UNHCR refugee settlement data available at https://data2.unhcr.org/en/geoservices/. Climate projection data available at https://esgf-node.llnl.gov/projects/cmip5/. Analysis and visualization by Van Den Hoek.

(Dai et al., 2018; Tabari, 2020). We do not need to wait decades to see the effects of climate change on refugee camps given varied examples of climate-related hazards observed in many of the world's refugee hotspots (Khadka, 2019) including heat stress in Pakistan (Buncombe & Waraich, 2009), Jordan (Taylor, 2015), Lebanon (Medecins Sans Frontieres, 2015), and Iraq (George, 2015); seasonal flooding in Bangladesh (Sengupta & Fountain, 2018), Ethiopia (Al Jazeera, 2014; Medecins Sans Frontieres, 2014), and Kenya (Save the Children, 2018); landslides in Thailand (Associated Press, 2018) and Uganda (Atuyambe et al., 2011); and winter storms in Lebanon (Al-Arian & Sherlock, 2019).

For refugee settlements facing extreme climate change effects, there are three potential options: the development of extensive on-site adaptation measures, migration to a third country or the origin country, or relocation within the host country. Building adaptive capacity is a challenge anywhere, but refugees are uniquely under-equipped to take actions to reduce their exposure and vulnerability to climate change. The clearing of forests and disruption of root structures to establish new camps can magnify flood, landslide, and drought risk, for example; dwellings in refugee camps typically do not offer sufficient protection from extreme weather; and most settlements are reliant on external support to meet even their most basic survival needs. Further, refugees routinely lack the right to citizenship, employment, and land ownership, are not registered in national censuses, and are excluded from national planning processes on climate change adaptation (for example, Green Climate Fund, 2019) and disaster risk reduction (DRR) (Peters et al., 2019). Restrictions on refugee movement, especially for warehoused refugees, further limits the potential for resettlement in a third country, repatriation, or integration in an urban region. These imposed constraints on adaptation and migration makes planned climate relocation a crucial option of last resort for refugees in camps exposed to climate change.

4.4 Looming refugee camp relocation

While refugee camps are not permanent solutions, the purpose of refugee camps is to offer security and a basic standard quality of life prior to finding an alternative durable solution (i.e., voluntary repatriation, integration, or resettlement). Where refugee camps no longer safeguard the human rights of refugees, they are no longer habitable. Climate change combined with human actions and inactions has the potential to create insecure living conditions in refugee camps and may necessitate relocating refugee camps to safer locations.

The case of Rohingya refugees in Bangladesh shows that climate relocation for refugee camps is not a thought experiment; it is an emergent reality. The Rohingya people were displaced due to gross human rights violations and extreme violence perpetrated by the Myanmar military, and approximately 750,000 people sought refuge in Bangladesh beginning in 2017 to join those previously displaced (International Rescue Committee, 2019). More than one million Rohingya refugees now live in refugee camps outside the coastal border town of Cox's Bazar, where they face overcrowding and regular risks of flooding and landslides. The Government of Bangladesh unilaterally proposed to relocate 100,000 of these refugees to the island of Bhasan Char in the Bay of Bengal, and has invested $350 million to prepare the island for refugee settlement, including building concrete housing and flood mitigation infrastructure (Beech, 2020). In December 2020, the government began transferring refugees to the island amidst concerns that at least some of the refugees were forced into relocation (Al Jazeera, 2020).

However, the low-lying silt island of Bhasan Char is extremely vulnerable to climate-related disasters, including flooding, tidal waves, and tropical cyclones, and the majority of the land mass is underwater for the entirety of the monsoon season from June to September (Banerjee, 2020). In fact, the free-flowing sedimentary island only formed in 2003 and may well disappear back into the ocean (Bremner, 2020). Rohingya refugees have opposed relocation to Bhasan Char in part due to the lack of high ground on the island, and human rights advocates have expressed concerns about its habitability. Located 37 miles from the coast, Bhasan Char is far from large mainland Bangladeshi settlements, which makes the logistics of humanitarian support even more challenging. The ongoing struggles faced by the Rohingya refugees makes it clear that swapping one set of risks for another during relocation does not provide safety and security and may trigger another humanitarian crisis for an already vulnerable population (Paul, 2020).

4.5 Planned climate relocation of refugee camps needs special consideration

Planned climate relocation in settings other than refugee camps is typically driven by and negotiated between local communities and national authorities in terms of collectively deciding to relocate, marshaling the resources and political will to enable a move, and building a resilient community not only in terms of physical infrastructure but also linking new locations with cultural, livelihood, and basic service needs. Planned climate relocation for refugee camps is a last resort, but we argue that it may be necessary when the following three criteria have been met:

1. The refugee camp is exposed to extreme effects of climate change (e.g., extreme weather conditions, extreme storms, and prolonged drought).
2. Mitigation efforts are not feasible or would not provide reasonable safeguarding for already vulnerable populations; the financial, ecological, and human costs of such efforts would outweigh those of relocation; or the political constraints prevent necessary measures (e.g., temporary shelters are not permitted to be made more "permanent" with durable materials and foundations).
3. It is not possible for refugees to leave the refugee camp by voluntarily returning to their countries of origin (due to protracted violence and/or home communities being rendered uninhabitable by climate change or other environmental processes); resettling in third countries in safer locations (due to political barriers in the international system); or integrating in urban regions in safer locations in the current host country (due to political barriers in the national system).

Even when these conditions are met, the already-complex processes of planned climate relocation become all the more challenging in refugee camps due to the demographic, social, institutional, and political barriers at play and the diverse stakeholders – including but not limited to refugees, host, and other local communities, national authorities, and the international community – involved in decision-making.

These additional layers of interests, needs, and stakeholders bring new challenges and advantages that must be navigated carefully with respect to the planned relocation of refugee camps (see Table 4.1). Refugee camps are embedded within multiple layers of conflicts even without considering relocation. For example, the influx of Rohingya refugees in Bangladesh has been accompanied by tensions between refugees and host communities largely over perceptions of unfairness in the distribution of resources (ACAPS, 2018; International Rescue Committee, 2019; Krehm & Shahan, 2019), as well as tensions between the Government of Bangladesh and the international community over what Bangladeshi Prime Minister Sheikh Hasina referred to as the "untenability" of Bangladesh managing the bulk of the refugee crisis and the world's single largest refugee camp population on its own (Associated Press, 2019; Tharoor, 2019). The topic of refugee camp relocation is likely to introduce new controversies that will need to be managed to meet the urgent as well as long-term needs of refugees living in increasingly uninhabitable camps.

4.6 Refugee relocation guided by social justice

The clear goal of climate relocation for refugee camps must be to establish places of refuge from the extreme effects of climate change, violence, and political persecution. The relocation of refugee camps (i.e., a secondary displacement due to climate change) must respect the human rights of refugees by (1) acknowledging the conditions refugees face in camps; (2) reducing and not replacing or redistributing

Table 4.1 Refugee camp characteristics, challenges, and advantages linked to planned climate relocation

	Theme	Refugee camp characteristics
Challenges	Demographic stability	Potentially unstable or growing population, with some refugees arriving and others leaving
	Social cohesion	Potentially lacking social cohesion due to refugees coming from different cultures, communities, and/or countries and potentially speaking different languages
	Governance structure	May not be included in governance systems, or systems of governance may be evolving, and may not have the means to advocate for themselves
	Resources	Largely or fully dependent on external (international and national) aid to meet basic needs
	Security in transit	Process of transit to a new location may expose refugees to new violence and insecurity
	Political sensitivity surrounding protractedness	Long-term existence of refugee camp likely to be a source of political tension and social resentment, and it may be extremely politically sensitive to consider openly protracted refugee scenarios
Advantages	Attachment to place	Unlikely to be attached to a sense of place in the refugee camp and have greater willingness to move to safer locations
	Resources	May have international resources and advocacy to draw from
	Rights	The host country has an obligation to protect refugees as signatory to the 1951 Refugee Convention

the risk of climate-related impacts and disasters and/or politically motivated violence or persecution; and (3) engaging with the full range of stakeholders to determine and pursue durable solutions.

At a basic level, relocation strategies would further benefit from social justice principles, which guide the fair and compassionate distribution of power and resources, "starting with the right of all human beings to benefit from a safe and pleasant environment" (UN, 2006). The concept of social justice has been well-discussed by social scientists, theorists, and philosophers, though it is a far-from-unified field owing to the diversity of perspectives on what constitutes a just world. For example, *distributive* justice (Rawls, 1971) raises questions about the fair distribution of benefits and costs; *procedural* justice (Thibaut & Walker, 1975) is concerned with fair process to guarantee fair outcomes; and *restorative* justice

(Eglash, 1977; Zehr, 2002) seeks to restore relationships and avoid stigmatization while fighting against injustice.

Drawing from these perspectives, a justice-based approach to relocation for refugee camps would include the following principles:

1. refugees and local communities have equitable and fair opportunities to relocate and benefit from the outcome of planned climate relocation, while national and international bodies bear an equitable distribution of the costs through *distributive justice*;
2. refugees and local communities meaningfully participate in the decision-making, implementation, and monitoring/evaluation processes associated with relocation in conjunction with national and international authorities through *procedural justice*; and
3. refugees and local communities are provided with opportunities for improved relationships, building of trust, and social integration/inclusiveness through *restorative justice*.

Distributive justice recognizes that the costs of relocation are not only financial, but also extend to environmental, political, and social domains. For example, the establishment and habitation of a refugee camp contributes to local landscape changes, particularly in the conversion of forests to agricultural land (Maystadt et al., 2020). Conversely, the abandonment of a refugee camp may reduce pressure on a local ecosystem. For this reason, a refugee camp may be seen as problematic or beneficial to local communities in terms of how it redirects resources to refugees or stimulates the local economy (Alix-Garcia et al., 2018). Where refugee camps are seen as undesirable by a host country for social, economic, or political reasons, the host country may push for relocation to more marginal areas, which may not be in the best interest of refugees. Similarly, refugee needs for relocation should be considered within the broader context of local communities that may express similar needs and desire for relocation but may lack the resources and advocacy to do so. Relocation planning can capitalize on positive-sum solutions that result in greater benefits than costs for all parties.

Decisions about whether and where to move refugee camps must take into consideration such costs and benefits in terms of type (e.g., financial and environmental), institutional scale (e.g., local or international), temporal scale (e.g., short-term or long-term), and magnitude (e.g., a matter of convenience or survival). Where benefits or costs are distributed disproportionately among refugees, local communities, and national and international groups, effort must be made to redistribute the burden while still adhering to the international commitment to provide safety and security to refugees. If relocating a refugee camp brings more costs than benefits to the involved parties, the situation may act as an incentive to dissolve a national camp policy and pursue the preferred options of integration into urban centers or resettlement in a new, third country. The spirit of distributive justice is reflected in the Global Compact on Refugees (signed in December 2018; UNHCR, 2018) that states that the broader international community should provide financial and other assistance so that the burden of hosting refugees does not fall unduly on host countries, which often lack the needed resources. The financing of relocation should draw from international resources as well.

Procedural justice requires that all stages of relocation are conducted through participatory processes with all relevant stakeholders. Of particular importance is the refugee community's full and informed consent for relocation and consultation and participation in relocation planning; solutions must not be imposed on refugees, host communities, or host countries. Participatory processes like these will help to prevent unilateral decision-making that prioritizes political objectives of governments that deem refugee camps undesirable and seek to marginalize or isolate them from the general population, and also help to ensure the equitable distribution of costs and benefits (i.e., distributive justice). Stakeholders, including refugees, host communities, and other local communities, must be provided with unbiased and transparent information that is easily accessible and distributed in local languages and appropriate channels. This includes information about immediate and anticipated climate hazards, impacts, costs and benefits of relocation. In turn, the authentication and endorsement of this information by outside parties, such as UNHCR, refugee advocacy groups, and scientific bodies, would build trust in the process and outcomes and assure refugees that their human rights and basic needs will continue to be supported in accordance with international law.

It is essential for participatory processes to include diverse representatives from communities and make efforts to capture important cross sections of vulnerable and classically underrepresented groups (e.g., refugees living with disabilities, youth, and women). Care must be taken to encourage the full participation of marginalized groups and to deliberately *even the playing field* in terms of power differences between them and other stakeholders. Bringing together diverse stakeholders in participatory processes can highlight tensions between priorities and interests, and conflicts can create new vulnerabilities. Hence, trusted mediators should carefully balance the interests and needs that emerge in participatory processes. Participatory processes should be understood as long-term dialogues rather than one-time events.

The redistribution of resources (i.e., distributive justice) through participatory processes (i.e., procedural justice) helps to break down social divisions and build meaningful relationships, leading to restorative justice. Restorative justice has typically been applied in alternative legal settings with perpetrators and victims of minor crimes, but it has also been used in cases involving genocide (e.g., the Gacaca courts in the wake of the Rwandan Genocide) and apartheid (e.g., the Truth and Reconciliation Commission after the end of the South African apartheid). Restorative justice is relevant to planned climate relocation through its values of nondomination, empowerment, respectful listening, equal concern for all stakeholders, accountability, and respect for fundamental human rights (Braithwaite, 2003). Restorative justice encourages opportunities for stakeholders to identify positive-sum solutions that support the empowerment and dignity of refugees and local communities, groups that are often pitted against each other through the inequitable distribution of resources and lack of representation in decision-making processes. Critically, refugee camps most often isolate refugee and local communities from each other, but contact and communication are needed to build relationships. These relationships can lead not only to positive social effects related to a reduction in local episodes of conflict, but they can also lead to superior collaborative strategies to address broader disaster risk and climate change. It is not just within-group (i.e., bonding) but also across-group (i.e., bridging) relationships that contribute to resilience (Islam & Walkerden, 2014). Restorative justice allows us to reimagine refugee camps as living structures that empower and connect the people who reside in them and the communities that host them. This approach may also dissolve the social and political barriers that divide refugee camps and host communities that are often codified into national policies, effectively transforming refugee camps into hybrid camp-integration experiments that forge pathways to truly durable solutions.

4.7 Conclusion

The planet is changing, and so must we. In fulfilling international obligations to protect the human rights of refugees, we must forge new pathways to address the unprecedented challenges associated with climate change. Refugees are vulnerable, but they are also resilient in part owing to the hardships they have surmounted (Uekusa & Matthewman, 2017) and must be considered a vital part of the solutions that affect them, including those related to planned climate relocation. The planned climate relocation of refugee camps may be a particularly difficult challenge, but it also presents an important step towards restoring the security and freedom for refugees that Tuan (1977) identified as being essential for any lived-in place. Refugee camps – as any human settlement – are more than physical infrastructure and represent spaces of relative stability, habitation, and hope for people; their long-term durability matters to achieving security from socially induced violence and climate-related risks. Where short- and long-term durability cannot be achieved in original refugee camp sites, planned climate relocation may be a necessary solution to avert a cascading humanitarian crisis where refugees are forcibly displaced from violence only to be forcibly displaced – or trapped – by the extreme effects of climate change in places of supposed refuge.

An approach to the planned climate relocation of refugee camps that leverages distributive, procedural, and restorative justice is likely to yield solutions that are durable from the perspective of refugees, host communities, and national and international policy communities, alike.

In planned climate relocation – regardless of the type of community to be relocated – the goal should be to "build back better" or "build back safer" and in ways that support community well-being and revival under current and projected climate impacts. The habitability of refugee camps must be judged by how well they protect refugees from the deleterious effects of climate change, violence, and political persecution as well as how they support the dignity and human rights of refugees and host communities. Human beings are not only defined by their vulnerabilities and victimizations but also their capacities and agency. The relocation of refugee camps – in terms of the re-siting and implementation process – can and must holistically engage with these intersectional vulnerabilities and capacities. Disasters – including those influenced by climate change – are socially constructed through vulnerabilities (Hewitt, 1983; Lewis, 1999; Wisner et al., 2004), so it is within our power to socially deconstruct them as well. Refugee camps should be built through risk-informed, conflict-sensitive, and justice-infused sustainable development strategies, and through their design, they can contribute to the creation of an inclusive, synthetic culture that mitigates the risks of climate-related hazards and disasters.

4.8 Recommendations

- Restrictions on voluntary migration of refugees from camps should be recognized as being detrimental to climate change mitigation and adaptation, and refugees should be granted explicit international assistance and protections from the effects of climate change;
- planning for climate relocation of refugees living within camps should be undertaken in refugee-hosting regions that are expected to become unlivable due to climate change;
- the process of climate relocation for refugee camps, from conceptualization to implementation, should take on a social justice orientation – including distributive, procedural, and restorative justice – to safeguard and restore the human rights and dignity of refugees; and
- further studies should be undertaken on the planned climate relocation of refugee camps, including the extent to which this adaptation strategy can mutually benefit refugees and host communities.

References

ACAPS. (2018). *Rohingya crisis: Host communities review*. https://www.humanitarianresponse.info/sites/www.humanitarianresponse.info/files/assessments/180131_host_communities.pdf

Al-Arian, L., & Sherlock, R. (2019, January 9). Heavy winter storm wrecks Syrian refugee camps in Lebanon. *National Public Radio*. https://www.npr.org/2019/01/09/683528148/heavy-winter-storm-wrecks-syrian-refugee-camps-in-lebanon

Al Jazeera. (2014, September 5). *Ethiopia refugee camp submerged by floods*. https://www.aljazeera.com/news/africa/2014/09/ethiopia-refugee-camp-submerged-floods-201495152914985589.html

Al Jazeera. (2020, December 29). *Bangladesh moves nearly 2,000 Rohingya refugees to remote island*. https://www.aljazeera.com/news/2020/12/29/new-group-of-rohingya-refugees-moved-to-bangladesh-remote-island

Alix-Garcia, J., Walker, S., Bartlett, A., Onder, H., & Sanghi, A. (2018). Do refugee camps help or hurt hosts? The case of Kakuma, Kenya. *Journal of Development Economics*, 130, 66–83.

Associated Press. (2018, September 17). *Refugee camp landslide in Thailand leaves 1 dead, 7 missing*. https://apnews.com/fa87bef560f14441ba686e08a196c28f

Associated Press. (2019, September 27). *Bangladeshi leader at UN: Rohingya refugee crisis worsening*. https://apnews.com/article/52d55c04eafa4ebdb3e3b30d62346607

Atuyambe, L. M., Ediau, M., Orach, C. G., Musenero, M., & Bazeyo, W. (2011). Land slide disaster in eastern Uganda: Rapid assessment of water, sanitation and hygiene situation in Bulucheke camp, Bududa district. *Environmental Health*, 10(38), 1–13. 10.1186/1476-069X-10-38

Banerjee, S. (2020). From Cox's Bazar to Bhasan Char: An assessment of Bangladesh's relocation plan for Rohingya refugees. *Observer Research Foundation*. https://www.orfonline.org/research/from-coxs-bazar-to-bhasan-char-an-assessment-of-bangladeshs-relocation-plan-for-rohingya-refugees-65784/

Beech, H. (2020, December 4). From crowded camps to a remote island: Rohingya refugees move again. *The New York Times*. https://www.nytimes.com/2020/12/04/world/asia/rohingya-bangladesh-island-camps.html

Braithwaite, J. (2003). Principles of restorative justice. In A. von Hirsch, J. Roberts, A. E. Bottoms, K. Roach, & M. Schiff (Eds.), *Restorative justice and criminal justice: Competing or reconcilable paradigms?* (pp. 1–20). Hart Publishing.

Bremner, L. (2020). Sedimentary logics and the Rohingya refugee camps in Bangladesh. *Political Geography*, 77, 102109. 10.1016/j.polgeo.2019.102109

Buncombe, A., & Waraich, O. (2009, May 16). Refugees' plight worsens in searing heat. *Independent*. https://www.independent.co.uk/news/world/asia/refugees-plight-worsens-in-searing-heat-1685724.html

Connolly-Boutin, L., & Smit, B. (2016). Climate change, food security, and livelihoods in sub-Saharan Africa. *Regional Environmental Change*, 16(2), 385–399.

Conway, D., Nicholls, R. J., Brown, S., Tebboth, M. G., Adger, W. N., Ahmad, B.,… & Said, M. (2019). The need for bottom-up assessments of climate risks and adaptation in climate-sensitive regions. *Nature Climate Change*, 9(7), 503–511.

Dai, A., Zhao, T., & Chen, J. (2018). Climate change and drought: A precipitation and evaporation perspective. *Current Climate Change Reports*, 4(3), 301–312.

Devictor, X. (2019, December 9). *2019 update: How long do refugees stay in exile? To find out, beware of averages*. World Bank Blogs. https://blogs.worldbank.org/dev4peace/2019-update-how-long-do-refugees-stay-exile-find-out-beware-averages

Eglash, A. (1977). Beyond Restitution: Creative Restitution. In J. Hudson & B. Galaway (Eds.), *Restitution in criminal justice*. Lexington Books.

Füssel, H. M. (2010). *Review and quantitative analysis of indices of climate change exposure, adaptive capacity, sensitivity, and impacts*. World Bank.

George, S. (2015, July 31). Iraq hit by heatwaves, making life for refugees even tougher. *The United Nations High Commissioner for Refugees*. https://www.unhcr.org/en-us/news/latest/2015/7/55bb414f6/iraq-hit-heatwaves-making-life-refugees-tougher.html

Green Climate Fund. (2019). *Climate resilient development in refugee camps and host communities in Kigoma region, Tanzania*. Concept note. https://www.greenclimate.fund/sites/default/files/document/21880-climate-resilient-development-refugee-camps-and-host-communities-kigoma-region-tanzania.pdf

Hewitt, K. (Ed.). (1983). *Interpretations of calamity from the viewpoint of human ecology*. Allen & Unwin.

International Rescue Committee. (2019). *Left in limbo: The case for economic empowerment of refugees and host communities in Cox's Bazar, Bangladesh*. https://reliefweb.int/sites/reliefweb.int/files/resources/finalircbangladeshlivelihoodspolicybrief.pdf

Islam, N., & Winkel, J. (2017). *Climate change and social inequality* (UN Department of Economic and Social Affairs Working Paper 152). New York: United Nations. 10.18356/2c62335d-en

Islam, R., & Walkerden, G. (2014). How bonding and bridging networks contribute to disaster resilience and recovery on the Bangladeshi coast. *International Journal of Disaster Risk Reduction*, 10(Part A), 281–291. 10.1016/j.ijdrr.2014.09.016

Khadka, N. S. (2019, December 10). Refugees at 'increased risk' from extreme weather. BBC. https://www.bbc.com/news/science-environment-50692857

Krehm, E., & Shahan, A. (2019). Access to justice for Rohingya and host community in Cox's Bazar. *International Rescue Committee*. https://www.rescue.org/sites/default/files/document/3929/accessingjusticeassessmentexternalfinalsmall.pdf

Lewis, J. (1999). *Development in disaster-prone places: Studies of vulnerability*. Intermediate Technology Publications.

Medecins Sans Frontieres. (2015, August 20). *Heat wave adds to the woes of Syrian refugees in Bekaa Valley*. https://www.msf.org/lebanon-heat-wave-adds-woes-syrian-refugees-bekaa-valley

Maystadt, J. F., Mueller, V., Van Den Hoek, J., & van Weezel, S. (2020). Vegetation changes attributable to refugees in Africa coincide with agricultural deforestation. *Environmental Research Letters*, 15(4), 044008. 10.1088/1748-9326/ab6d7c.

Medecins Sans Frontieres. (2014, September 4). *Refugees attempt to survive in flooded camps*. https://www.msf.org/ethiopia-refugees-attempt-survive-flooded-camps

Nadeau, C. P., Urban, M. C., & Bridle, J. R. (2017). Climates past, present, and yet-to-come shape climate change vulnerabilities. *Trends in Ecology & Evolution*, 32(10), 786–800.

National Oceanic and Atmospheric Administration. (2020). *State of the climate: Global climate report for annual 2019*. https://www.ncdc.noaa.gov/sotc/global/201913

Paul, R. (2020, February 26). Bangladesh rethinks plan to move Rohingya refugees to island: Minister. *Reuters*. https://www.reuters.com/article/us-myanmar-rohingya-bangladesh/bangladesh-rethinks-plan-to-move-rohingya-refugees-to-island-minister-idUSKCN20K29G

Peters, K., Eltinay, N., & Holloway, K. (2019). Disaster risk reduction, urban informality and a 'fragile peace': The case of Lebanon. *The Overseas Development Institute*. https://www.odi.org/sites/odi.org.uk/files/resource-documents/12911.pdf

Rawls, J. (1971). *A theory of justice*. Harvard University Press.

Save the Children. (2018, April 20). *Refugee camp submerged as flash floods hit northern Kenya*. https://reliefweb.int/report/kenya/refugee-camp-submerged-flash-floods-hit-northern-kenya

Schmidhuber, J., & Tubiello, F. N. (2007). Global food security under climate change. *Proceedings of the National Academy of Sciences*, 104(50), 19703–19708.

Sengupta, S., & Fountain, H. (2018, March 14). The biggest refugee camp braces for rain: 'This is going to be a catastrophe'. *The New York Times*. https://www.nytimes.com/2018/03/14/climate/bangladesh-rohingya-refugee-camp.html

Tabari, H. (2020). Climate change impact on flood and extreme precipitation increases with water availability. *Scientific Reports, 10*(1), 1–10.

Taylor, A. (2015, August 3). In the Middle East's largest refugee camp, it's so hot you can fry an egg. *The Washington Post.* https://www.washingtonpost.com/news/worldviews/wp/2015/08/03/in-the-middle-easts-largest-refugee-camp-its-so-hot-you-can-fry-an-egg/?noredirect=on

Tharoor, I. (2019, September 29). The Rohingya crisis can't stay Bangladesh's burden, prime minister says. *The Washington Post.* https://www.washingtonpost.com/world/2019/09/30/rohingya-crisis-cant-stay-bangladeshs-problem-prime-minister-says/

Thibaut, J., & Walker, L. (1975). *Procedural justice: A psychological analysis.* Lawrence Erlbaum Associates.

Tuan, Y. F. (1977). *Space and place: The perspective of experience.* University of Minnesota Press.

Uekusa, S., & Matthewman, S. (2017). Vulnerable and resilient? Immigrants and refugees in the 2010–2011 Canterbury and Tohoku disasters. *International Journal of Disaster Risk Reduction, 22,* 355–361. 10.1016/j.ijdrr.2017.02.006

United Nations (UN). (2006). *Social justice in an open world: The role of the United Nations.* United Nations Publications.

United Nations High Commissioner for Refugees (UNHCR). (2018). Global compact on refugees. *UNHCR.* https://www.unhcr.org/5c658aed4

United Nations High Commissioner for Refugees (UNHCR). (2020). Global trends: Forced displacement in 2019. *UNHCR.* https://www.unhcr.org/en-us/statistics/unhcrstats/5ee200e37/unhcr-global-trends-2019.html

United Nations High Commissioner for Refugees. (n.d.). *Alternatives to camps.* https://www.unhcr.org/en-us/alternatives-to-camps.html

U.S. Committee for Refugees and Immigrants (USCRI). (2019). Lives in storage: Refugee warehousing and the overlooked humanitarian crisis. https://reliefweb.int/sites/reliefweb.int/files/resources/USCRI-Warehousing-Dec2019-v4.pdf

Wisner, B., Blaikie, P., Cannon, T., & Davis, I. (2004). *At risk: Natural hazards, people's vulnerability and disasters* (2nd ed.). Routledge.

Xu, C., Kohler, T. A., Lenton, T. M., Svenning, J. C., & Scheffer, M. (2020). Future of the human climate niche. *Proceedings of the National Academy of Sciences, 117*(21), 11350–11355.

Zehr, H. (2002). *The little book of restorative justice.* Good Books.

Interlude 1 *Origins of Limestone*

Martha Lerski

Essentially, marble is long-compressed limestone, which is stone formed of marine life. Seeing evidence of ocean life in the sediments of places now far from seashores extends an observer's view into prehistoric geographies and time periods – forcing a truly long-term perspective.

Human life did not always exist on earth, nor is its foothold guaranteed other than as layers in paleontological strata (Kolbert, 2014). Decisions taken by the dominant species of the current epoch, the Anthropocene, have led to the demise of many concurrent life forms. While the start of the current epoch is a subject of debate (Oxford University Press, 2020), the accelerating and intensifying environmental damage during this Anthropocene era is widely documented. Impacts of deforestation; commercial whaling and fishing; and industrial and individual outputs of carbon, myriad chemicals, and plastic debris pollution threaten irreparable harm.

This sculpture's form is reminiscent of a wave, conveying history's movement in geologic scale. Though an abstraction rather than a depiction, this stone carving's movement and shape were inspired by the ocean. I was born and raised in San Francisco, where the sea is central. The thunderous saline surf of Northern California's Pacific Ocean continues to ring through my ears, course through my veins and inform my field studies. I examine vulnerabilities and resilience mechanisms of cultural heritage in response to flooding and drought. Pondering the specific processes whereby shells and marine life morph into stone, one might ask this question: Is mankind capable of acknowledging and effectively responding to its own vulnerability?

DOI: 10.4324/9781003141457-101

Origins of Limestone 63

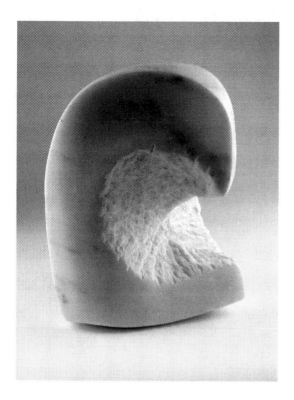

Figure Int 1.1 Origins of limestone, marble (view (a) of the sculpture by Lerski); photos by Javier Agostinelli.

Figure Int 1.2 Origins of limestone, marble (view (b) of the sculpture by Lerski); photos by Javier Agostinelli.

References

Kolbert, E. (2014). *The sixth extinction: An unnatural history*. Henry Holt and Company.
Oxford University Press (September 2020). Anthropocene, n. and adj. In *OED Online*. Retrieved October 5, 2020 from www.oed.com/view/Entry/398463

Part II
Shifting lands, resistance, and acceptance

5 Breaking the borderscape: Migration, resettlement, and citizenship on the Anthropocene Brahmaputra

Kevin Inks

5.1 Introduction

The Brahmaputra River, which runs through the Northeast Indian state of Assam, is a site of immense geophysical and cultural dynamism. The communities built along its banks endure intense seasonal flooding and periodic displacement due to erosion. Government disaster management offers relief for the displaced, but only provides formal resettlement assistance if the ownership of lost land can be proven using outdated government surveys. In the context of a river system which consumes and produces land every year, much of which is never surveyed, the result is innumerable displaced people resorting to a semi-migratory lifestyle between lands which are not recorded in formal surveys.

Although the Assam state government has recognized this crisis, it has simultaneously experienced social upheaval in response to continuing climate-induced migration from Bangladesh. Undocumented Bengali migrants, like people displaced within Assam, are often forced to settle in vulnerable, unsurveyed lands near the river's banks. After the creation of a pathway to citizenship for non-Muslim migrants in 2019, protesters in Assam expressed fear that Bengali migrants would occupy Assamese land and dilute Assam's heritage. These two migration regimes – periodic displacement of communities along the Brahmaputra and climate-induced migration from Bangladesh – are distinct, but similarly drive migrants into positions of precarity on unsurveyed lands vulnerable to flooding, erosion, and further displacement. To confront this crisis, government land and disaster managers must move beyond a cartographic paradigm in which the Brahmaputra's shoreline is an unchanging demarcation between land and water. Similarly, cadastral surveys which enforce boundaries between private and unowned land, as well as the national borders which separate Assam from Bangladesh, are unsuitable for the dynamics of a changeable, braided river. As climate-induced migration, internal displacement and subsequent tensions increase, disaster managers should leverage local resilience and migration strategies to break the conceptual borderscapes of the Brahmaputra and embrace more fluid understandings of the landscape.

DOI: 10.4324/9781003141457-5

5.2 Displacement and boundaries

On 9 August 2018, Assam Chief Minister Sarbanada Sonowal visited the remote island district of Majuli to christen a fleet of new ferries. Majuli, an island in the Brahmaputra, is widely understood as a site of immense cultural and religious heritage. It is famous for its *satras*[1] and its Indigenous forms of dance, weaving, and craft. It is also, unfortunately, famous for its rapidly eroding coastline and the near-annual displacement of its riparian communities. The Chief Minister's speech, which focused on the importance of the Brahmaputra as an inland waterway and the economic development promised by a new ferry system, included a portion addressing the concerns of displaced people on the island. The government, he said, would be offering *pattas*[2] to a handful of households informally settled on state-owned land.

The remark was well received. No discussion of the future of Majuli, regardless of ferry service, can ignore the dozens of small tarpaulin shelters which line the road from Kamalabari to the ferry docks, nor the fact that the displaced residents of those shelters were forced to build them because they lacked patta of their own. Chief Minister Sonowal's speech seemed to suggest that the new ferry system represented symbolic domination of a river which had caused unimaginable human misery among the people of the island. Undergirding the distribution of patta, however, was an implicit admission: That displacement and migration along the banks of the Brahmaputra, while prompted by the motion of land and water, are also driven by the cartographic technologies employed by the state to tame that motion.

The crisis of environmental displacement on Majuli is often framed as a "natural" disaster, or an inevitable consequence of life on what one public official described to me as "the most violent river in the world" (Inks, 2020). The Brahmaputra, which rises in Tibet and gathers up its tributaries in Northeast India before joining with the Ganga in Bangladesh and emptying into the Bay of Bengal, has a strong claim to that dubious distinction. The Brahmaputra River Valley in Assam is one of the most seismically active regions on the planet, and a combination of high seasonal rainfall, Himalayan meltwater, and large loads of exceedingly fine sediment have made the river prone to drastic channel avulsion and frequent flooding (Sarma, 2005).

The Brahmaputra's geophysical dynamism is expressed in the frequent creation of *saporis*[3] through sediment deposition as well as massive erosion which consumes homes, grazing pastures, and incalculable *bighas*[4] of agricultural land. Erosion is especially prominent on Majuli, which sits at the confluence of the Brahmaputra and its tributaries in the Kherkutia Xuti and Subansiri Rivers. Dutta et al. (2010) estimate that the island has shrunk by an average of 1.87 square kilometers annually between 1975 and 2008. Although Majuli is shrinking due to erosion, this process is not linear. The Brahmaputra deposits sediment even as it whisks it away, creating new banks and river islands. The size of the island increases during some years and decreases during others. Disaster management and resettlement are, as a consequence, primary points of interaction between the people of Majuli and the state.

Breaking the borderscape 69

Disaster management policy on the Brahmaputra, coordinated by the Assam State Disaster Management Authority (ASDMA), can be divided into two broad categories: *Immediate relief* and *restitution*. Short-term, immediate relief is coordinated from the ground up, with community representatives communicating their particular needs to district and state representatives who then commission and deliver food aid. This relief, I was told by a high ranking anonymous official in the ASDMA, "has been administered since time immemorial" in one form or another (Interview by Author, 17 August 2018, Guwahati). The system's decentralized approach allows it to meet the most pressing needs of particular communities at a granular level while coordinating the logistics of large-scale aid through a series of federalized regions and a "network of village heads." Although most communities receive only a few days' worth of food and supplies, the system's swiftness and flexibility is generally seen positively by residents impacted by flooding and inundation. "We face danger each and every day of our lives," I was told by a farmer living near the river in Sumoimari Ghat, "but we like the local relief given by number living in [the] household" (Interview by Author, 9 August 2018, Sumoimari Ghat).

The same cannot be said for disaster restitution. While short-term, immediate relief is often dexterous enough to provide food and shelter to people temporarily displaced by flooding, it offers few long-term solutions for those who have lost land, family members, animals, or homes to erosion. Nominally, legally recognized owners of land are entitled to restitution and resettlement assistance after displacement. In practice, many displaced people seeking restitution and resettlement assistance are unable to prove ownership of their land and therefore receive nothing from state disaster managers. Disaster restitution and resettlement assistance is, for these displaced people, a bureaucratic fiction. Without patta, receiving restitution for lost land and homes is largely impossible. "I have applied for proper land to live," said another previously displaced resident of Sumoimari Ghat, "and have gotten nothing" (Interview by Author, 10 August 2018).

5.3 Disappearing islands

Although Chief Minister Sonowal handed out several pattas during his visit to Majuli, the standard process is more circuitous. Pattas signify ownership of a parcel of land, but these parcels must be cadastrally surveyed and titled by the state before patta can be distributed. As the Brahmaputra undulates over time, eroding away existing land and creating new land through sediment deposition, the cadastrally surveyed and titled lands recognized by state disaster managers are consumed and replaced by unsurveyed land. This newly created land remains untitled, unrepresented in government records, and *de facto* unowned until a state cadastral survey is performed in order to map, parcelize, and distribute it.

These ephemeral lands, on which all settlement is necessarily informal, are referred to as *unmapped lands* (Inks, 2020). Their residents, whether in Assam,

West Bengal, or Bangladesh, generally suffer from unequal access to services and infrastructure in addition to being largely outside of the protection of disaster management (Lahiri-Dutt & Samanta, 2013). As the geophysical dynamism of the river continues to consume surveyed land and produce unmapped land, the proportion of unmapped land along the river increases along with the proportion of residents unable to access restitution or resettlement assistance. Informally settled communities are, as a result, compelled into a perpetual state of semi-migration in the absence of a new cadastral survey. "We were shifted a few years ago to government land," I was told by a farmer in Bhakat Sapori, "so we couldn't show proof, and got nothing" (Interview by Author, Bhakat Sapori, 11 August 2018). Rather than "natural" disasters, the differential impacts of flooding and erosion are mediated by a cartographic imagination which simply cannot see migrants and displaced people on its maps. In the Anthropocene, climate change-induced increases to erosion is likely to worsen the displacement crisis (Das, 2015).

The most recent survey of Majuli prior to 2019 was carried out in 1969. During the subsequent half-century of sediment deposition, an annually increasing number of communities settled on vulnerable unmapped lands were left unserved by disaster managers. In the interior of Majuli, where displacement is a less pressing concern, the land parcels created by the 1969 survey are still in existence. Landowners holding private title in the interior are able to receive disaster restitution and resettlement assistance, while communities informally settled near the river are unable to. The households most in need of resettlement assistance are, as a consequence, the least likely to receive it. The disaster management system, based as it is on rigidly cadastral understandings of a shifting landscape, reproduces the conditions which it seeks to alleviate.

Discourse surrounding erosion on Majuli often frames it as a vanishing landscape, both physically and culturally. Many communities feel that the failure of disaster managers to protect Majuli itself – and not just its residents – from erosion represents a profound betrayal of an archetype of Assamese culture. "There is no other place better than Majuli in this world," one farmer told me as we sat on his front porch in Mazgaon. "I want to save Majuli. Majuli is rich in culture, satra culture" (Interview by Author, 31 July 2018, Mazgaon). The island is understood as a discrete body, and the gradual loss of its towns, shorelines, and roadways has led to a narrative of tragedy and almost-carnal violence surrounding the island. None of these, however, are cited as often by concerned residents as the disappearance of Majuli's satras.

In August of 2018, a few days after Chief Minister Sonowal's speech, I visited the Auniatri Satra near Kamalabari to interview perhaps the most influential living figure in the Neo-Vaishanavite movement, Dr. Pitambar Dev Goswami. A prolific author and political power in his own right, Dr. Goswami is a critic of government disaster management and a vocal leader in the movement opposing the migration of Bengali people into Assam. During our conversation, he described Majuli as the "nerve center of satra culture and Assamese culture." While criticizing the lack of government action to prevent erosion, he told me that 32 or 33 satras had already been lost to erosion. "If the Satras are lost," he said, "then Majuli is finished and Assam is empty" (Interview by Author, Auniati Satra, 11 August 2018). The perception that the shifting physical and cultural landscape represents an existential threat to Assam itself bleeds into Dr. Goswami's opposition to migration from Bangladesh.

5.4 Climate migration and borderscapes

Migration into Assam from Bangladesh has been a contentious issue in the state for decades, and antimigrant sentiment continues to periodically erupt into violence. Millions of residents of Bangladesh, widely understood to be among the most vulnerable countries on the planet to the impacts of climate change, are likely to make the decision to migrate in the coming decades (Hassani-Mahmooei & Parris, 2012). Recent scholarship has identified migration as an important strategy for adaptation to environmental, economic, and climate change among at-risk people in rural Bangladesh, and Assam remains a prominent destination for Bangladeshis migrating for climate-related reasons (Amit, 2016; Martin et al., 2014; Stojanov et al., 2016).

Anti-Bengali sentiment reached a crescendo in December of 2019 with the passage of the Citizenship (Amendment) Act (CAA) in the Parliament of India. The CAA offered a path to citizenship for Hindus, Jains, Buddhists, Sikhs, Christians, and Parsis who had migrated from Afghanistan, Bangladesh, or Pakistan before December of 2014. While opposition to the CAA in mainland India and from the international community saw it as an attack on the rights of Muslims living in India, opposition in Assam was largely based on its perceived leniency toward Bangla-speaking Hindu migrants. Dr. Goswami was vocal during the protests, arguing that the CAA would "pollute the Assamese community and language" (Press Trust of India, 2019). For Dr. Goswami, anxieties around the diminution of Majuli's physical boundaries mirror anxieties surrounding the perceived diminution of its culture. The conceptual boundaries weakened by the CAA, like the cadastral boundaries of the island produced by surveying, are perceived to be slowly fraying. In the words of one resident: "If it continues there may be no Majuli. The signature of Majuli will be erased from the map" (Interview by Author, Bekuli Mari, 8 August 2018).

The formalization of the riverine landscape within the cartographic imagination of the state mirrors the formalization of the national borders which bisect the river and which define the citizenship – or lack of it – of people living on the Brahmaputra. Borrowing Mezzadra and Neilson's (2013) understanding of borders as cartographic entities which not only mark national territories but also permeate subnational space, we may begin to articulate the extensive hidden boundaries, produced by state surveying policy, which criss-cross the Brahmaputra and which dictate much of the ongoing crisis of displacement and resettlement. On the whole, we may describe the state's cadastral rendering of the Brahmaputra as what Perera (2007) would describe as a *borderscape*: That is, a landscape in which cartographies are utilized to demarcate spaces of ownership, belonging, and exclusion.

If the imaginary Brahmaputra ossified in land records and on the pages of atlases has been consistent in shaping the risks faced by residents of unmapped lands, the forms of settlement practiced on these lands varies widely. While many unmapped lands are occupied by previously displaced Assamese communities, they are also commonly settled by Mising people who understand themselves as semimigratory. Bangla-speaking people, too, are often compelled to settle on unmapped lands. Many of these communities are composed of undocumented migrants who have made the choice to migrate both to avoid climate catastrophe and erosion in Bangladesh and to seek opportunities for an improved life in Assam. Unmapped lands, given that they are without formal title and more likely to be available for settlement than surveyed lands, are often the best or only resettlement opportunity for undocumented Bangladeshi migrants.

Undocumented migrants and Bangla-speaking immigrants, like displaced people on Majuli, struggle to find land which is suitable for long-term prosperity. Many at-risk residents believe that there is no safe place for resettlement left on Majuli. "We will live wherever we are given land," I was told by a farmer in Hunari Bari. "I don't think there are any safe places in Majuli" (Interview by Author, Hunari Bari, 7 August 2018). The purchase of land in the interior of the island, which remains fully surveyed and titled, is beyond the means of nearly all displaced people and especially undocumented migrants.

5.5 Surveys and resettlement

Unmapped land along Majuli's shorelines is dangerous, and informally settled people on these lands generally doubt that their families will be able to remain in their homes for long. "It is not possible for [my] grandchildren to live here," said one farmer in Bhakat Sapori when I asked him about long-term prospects of life on the island. "If the water breaks the embankment, this place will be destroyed" (Interview by Author, Bhakat Sapori, 5 August 2018). Many young people have migrated away from Assam both for work and to avoid displacement. "Most of the young boys leave for work in Hyderabad, Goa, Bangalore," I was told by a man in Salmora. "[It is] not possible to be living here" (Interview by Author, 5 August 2018, Salmora). Some households have embraced a strategy of managed retreat, saving money to purchase land far from the river in another district. Movement away from the river is not an option for most at-risk residents, however. One resident of Khorahala by the river estimated that they would need to save for "10 or 12 years" to buy land which is safe for settlement (Interview by Author, 7 July 2018).

Discussions of optimal resettlement locations are, for most people at risk of displacement on Majuli, an exercise in theory. Those that lose agricultural land often become laborers. This progressive change in the material conditions of residents has deepened inequities on the island, pushing displaced people to settle on precarious unmapped lands at the greatest risk of further erosion. The result is continuous stratification of social relations between title holders and informally settled people. Some displaced residents seek out newly produced unmapped land which remains unsettled, and various NGOs provide financing and materials for informally settled people to build new homes. Although displaced people exercise agency in adapting to and preparing for displacement due to erosion, the process of resettlement itself is mediated by both visible and implied state power. The roadsides of Majuli, generally built on high ground to avoid flooding, are lined with tarpaulin shelters built by displaced people. "Obviously," I was told by a woman in Pohardia when I asked what her family's resettlement strategy was, "we must go to the roadside because we don't believe the government will help" (Interview by Author, Phardia, 6 August 2018). Her family's home had been destroyed once before – by a rampaging elephant rather than the river – and they had been forced to stay in the roadside. They anticipate needing to do so again in the near future.

Although they do not exist in the cartographic imagination of the state, unmapped lands are not outside of state power. Informal settlement has served as a pretext for state violence against people assumed to be undocumented migrants, including more than 600 Bengali Muslim families who were forcibly evicted from unmapped land in Karbi Anglong in 2019 (Azad, 2019). Some displaced residents of Majuli report being forced away from roadsides and onto vulnerable unmapped lands by police force. "We got informed we were getting a home," I was told by a farmer in Bekuli Mari, "but we have heard nothing… we have been shifted by police because of dangerous soil erosion" (Interview by Author, Bekuli Mari, 8 August 2018). The machinery of resettlement on Majuli, as elsewhere along the Brahmaputra, is built on cadastral surveys and lines on a conceptual map. Undocumented migrants and people displaced from unmapped land are forced to navigate this machinery while remaining unseen and unheard by the bureaucracy which built it.

Dr. Pitambar Dev Goswami, in his battle against both displacement in Majuli and the free movement of Bengali people in Assam, obscures the fact that these processes are mutually reinforced by the taxonomization and borderization of the Brahmaputra. These multiplicative boundaries, produced by mapping technologies and made real by governmental policy, have become the axis on which "natural" disaster swings. In a region increasingly marked by environmental displacement and climate migration from Bangladesh, the creation of cartographic demarcations – between cadastral land and unmapped land, between citizen and foreigner, and between the settled and the migratory – produces and reinforces the differential consequences of life in a shifting landscape. The Brahmaputra borderscape represents the profound inversion of a classic axiom: The map manufactures the territory.

Public officials in Majuli are aware of the link between surveying and disaster management. In June of 2019, after years of lobbying the state government for a new survey of the island, surveyor teams arrived on the island ("The Assam Sentinel," 2019). Beginning in Ahatguri Mouza,[5] the surveying teams began working to re-calibrate the cartographic imagination of the state. The new survey promised a reset of the island's disaster management apparatus, and suggested that Chief Minister Sonowal would soon be able to deliver far more than the handful of pattas to landless people he had in August of 2018. The disaster management system based on cadastral surveying, which has been largely unable to prevent the most severe consequences of flooding and erosion in riverine communities, remained intact.

The 2019 survey is indisputably a positive development. Even an equitable and thorough land survey, however, would only temporarily reset the creeping disjunction between the map and the territory of Majuli. Cadastral land is consumed and unmapped land produced during every rainy season. Barring new surveys every year, the cycle of resettlement on unmapped land and subsequent displacement will continue. Yearly surveys, although technically within the power of the state, are likely impossible given state budgets and a difficult riparian terrain. One may also be wary of solutions which attempt to resolve the contradictions within technocratic disaster management by applying increasingly technocratic solutions. That periodic re-surveying is necessary itself suggests that cadastral understandings of space are not suited to the Brahmaputra's shifting landscape, and that more radical and imaginative transformation is required.

5.6 Breaking the borderscape

Surveying alone will not serve to recognize the rights of unlanded migrants fleeing rising sea levels in the Bay of Bengal during the Anthropocene, or will it undo the processes driving these migrants onto vulnerable unmapped lands. Increased meltwater from Himalayan glaciers is likely to increase climate migration and displacement along the length of the Brahmaputra. Migration into Assam from Bangladesh, as well as the movement of internally displaced peoples along the Brahmaputra itself, demands a radical reimagining of disaster management in shifting landscapes. Just as imperial cartographers imagined a static Brahmaputra doomed to gradually consume itself, so too can disaster managers and cartographers in the Anthropocene imagine a Brahmaputra of land-as-motion.

This reimagining demands that disaster management be untethered from the parcelization of land imposed by cadastral surveying and titling. A landscape taxonomized into a patchwork of the legally recognized and the unmapped can only serve to taxonomize the residents of these lands along similar lines. Bangla-speaking Indians, undocumented migrants, and displaced people are sorted onto unmapped land by the action and inaction of the state, and are therefore forced to migrate perpetually from danger to danger. Rather than try to bring informally settled people into the fold of cadastral land tenure through surveying, disaster managers along the Brahmaputra could embrace Indigenous understandings of land-as-motion. Doing so would necessitate breaking the conceptual linkages between private titling and disaster relief which reproduce the borderscape after cadastral surveying, offering restitution and resettlement assistance in much the same way that short-term relief is offered: Through a bottom–up and community-led recognition of need. Aid could be offered to displaced people regardless of ability to produce private land title, and resettlement assistance could be coordinated with a full and honest accounting of the riparian landscape rather than simply those lands which appear in the latest survey.

Rather than a map of a calcified river frozen in time, the cartographic imagination of the state could see the Brahmaputra as a constantly moving assemblage of lands and waters. This reimaging would create new possibilities for the resettlement and restitution of all displaced people rather than just those holding patta. Instead of forcing displaced people from roadsides and onto vulnerable unmapped lands, scientific, and Indigenous understandings of the river could be entwined to guide resettlement toward lands at the least risk and which best suit the community's needs and ambitions.

A cartographic imagination of the Brahmaputra in which unmapped lands and their residents are recognized as a part of the landscape would also do much to protect Bangla-speaking peoples, both migrants and Indian-born, from the threat of displacement and violence. Muslim Bengalis who are not citizens or are unable to prove their citizenship, such as those displaced in Karbi Anglong in 2019, are at particular risk of displacement from unmapped land by state force. Non-Muslim Bengali migrants, offered a path to citizenship under the CAA which was denied to them by the Assam Accord, may also be targeted for violence as they were during the 1983 Khoirabari Massacre in Bodoland.[6] More broadly, however, the production of unmapped land along the Brahmaputra borderscape creates spaces in which climate migrants and environmentally displaced people are made vulnerable by the taxonomy of the lands which they are compelled to settle. Undoing the contour lines which govern the differential violence experienced by these groups is vital to a just and equitable disaster management system on the Anthropocene Brahmaputra.

On 28 January 2020, Chief Minister Sarbananda Sonowal returned to Majuli. He was there to distribute pattas and land allotment letters to 4,513 landless households, a direct result of surveying efforts in 2019 and a signal of commitment to informally settled people unable to access disaster restitution. As his convoy passed through the town of Garmur, however, it was met with a series of black flags raised in protest against the CAA ("The Assam Sentinel," 2020). Dr. Pitambar Dev Goswami, concerned that the Act would welcome Bangla-speaking migrants damaging to the cultural and religious integrity of Assam, had organized the demonstration. The Chief Minister's subsequent speech, coming as it did on the tail of violent protests throughout the state, focused less on concern for Majuli's disappearing riverbanks than it did on the perception of its disappearing culture. He promised to protect the Indigenous art, language, and religion from the effects of migration into Assam, and emphasized the island's history as an oasis of relative peace and ethnic cooperation in a state historically marred by communal violence.

Unsaid in the Chief Minister's speech was the inevitability of the Brahmaputra's continued motion and the state's contribution to the perpetual production of a borderscape of unmapped lands. Unsaid, too, was the influence of that production on the well-being and safety of climate migrants, whether Bengali, Assamese, or Adivasi, during the Anthropocene. As the Chief Minister spoke and Dr. Goswami's black flags waved in Garmur and Kamalabari, state cartographies continued to deepen their conceptual grid lines across the undulating island. Boundaries and borders continued to be drawn with the motion of the river, demarcating the land and the water, the real from the imagined, and spaces of inclusion from those subject to the violence of being left off of the map.

5.7 Recommendations

- Create restitution opportunities for residents of informally settled land, regardless of legal ownership
- Embrace alternative methods for establishing residency besides cadastral surveying and titling, including remote sensing of riverbank erosion and community-based recordkeeping
- Embrace and encourage Indigenous resiliency and migration strategies, allowing easier movement of people with the river

Notes

1 Monastic centers of the Neo-Vaishnavite religious movement founded by the Saint Sankaradeva.
2 A private land title granted to a landowner after cadastral surveying.
3 Sandy river islands which are often ephemeral, also called *chars* in Lower Assam and Bangladesh.
4 1,340 square meters in Upper Assam, although the size of the unit of measure changes by location.
5 The Southwestern extremity of Majuli, and the part of the island most affected by erosion.
6 A pogram in which as many as 500 Hindu Bengalis were killed for their perceived migrant status.

References

Amit, R. (2016). Migration from Bangladesh: Impulses, risks, and exploitations. *The Round Table*, 105(3), 311–319.
Azad, A. K. (2019). In India's Assam, Muslim families evicted weeks before elections. *Al Jazeera*, 25 March.
Das, D. (2015). Changing climate and its impacts on Assam, Northeast India. *Bandung: Journal of the Global South*, 2(1), 1–13.
Dutta, M. K., Barman, S., & Aggarwal, S. P. (2010). A study of erosion-deposition processes around Majuli island, Assam. *Earth Science India*, 4(3), 206–216.
Hassani-Mahmooei, B., & Parris, B. (2012). Climate change and internal migration patterns in Bangladesh: An agent based model. *Environment and Development Economics*, 17(6), 763–780.

Inks, K. (2020). Reimagining a violent landscape: Disaster, displacement, and cartographic imagination in the Brahmaputra River Valley (unpublished manuscript submitted to *Annals of the Association of American Geographers*).

Lahiri-Dutt, K., & Samanta, G. (2013). *Dancing with the river: People and life on the Chars of South Asia*. Yale University Press.

Martin, M., Billah, M., Siddiqui, T., Abrar, C., Black, R., & Kniveton, D. (2014). Climate-related migration in rural Bangladesh: A behavioural model. *Population and Environment*, 86(1), 85–110.

Mezzadra, S., & Neilson, B. (2013). *Border as method, or, the multiplication of labor*. Duke University Press.

Perera, S. (2007). A Pacific zone? (In)security, sovereignty, and stories of the Pacific borderscape. In P. K. Rajaram & C. Grundy-Warr (Ed.), *Borderscapes: Hidden geographies and politics at territory's edge*. University of Minnesota Press.

Press Trust of India. (2019). Citizenship Act protests: Assam's Satras urge people to hoist black flags outside houses. *The New Indian Express*, 13 December.

Sarma, J. N. (2005). Fluvial process and morphology of the Brahmaputra River in Assam, India. *Geomorphology*, 70(3-4), 226–256.

Sentinel Digital Desk. (2019). After 55 long years, land survey undertaken by Assam government in Majuli. *The Assam Sentinel*, 12 June.

Sentinel Digital Desk. (2020). Chief Minister Sarbananda Sonowal faces black flag protest in Majuli. *The Assam Sentinel*, 28 January.

Stojanov, R., Kelman, I., Ullah, A., Duží, B., Procházka, D., & Blahůtová, K. (2016). Local expert perceptions of migration as a climate change adaptation in Bangladesh. *Sustainability*, 8(12), 1223–1238.

6 Losing ground: Rethinking land loss in the context of managed retreat

Maggie Tsang and Isaac Stein

6.1 Introduction

Many localities across the United States are facing increasing flood risk. Recent data have shown that nearly 14.6 million properties across the nation may be at risk from a 100-year flood due to climate change and sea level rise (Flavelle, 2020). The common narrative around this climate crisis describes innumerable losses: Loss of land, loss of property, loss of livelihood, and loss of value. This chapter examines the meaning of "loss of land" and its relationship with property rights, environmental histories, and processes of urbanization. Through two case studies, this chapter draws attention to the ways in which the material properties of land upend underlying assumptions about (1) the stability and immutability of private property; (2) the location and identification of flood risk; (3) planning and policy approaches to hazard mitigation, adaptation, and managed retreat.

First, we examine the erosion of barrier islands and property rights along the North Carolina coast. Second, we explore the history of urbanization on present-day flood risk in North Miami. And finally, we demonstrate the benefits of reorienting planning and financial tools to produce adaptive land management strategies that reframe "land loss" as a public benefit.

While the trauma that results from loss of land is immeasurable and should not be undermined or diminished, this chapter presents these case studies to distinguish the concept of land, as a physical matter, from its imbued values: Economic, cultural, and personal. By separating these concepts, and by focusing on the physical and environmental processes the produce land loss, this chapter provides additional context needed for the discovery of new forms of mitigation, adaptation, and retreat (Figure 6.1).

6.2 Land loss and property

Land loss is often conflated with loss of property. The two, however, are not always synonymous. This tension is nowhere more evident than on barrier islands of the eastern seaboard of the United States. Here, the fixed definition of property contradicts the mutable condition of the ground.

DOI: 10.4324/9781003141457-6

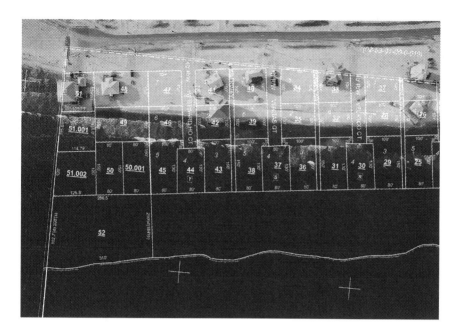

Figure 6.1 Property on barrier islands is frequently literally underwater. (Image by authors).

Hatteras Island, North Carolina is an archetype of this phenomenon. The island itself is comprised of quartz sand eroded from the Appalachian Mountains that has been shaped by coastal forces such as wave action, longshore drift, tides, sea level rise, and storms. Having only been formed in the last 3,000 years, in relative geologic time, Hatteras is young ground (the Appalachian Mountains are half a billion years old), and as such it is constantly moving and migrating in response to the "endless interplay of the sea" (Kaufman & Pilkey, 1983). The barrier island is also true to its name. It is a barrier that protects inland North Carolina coastal communities from storm surge (Dolan & Lins, 1986). Tropical storms and nor'easter cause dramatic shifts in the land: New passes, new spits, and sediment overwash. Everyday waves, tides, and incremental sea level rise push the island landward. Coastlines retreat at rates of up to 15 feet per year (North Carolina Division of Coastal Management, 2019).

Although it has one of the highest shoreline erosion rates in the country, Hatteras Island is also defined by its desirable beachfront properties. Of the 4,000 homes on Hatteras Island, 97% are vacation homes; and following storms or extreme weather events, these properties and their services are maintained, reconstructed, and fortified despite the costs. For example, since 2012, the North Carolina Department of Transportation has invested nearly a billion dollars in the reconstruction of NC-12, the island's main road, to maintain

80 Tsang and Stein

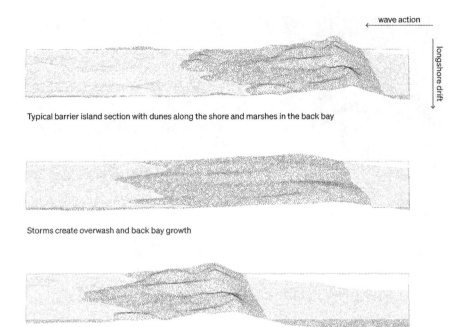

Typical barrier island section with dunes along the shore and marshes in the back bay

Storms create overwash and back bay growth

Continued wave action and longshore drift allow the island to migrate

Figure 6.2 Undisturbed, barrier islands naturally migrate landward through a process of erosion and overwash. (Image by authors).

access to the few vacation homes along the coast (North Carolina Department of Transportation, 2020).

In this context, land loss is predominantly viewed as a threat to private property, real estate markets, municipal budgets, and financial value. This economic approach fails, however, to acknowledge the broader implications of land loss as a physical process. Through sediment transfer and land migration, land loss on the seaward side of barrier islands does not mean permanent land loss, but rather land redistribution. On a healthy barrier island, sediment is supposed to overwash and accrete on the bayside of the island to facilitate a natural "rolling" process. Preventing seaward erosion ultimately deprives sediment transfer across the island and diminishes the capacity of the barrier island to protect the larger mainland population from flooding and other environmental risks (State of North Carolina Department of Environmental Quality, 2020) (Figures 6.2 and 6.3).

And yet, localities are preoccupied with saving beachfront property. "Resilient" plans and studies have assured the stabilization of the coastline and the future viability of these private properties. But ultimately, these "resilience" measures promise the resilience of property and economic stability, not the resilience of ecological systems and natural resources (Holling, 1973). In a

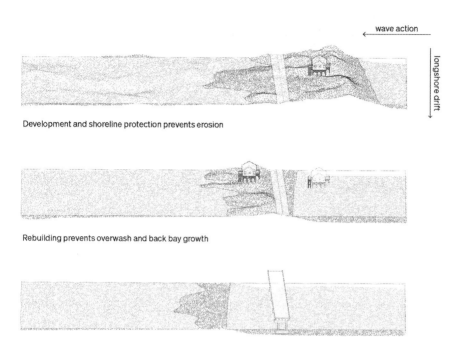

Figure 6.3 With continued development and fortification of the coastline, barrier islands are deprived of sediment transfer and gradually winnow and disappear. (Image by authors).

different light, if we disentangle land loss from property loss, we are able to see that the retreating coastline is in fact a natural form of coastal defense that functions to protect the coastal plain of North Carolina from flooding and other environmental risks. By stabilizing properties, the nature of the barrier island and its ability to serve its barrier function is lost (Tsang & Stein, 2019). This land-based approach ultimately demonstrates the short-sightedness of planning for the resilience of property without acknowledging underlying environmental and geological systems in motion.

6.3 Land loss as environmental history

Flood maps, such as those created by the Federal Emergency Management Agency (FEMA) or online platforms like Flood Factor (Flood Factor, 2020), measure and determine flood risk by synthesizing various data points such as rainfall, projected sea level rise, and elevation. While these digital tools are useful for visualizing information, they provide a limited snapshot based on a present-day condition and tend to overlook the history of development that has contributed to flood risk in the first place. Flood risk is not happenstance or random. Rather, it is a result of human-made infrastructure, development, and engineering; urbanization produces flooding.

In North Miami, for example, a municipality in greater Miami-Dade County, after a typical heavy rain event, the city's stormwater pipes reach their capacity and standing water can occupy entire blocks and remain for days as it slowly drains. Houses become islands. Cars wade through submerged streets with their wakes lapping onto front porches and into living rooms. Water is always present, even on sunny days and in dry seasons: A sample excavation on a residential lot in North Miami, several miles west of the coastline, reveals that the water table is only inches below grade.

In fact, there are as many as 78 properties within North Miami that experience repeat flooding damage, a phenomenon called "repetitive loss properties." FEMA defines repetitive loss properties (RLPs) as "any insurable building for which two or more claims of more than $1,000 were paid by the National Flood Insurance Program (NFIP) within any rolling ten-year period since 1978." Viewed within this narrow scope, "repetitive loss properties" appear random or abstract.

But RLPs are tied to the legacy of land itself. A large portion of the City of North Miami lies within the historic watershed of the Arch Creek Basin. This former waterway is known as a Transverse Glade, a hydrological formation that naturally channels water from the Everglades to Biscayne Bay. Today, however, the basin has been infilled and re-routed like much of South Florida. Development and engineering have significantly degraded the natural function of the creek. Water is regulated and controlled through canals, pumps, reservoirs, and flood gates. Impervious surfaces such as roads, parking lots, and building roofs reduce the area that can absorb stormwater. And finally, sea level rise has elevated the water table across the region, limiting space for infiltration and drainage (Urban Land Institute, 2016).

The ground has been fundamentally altered to produce flooding for residents.

The legacy Arch Creek Basin remains. Almost all 78 "repetitive loss properties" in North Miami fall within the creek's historic imprint. Here, "repetitive loss" is simply an indicator of a prior condition of land that was once watershed and was designed to produce flood conditions. Rarely does planning and policy acknowledge this connection between present-day flooding and past environmental histories. Instead, flood risk management is primarily focused on the protection of property and economic value. Land "lost" to flooding may not be suitable for buildings, but it is not truly lost. In fact, flooded land provides multiple benefits: Parks and open space for residents, wetland habitats, and green/blue stormwater infrastructure. Again, conflating loss of land with loss of value creates blind spots. (1) It overlooks the impacts of urbanization and infrastructure as a primary cause for flooding; (2) It minimizes the importance of building a social infrastructure that engages community members and residents in a dialog about the relationship between flooding and urban development; (3) It forecloses on new ways of thinking about flooding itself as an infrastructure. Ultimately, we need to address these blind spots to better understand how communities can gradually implement managed retreat (Figure 6.4).

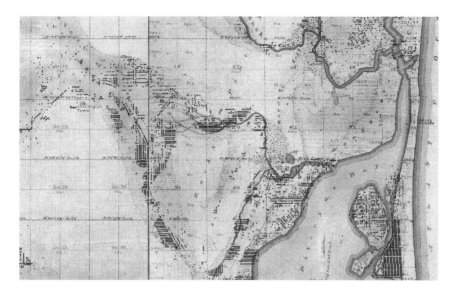

Figure 6.4 A map of projected "repetitive loss properties" with 1.5 meters of sea level rise reveals the figure of the historic Arch Creek Basin. (Image by authors).

6.4 Land loss as potential public benefit

In North Miami, as on Hatteras Island, when land loss is conflated with loss of property, it is commonly perceived as a loss of value. As flooding threatens more properties, local governments must contend with the effect loss of property has on their property tax revenue.

RLPs that generate multiple flood claims are costly for owners and localities to maintain. To help recoup some of the monetary value of flooded properties, FEMA funds buyouts for RLPs. Homeowners volunteer to have their property purchased by a public agency. The home is acquired at market value and may include relocation costs. Ultimately, ownership of flooded land is transferred from the private sector to local or state governments. In some cases, it is entrusted to nonprofit land trusts. By law, these properties cannot be sold or developed again (FEMA, 2005).

Currently, these government-funded buyouts are the primary mechanism for facilitating managed retreat. Properties that are a burden and liability to homeowners become an asset to flood mitigation in urbanized coastal areas. Rather than lay vacant, the land acquired through buyouts should be leveraged as a public amenity that can give back to the community. Though devalued for the purposes of real estate, the inherent spatial, physical, and material qualities of flood-prone land are uniquely suited for flood mitigation that is feasible, scalable, and localized.

84 *Tsang and Stein*

For example, we were involved in the design and construction of Good Neighbor Stormwater Park in North Miami, which transformed a vacant RLP into a community space to reduce flooding for the neighborhood. Situated on a half-acre lot in a medium density neighborhood, Good Neighbor Stormwater Park features a large retention basin as well as a new walking path through native plant communities that thrive in flooded conditions. In the retention basin, elevation markers are designed as a form of public artwork that also indicates fluctuating water levels. The visibility of the water was an important point of conversation with residents and local stakeholders who were initially wary of the exposed water. However, during the initial phases of the design, we hosted an onsite event where we dug a hole with some neighbors to show that water was just a few inches below grade; at the same time, we also shared a historic map of the area to show that the park was situated in a historic creek bed. Ultimately, these actions helped to convey the presence of water just beneath the ground and that the stormwater park was a place where water levels were made legible and visible to the public as a way of raising awareness of the environment and its history. Unlike typical utilities and infrastructure which are buried underground and out of site, Good Neighbor Stormwater Park shows that the visibility of infrastructure can be an important aspect of communicating flood risk (Harris, 2019) (Figure 6.5).

Converting just one lot into a stormwater retention basin can reduce flooding for a small watershed of homes, but creating multiple stormwater parks has compound benefits: Reducing flooding at the neighborhood scale, protecting municipal water supply from saltwater intrusion, and improving livability.

Lot-by-lot, even in a "checkerboard" pattern, a network of stormwater parks can grow into a city-wide green infrastructure that reframes "land loss" as an opportunity to give back to the community. This land-based approach has multiple benefits: First, it reduces stress on existing systems and services. Second, it increases stormwater capacity at a larger scale which helps recharge the city's freshwater supply. Third, it addresses multiple activity areas within FEMA's Community Rating System thereby reducing flood insurance costs for city residents (FEMA, 2020).

Ultimately, by looking beyond the financial burden of flooded property and examining the spatial and material realities of flood-prone land, so-called "lost land" can be repurposed as a public amenity. Especially in areas where residents cannot necessarily afford to relocate, strategies are needed that provide interim and intermediate relief from flooding while strengthening community resilience and reducing collective risk. The value of land – economic, social, cultural – is imposed on the land and can be changed; loss can be reframed as opportunity.

6.5 Conclusion

> *"The American penchant for owning nature shows in one specific and enormously important way how absurd and contradictory life in the modern world has*

Figure 6.5 The Good Neighbor Stormwater Park transforms a vacant "repetitive loss property" into a local stormwater park, reducing flooding for neighbors and raising awareness through signage and environmental monitoring. (Photo by authors).

become. There is no denying the whimsy and confusion of a culture that has tried to impose capitalist logic on the seemingly non-ideological matter-in-motion we call nature." (Steinberg, 1995)

If typical approaches to managing land loss focus primarily on economic value, then this chapter has proposed a counter-narrative that directly addresses "matter-in-motion."

Through case studies in coastal North Carolina and South Florida, where sea level rise and flood risk endanger residents and localities, this chapter argues for a more thorough exploration of land loss as material and physical process. Departing from traditional economic and policy approaches, these case studies draw attention to the causes and consequences of land loss – from patterns of sedimentary deposition to municipal stormwater management. Theodore Steinberg's "matter-in-motion" ultimately provides inspiration for a reorientation away from "land loss" and toward an understanding of land as matter, both subject and responsive to natural forces as well as human interventions (Steinberg, 1995). While the economic, social, and cultural implications of land loss are not to be overshadowed, this focus on the material and spatial, looks beyond the often paralyzing narrative of loss and crisis and instead gives us an opportunity to imagine how ground itself can be leveraged for new forms of mitigation, adaptation, and retreat.

6.6 Recommendations

- Planning and funding agencies should look toward community-oriented, neighborhood-scale infrastructure strategies to reduce flooding that consider public benefits, such as programming, education, and environmental stewardship.
- Engaging community members and residents of flood risk areas about the history and legacy of their local watersheds and drainage systems creates broader awareness. This social infrastructure and community buy-in is essential to minimizing risk and facilitating retreat.
- In addition to near-term fixes and long-term plans, intermediate strategies are needed: What is known as the "checkerboard effect" of buyouts should not be viewed as inherently negative. This pattern of buyouts should be leveraged to retain value, mitigate flood risk, add to the urban fabric and public realm, and grow public awareness and agency.

References

Dolan, R., & Lins, H. F. (1986). *The outer banks of North Carolina*. U.S. Geological Survey professional paper; 1177-B.

Federal Emergency Management Agency. National Flood Insurance Program: Frequently asked questions Repetitive loss. (2005). Retrieved June 21, 2020 from https://www.fema.gov/txt/rebuild/repetitive_loss_faqs.txt.

Federal Emergency Management Agency. National Flood Insurance Program Community Rating System. (2020). Retrieved June 21, 2020 from https://www.fema.gov/national-flood-insurance-program-community-rating-system.

Flavelle, C. (2020, June 29). New data reveals hidden flood risk across America. *The New York Times*. Retrieved July 13, 2020 from https://www.nytimes.com/interactive/2020/06/29/climate/hidden-flood-risk-maps.html

Flood Factor. (2020). Methodology. Retrieved November 14, 2020 from https://floodfactor.com/methodology

Harris, A. (2019, September 27). North Miami bought her flooded home. Now it's going to become a park to fight sea rise. *Miami Herald*. Retrieved July 13, 2020 from https://www.miamiherald.com/news/local/environment/article235403232.html

Holling, C. S. (1973). Resilience and the stability of ecological systems. *Annual Review of Ecology and Systematics*, 4, 1–23.

Kaufman, W., & Pilkey, O. H. (1983). *The beaches are moving: The drowning of America's shoreline*. Duke University Press.

North Carolina Division of Coastal Management. (2019). *North Carolina 2019 oceanfront setback factors & long-term average annual erosion rate update study: Methods report*.

North Carolina Department of Transportation. (2020). High-profile projects & studies: Dare County. Retrieved July 13, 2020 from https://www.ncdot.gov/projects/

State of North Carolina Department of Environmental Quality. (2020). *Climate risk assessment and resilience plan: Impacts, vulnerability, risks, and preliminary actions. A comprehensive strategy for reducing North Carolina's vulnerability to climate change.*

Steinberg, T. (1995). *Slide mountain or the folly of owning nature.* The University of California Press.

Tsang, M., & Stein, I. (2019). *Lines in the sand: Rethinking private property on Barrier Islands.* Harvard Graduate School of Design.

Urban Land Institute. (2016). *Arch Creek Basin, Miami-Dade County, Florida.* Advisory Services Panel Report.

7 Resistance, acceptance, and misalignment of goals in climate-related resettlement in Malawi

Hebe Nicholson

7.1 Introduction

Climate change is causing the conventional weather patterns of countries to change (IPCC, 2014). This is particularly detrimental to countries where much of the population is dependent on the environment for their livelihoods, such as those heavily reliant on agriculture (IPCC, 2014; Suckall et al., 2015). Resettlement, otherwise called planned retreat or planned relocation, is a climate change adaptation strategy that has received much discussion in recent years (Arnall, 2018). Four international organizations: The World Bank, the Norwegian Refugee Council, UNHCR, and Displacement Solutions, have produced guidelines for resettlement as a form of climate change adaptation. These organizations have been producing guidelines on resettlement more broadly since the 1990s but it is only in the past 10 years that resettlement as climate change adaptation has been considered. Similar to the broader guidelines on resettlement, these climate change adaptation guidelines highlight how and why resettlement processes must be voluntary and participatory: Fully involving those who are expected to relocate (Correa et al., 2011; Displacement Solutions, 2013; Norwegian Refugee Council, 2011; UNHCR, 2014). However, moving an established community, especially one that is dependent on the land, involves numerous considerations, challenges and resources to make it a satisfactory move for the participants (Dun, 2011; Stal, 2011). In this chapter, I draw on empirical study from Lower Shire Region of Malawi to illustrate the challenges associated with acceptance and resistance to planned resettlement.

I structure this chapter as follows, first, I set out why participation is important in resettlement. Next, I provide some brief context on Malawi and my research methods. Then in the results sections I go through each community to explore the participation occurring, and how community members influence the resettlement process. I conclude with three key recommendations.

7.2 The importance of participation in resettlement

Owing to the threat of climate change, some governments, such as in Malawi, are planning to proactively move people out of areas vulnerable to environmental change. To require a community to move home due to a threat that may not yet be imminent, requires clear communication and deep understanding and acceptance from communities at risk. The four major international guidelines on resettlement as a form of adaptation to climate change state that proactive resettlement processes must be participatory (Correa et al., 2011; Displacement Solutions, 2013; Norwegian Refugee Council, 2011; UNHCR, 2014). There are two key reasons for this: First, moving home is a personal process; and second, resettlement can be used by the government to serve broader developmental agendas (Arnall, 2018). For example, resettlement in China, involved moving rural communities to urban areas to encourage an urban economy (Rogers & Wilmsen, 2019). This approach has been critiqued by resettlement scholars (Rogers & Wilmsen, 2019).

Participatory approaches are difficult to enforce in practice. Participation is meant to address power imbalances, but some argue that it can still reinforce them (Cooke & Kothari, 2001). In their seminal work, *Participation: The New Tyranny?* Cooke and Kothari (2001) highlight how those in charge of the funding are often viewed as having greater influence over the output of a development project than those for whom the project is meant to help. The critiques of participation suggest that if a development project is initiated by a government, an international organization, or NGO, it can be difficult for it to lose its top–down origins regardless of the platforms for participation provided (Cooke & Kothari, 2001; Jakimow, 2013; Kapoor, 2005). In these instances, there is a danger that participation could be ill-used within the resettlement process, potentially making it maladaptive (Miller & Dun, 2019). However, when done appropriately, with adequate inclusion of all parties involved from the start, it can be incredibly empowering and transformative (Hickey & Mohan, 2008). Therefore, it is important to critically analyze the participation occurring in the resettlement process, to highlight the inclusivity, respect and any ulterior agendas involved. Whilst there is established literature exploring the participatory nature of resettlement related to development projects (Goebel, 1998; Mcdonald et al., 2008), there is burgeoning research exploring the participatory nature of proactive, climate-related resettlement (Arnall, 2014). In this chapter, I attempt to add to that by providing a case study highlighting the role of participation within resettlement processes in the Lower Shire Region of Malawi.

7.3 The study area: The Lower Shire Region of Malawi

This study was conducted in the Lower Shire Region of Malawi. Malawi is a small, densely populated country in South East Africa. It is relatively poor, with 70.3% of the population living below the poverty line, on less than $1.90 per day (The World Bank, 2018). The life expectancy at birth in 2017 was 63.3 (The World Bank, 2018). Additionally, 85% of the population rely on agriculture for their income (The World Bank, 2016), making the state of the environment crucial to the Malawian economy. Malawi is split into 26 districts, with the Lower Shire Region containing the two most southern districts: Chikwawa and Nsanje.

The Lower Shire Region is the valley base of the large Shire River. This river floods annually and is the most flood-prone area of Malawi (Government of Malawi, 2019). Malawian government data suggest that for the past two decades the flooding is getting increasingly severe and unpredictable due to changing weather patterns. Temporary resettlement, whereby communities move out of flood-prone areas in times of flooding, is a common strategy used by locals. The government is interested in formalizing this to permanently resettle communities particularly vulnerable to flooding. However, this interest in resettlement has only surfaced within the past 5 years, since the heavy flooding in 2015. It therefore makes it a particularly interesting time to explore how the climate-related resettlement process is being established in Malawi.

The government resettlement within the Lower Shire Region fits well with broader disaster management. Malawi has different forms of governance framework and agencies that manage disasters. There is the Department of Disaster Management Affairs (DoDMA) at the national level and the district disaster officer at the local level within the district council. Additionally, within each district there are two forms of customary governance. Each district is made up of several Traditional Authorities (TAs), which are presided over by a TA chief and within each TA there are several villages, which are led by a village head. To coordinate between the different levels of governance within each district there are committees. Disasters and resettlement due to flooding fit within the civil protection committees. At the most local level, there is the Village Civil Protection Committee (VCPC), which includes prominent people from the village. Next, there is the Area Civil Protection Committee (ACPC), which contains the village heads from within the TA and the TA chief. Finally, there is the District Civil Protection Committee (DCPC), which includes the TA chiefs, the district disaster officer and other relevant district council officials, and also stakeholders from relevant NGOs that work within the district. There are then further national level committees dealing with different foci of disaster management which the DCPC feeds into, and which ultimately feed back into DoDMA. Participation within the resettlement process is meant to occur through these committees, with communication flowing both top–down and bottom–up. My research suggests that it is difficult to reach a consensus on resettlement processes, through these committees or otherwise. Community priorities do not appear to align with government priorities and this is causing some resistance to government resettlement from those in communities.

This chapter examines three communities with three distinct views or positions on resettlement. These can be seen on the map in Figure 7.1. The community at Jombo, in Chikwawa district, resettled after the severe flooding of 2015. This resettlement was initiated by members of a community from a neighboring TA, around 20 km from Jombo, who had a particularly traumatic experience with flooding. Jombo is an area relatively safe from severe flooding. The resettlement was assisted by the government and civil society. However, it

Resistance and acceptance 91

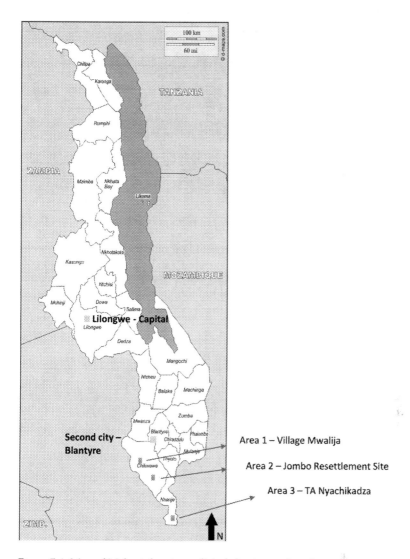

Figure 7.1 Map of Malawi showing political districts and study site.

was not a whole village that resettled, only around 60 households. Many, along with their previous village head, preferred to stay in their previous location. Their cultivation lands are also in this area and they regularly return to it for cultivation purposes. However, they now reside and are part of the host village of Jombo. The community at Village Mwalija, in Chikwawa district, were undecided whether to resettle, at the time of study. They consist of around 250 households. An NGO promised to help them with relocation if they considered it a viable option. However, there seemed to be several impediments to their resettlement. They are in a vulnerable area that had not had severe flooding at the time of study but has since, in March 2019, been devastated by flooding. Finally, those in TA Nyachikadza, in Nsanje district, were unwilling to relocate, despite a strong government focus aimed at resettling the community. This TA consists of several villages and around 1,000 households in total. It is situated on an island in a marsh area of the Shire River, making it a very fertile area but also highly prone to flooding. In 2012, the government declared TA Nyachikadza as not fit for human habitation, essentially making it a "no-go zone" for government and NGO services. There were no schools past the age of eight and no health facilities. To access these facilities, residents need to cross a river inhabited by hippos and crocodiles and travel several kilometers by foot.

I conducted my fieldwork between August and November 2017. I undertook 16 interviews and two focus groups for each of the three flood-prone communities described earlier. Both the interviews and the focus groups consisted of men and women in equal proportion, with a wide range in ages. Additionally, I conducted 21 interviews with stakeholders in government and NGOs that were involved in these communities and their resettlement plans. The interviews were conducted through the help of a translator. I transcribed and analyzed the results. My findings are presented in the following sections. I use pseudonyms to provide anonymity to the participants.

7.4 Expressing community needs within the resettlement process

This section illustrates how the community members made their voices heard within the resettlement process and caused those in government and NGOs to question how best to undertake resettlement projects. This is most apparent in Nyachikadza but is also present at village Mwalija and to some extent at Jombo resettlement site.

7.4.1 Nyachikadza

Nyachikadza provides a key example of the strength of a community against a coercive government-led resettlement strategy. The government has long been trying to get Nyachikadza residents to resettle. In 2012, the government declared Nyachikadza as not fit for human habitation because of the severe flooding that occurs in the area. They have restricted the humanitarian assistance and government services in this area to try and persuade people to move. Instead of those at Nyachikadza being disheartened by the lack of services and assistance available to them, there is a sense of pride and autonomy, as shown by Nyachikadza community member, Aliva:

> "In Nyachikadza, we do not depend on assistance, but in TA Ndamera we would depend on the government and NGOs for everything ... So, we are content, we have whatever we need" (Aliva, Nyachikadza, 2017).

Aliva highlights the land as the key priority for this community and indicates the limited influence of the government on them. The forceful measures used by the government, and the proud resistance demonstrated by Nyachikadza residents, indicates a mismatch in priorities and goals between those in government and those residing in Nyachikadza.

Moreover, my research suggests that those at Nyachikadza use their immobility as a bargaining tool in the resettlement process. This is how they communicate to the government their dissatisfaction with the current suggestion of resettlement. The government has offered to resettle those in Nyachikadza to land further upland, but the community members want to move close to their current area of cultivation. The cultivation land in Nyachikadza is some of the most fertile in the country. The community members said that they cannot move without the government providing them with a place to go close to their cultivation lands. This means they will refuse to move until this goal of having access to their source of livelihood, fulfilled.

The government's goal is to move Nyachikadza residents out of the flood-prone area and, to achieve this, the government does appear to now be listening to the needs of those in Nyachikadza. It considers temporary resettlement to the upper land closest to Nyachikadza during the rainy season, with access to the vulnerable area for farming, as the best compromise. This is expressed by National Government DoDMA employee, Mtafu:

> "Yeah, I think that one [temporary resettlement] at least the reluctance would be lower, because you can say, ok I can still use this land while saying you should move upland just during the rainy season to save lives and property, but when the rains have gone maybe in March, you can come down and do your winter cropping because that is a critical thing for them to do and that is their livelihood. And maybe later you can say to move permanently away, but just to kind of maintain the buy in, to motivate them we can still be using this one, to say no you are still entitled to this land but just use it for the production of crops, but invest in your permanent home, maybe upland or somewhere safe" (Mtafu, Government official, 2017).

Mtafu's acknowledgment that temporary resettlement would mean the "reluctance would be lower," suggests he is aware that the community do not currently want to resettle. However, the way Mtafu discusses resettlement also illustrates that the approach of temporary resettlement is used to ultimately lead to permanent resettlement. He says that temporary resettlement is used "to maintain buy-in" of residents to the resettlement process and implies they may be being coerced to fulfill the broader government agenda of permanent resettlement later.

7.4.2 Village Mwalija

The community at Mwalija is not directly opposed to government resettlement plans. They were offered assistance with the resettlement. An NGO has committed to provide assistance to the most vulnerable households and a borehole has been drilled in the place allocated for resettlement. Yet, there are impediments to their resettlement. Cultural perception and beliefs are important in the resettlement decisions. For example, there is a wide belief that if any community member were to move before the village head they would be cursed, and in this village the village head is reluctant to move. Thus, the community members have refused to resettle. Furthermore, the community members have complained about limited consideration of community views in the planning of the resettlement. There is particular grievance about the lack of consultation in the positioning of the new boreholes in the area where they are to resettle. This highlights another misalignment in views, this time between a community and the NGO assisting the community.

The resistance of the people in the village Mwalija community to resettle has led to some changes to the resettlement proposal. Indeed, the NGO worker involved, Jake, suggests that his opinion has changed from a permanent move to a more flexible form of resettlement:

> "Always I have been thinking, because they have not been contacting us ... You need to develop ways of how can we move forward with this approach? How can we understand each other with this approach? So that the permanent resettlement area is where their property and everything can be. And then this area is an area where they can still practice whatever, but during the rainy period we sensitize them and they go back to their permanent structures. Those are just kind of an office thing that I am trying to develop, because maybe we cannot understand. And we cannot push, but we can at least balance up to solve the situation at hand, which is flooding" (Jake, NGO representative, 2017).

Jake illustrates how the silence of the community on proceeding on the resettlement was a way of communicating their resistance. It has led him to think further on how he can change the approach to resettlement to make it more acceptable. It thus shows an adaptability in the governance of resettlement and, linking back to the Nyachikadza experience, reiterates how government perception is changing to appreciate the importance of continuing use of existing cultivation lands. Cultivation lands are particularly important in the Lower Shire Region because the annual flooding makes it very fertile and much of the livelihoods of the community members in this area depend on the produce from their lands.

7.4.3 Jombo

In Jombo, resettlement has taken place and is often presented as a success. However, the community at Jombo still perceive themselves as separate from their host community. This is seen most clearly in their disputes over leadership. It is also visible in their constant return and contact with their previous area, where their cultivation lands are situated.

There is a deep connection with their previous location which inhibits those at Jombo resettlement site from integrating into the new community. Moreover, as it is suggested that future resettlement should allow for return to the previous area for cultivation, it is likely that integration into a host area may continue to be difficult. Kenny, Jombo resettled community member, highlights how strong social, economic and cultural ties to a previous area can be:

> "I am worried that I have grown up under [village head] Champanda, and Champanda, I regard him as my father, and I have come here and now I am under a new father who adopted me. I felt like Champanda loved me unlike this one. That is why sometimes when items or food come to be distributed to this area, those from Lighton Jombo would prefer to write names from his village, and these ones [those resettled] become the last" (Kenny, Jombo community member, 2017).

Kenny illustrates the deep relationship possible with village heads. He uses the metaphor of family to discuss the connection, or lack of it, with a village head for those at Jombo resettlement site. The role of the village head appears to impact their sense of belonging and identities, seen through the hostilities related to aid relief. This makes life particularly hard for this community. Moreover, it is common for these hostilities to escalate into altercations. The village head is in charge of deciding who is on the VCPC and none of the resettled communities are part of the VCPC, making it difficult for them to express their concerns to the district council. This can lead the resettled community to feel desperate and hopeless about their situation.

However, the interviews with the government suggest that the difficulties experienced at Jombo have raised awareness of the need to include the host community in the resettlement process. Local government official Charles discusses the strategy of inclusion of the host community.

> "The host community is also ... maybe interested in the maybe social support that is maybe being channeled to the internally displaced people who have maybe possibly been resettled or integrated in the existing villages. So, the approach now, has been that whenever we are bringing an intervention in that particular community we should also make sure that, even some members of the host community should also be benefitting, which is even making it very expensive because you can not only maybe think of the vulnerable households which were possibly affected by the floods, but even considering people who were not affected by floods" (Charles, Government official, 2017).

Charles suggests that the government is aware that the involvement of a host community and the need for integration of one community into another can be problematic. The fact that he says "the approach now" implies that it may not have been the approach before and, as the situation at Jombo was being discussed shortly before this in the interview, it is likely that the change in approach is due to the lessons learned from the Jombo resettlement.

7.5 Conclusions

This chapter highlights how community views, goals, priorities, and inter-community relations between resettling and host communities shapes acceptance and resistance to planned relocation. In all three communities (Jombo, village Mwalija and TA Nyachikadza) examined in Malawi, access to cultivable and fertile land and sociocultural ties played a key role in determining communities' willingness to resettle. Each case highlighted different tensions within the suggested or occurred resettlement: At Nyachikadza, it was access to fertile land; at village Mwalija, it was chiefly the role of the village head; and at Jombo, it was the integration into the host community. For each of these communities the misaligned priorities between those organizing and those undergoing the resettlement led to community members either refusing to resettle or not successfully resettling. The government and NGO workers involved appeared to be responding to this. However, those in government and NGOs organizing resettlement should ideally be aware of community views, priorities and goals at the designing stage of the resettlement process not at the desired implementation stage, as was the case for those at village Mwalija and Nyachikadza, or post resettlement, as was the case at Jombo. Moreover, it is not clear whether community priorities were truly being respected in these cases or whether they were being included as a necessary step for facilitating resettlement and, eventually, a government agenda of permanent resettlement. This reiterates the need for inclusivity and participation in resettlement, as a more inclusive and participatory process could help to align priorities earlier. It highlights how more critical research is needed to ensure planned resettlement is participatory and adaptive, and not maladaptive.

7.6 Recommendations

- To better prepare for adaptive resettlement, there should be further research on potential priorities of those experiencing resettlement, such as those which came out of this research: The importance of land, the role of the village head, and the host community.
- To better align priorities of those involved in resettlement there should be greater emphasis placed on communication at the earlier stages of the resettlement process. For the case of Malawi, as the committee structure is already present this could be done through the existing committee structure.

- To ensure the alignment of priorities within the resettlement process a draft plan and schedule of the resettlement should be drawn up between those assisting with the resettlement (those in government and NGOs) and those undertaking the resettlement.

References

Arnall, A. (2014). A climate of control: Flooding, displacement and planned resettlement in the Lower Zambezi River valley, Mozambique. *The Geographical Journal*, 180(2), 141–150. doi:10.1111/geoj.12036

Arnall, A. (2018). Resettlement as climate change adaptation: What can be learned from state-led relocation in rural Africa and Asia? *Climate and Development*, 11, 1–11. doi:10.1080/17565529.2018.1442799

Cooke, B., & Kothari, U. (2001). *Participation: The new tyranny?* Zed Books.

Correa, E., Ramírez, F., & Sanahuja, H. (2011). *Populations at risk of disaster: A resettlement guide*. World Bank.

Displacement Solutions. (2013). *The peninsula principles: On climate displacement within states*. Displacement Solutions.

Dun, O. (2011). Migration and displacement triggered by floods in the Mekong Delta. *International Migration*, 49, e200–e223. doi:10.1111/j.1468-2435.2010.00646.x

Goebel, A. (1998). Process, perception and power: Notes from 'participatory' research in a Zimbabwean resettlement area. *Development and Change*, 29(2), 277–305. doi:10.1111/1467-7660.00079

Government of Malawi. (2019). *Malawi 2019 floods post disaster needs assessment report*. Government of Malawi.

Hickey, S., & Mohan, G. (2008). *Participation: From tyranny to transformation: Exploring new approaches to participation in development*. Zed Books.

IPCC. (2014). Summary for policymakers. In C. B. Field, V. R. Barros, D. J. Dokken, K. J. Mach, M. D. Mastrandrea, T. E. Bilir, M. Chatterjee, K. L. Ebi, Y. O. Estrada, R. C. Genova, B. Girma, E. S. Kissel, A. N. Levy, S. MacCracken, P. R. Mastrandrea, & L. L. White (Eds.), *Climate change 2014: Impacts, adaptation, and vulnerability. Part A: Global and sectoral aspects. Contribution of Working Group II to the Fifth Assessment Report of the Intergovernmental Panel on Climate Change*. Cambridge University Press.

Jakimow, T. (2013). Spoiling the situation: Reflections on the development and research field. *Development in Practice*, 23(1), 21–32. doi:10.1080/09614524.2013.753411

Kapoor, I. (2005). Participatory development, complicity and desire. *Third World Quarterly*, 26(8), 1203–1220.

Mcdonald, B., Webber, M., & Yuefang, D. (2008). Involuntary resettlement as an opportunity for development: The case of urban resettlers of the Three Gorges Project, China. *Journal of Refugee Studies*, 21(1), 82–102. doi:10.1093/jrs/fem052

Miller, F., & Dun, O. (2019). Resettlement and the environment in Vietnam: Implications for climate change adaptation planning. *Asia Pacific Viewpoint*, 60(2), 132–147. doi:10.1111/apv.12228

Norwegian Refugee Council. (2011). The Nansen Conference: Climate change and displacement in the 21st century. Retrieved from https://www.nrc.no/globalassets/pdf/reports/the-nansen-conference---climate-change-and-displacement-in-the-21st-century.pdf

Rogers, S., & Wilmsen, B. (2019). Towards a critical geography of resettlement. *Progress in Human Geography*, 44, 0309132518824659. doi:10.1177/0309132518824659

Stal, M. (2011). Flooding and relocation: The Zambezi River Valley in Mozambique. *International Migration*, 49, e125–e145. doi:10.1111/j.1468-2435.2010.00667.x

Suckall, N., Fraser, E., Forster, P., & Mkwambisi, D. (2015). Using a migration systems approach to understand the link between climate change and urbanisation in Malawi. *Applied Geography*, 63, 244–252. doi:10.1016/j.apgeog.2015.07.004

The World Bank. (2016). Malawi. *Data*. Retrieved from http://data.worldbank.org/country/malawi

The World Bank. (2018). Malawi. Retrieved from https://data.worldbank.org/country/malawi

UNHCR. (2014). Planned relocation, disasters, and climate change: Consolidating good practices and preparing for the future. Retrieved from http://www.unhcr.org/54082cc69.pdf

8 Land is life: A poem of the Philippines Lumad

Nikki C.S. Dela Rosa

Mindanao is in turmoil, and its Indigenous peoples, the Lumad, are experiencing the burden as they find themselves unhoused and uprooted. This poem is a compiled narrative from the Lumad people and the author's personal experience as she listened to their story of forced relocation and their fight for justice. The combination of the harsh realities and poetic metaphor brings forth the callous nature of their relocation that was organized by the Philippines military. This ongoing struggle against the Philippines government and foreign mining corporations shapes the lived experience of the Lumad and their undying hope of going back to their ancestral land to maintain their rich history.

Land is life:
A poem of the Phillippines Lumad

>Fearless, relentless —
>In the struggle for land and life.
>The Lumad of Mindanao
>Sing their songs of resistance.
>
>Mindanao, Land of Warriors,
>Where Mount Apo rests his soul,
>And the milkfish grilled al fresco.
>This is Lumad ancestral home.
>
>Mindanao, her blood flows regal blue,
>But her copper veins bleed rustic red.
>Her body was reaped more than mortals sow,
>And more than her barren scalp can regrow.
>
>Before the rest of the world took notice
>Of all the secrets buried below her bosom,

Mindanao preserved bounty for her children alone.
But visitors came and wanted to take it all.

The visitors had foreign names and shiny dimes.
They came from the North, the West,
And somewhere cold, and somewhere dry.
They brought their privileges and began to mine.

The Lumad saw the visitors arrive
With their cargo ships that hauled giant devices.
Mindanao was stripped naked, flattened, and stabbed.
The visitors were a new kind of monster,
Untold by their ancestors' past.

Mindanao became depleted and tired,
The unwanted visitors will not stop
They plundered and grabbed
But she stays awake, fearing the rifles fired.

When sundown comes in this Southern land,
The infantry battalion took up loaded arms —
Forcefully marched the Lumad out of their home,
Down from riverbanks and mountain tops.

The Lumad of Mindanao
Sing their songs of resistance
As they come down to the land unknown.
Relocated to a place made out of stone.
It is cold and foreign on their barefoot.

As the red sun rises in the East,
The Lumad remain fearless and relentless
To go back home and save their schools.
Indigenous at its core, their citadels of knowledge
Where they share generations of Earth appreciation.

The Lumad without their land,
Without their schools,
The soil will soften, unable to hold a root.
The river parched and drained will yield no food.

As the world comes together and heals the wound,
Deforestation of looted inherited ground is still afoot.
Let us bear witness to the Lumad song:

Makiisa sa pakikibaka ng mamamayang Mindanao
Laban sa militarisasyon at pandarambog.[1]

The women mobilize in the streets
With picket signs, dressed in woven sheets.
Voicing their innate desire for justice.
A forced evacuation to a gymnasium
Is not a home, this is sure.

The scorching heat of the high noon sun will come,
So will the courage of displaced sons
And the strength of Mindanao daughters.
Fearless, relentless —
In the struggle for land.
Hopeful in their zest for life.

The Philippines has an estimated $850 Billion worth of untouched mineral resources such as copper and gold, the majority of which can be found in the southernmost island of the archipelago, Mindanao (Mining Technology, 2020; Philippines Mining Bureau, 1915). The Philippine Mining Act of 1995 allowed foreign multinational mining corporations to operate open-pit mining in Mindanao (Republic of the Philippines, 1995). The mining operations caused severe environmental degradation including siltation, poisoning of rivers from toxic chemical deposits, and deforestation – leaving communities prone to flooding.

The Lumad, a group of Austronesian Indigenous people found in Mindanao, are most affected by the mining operations because they occupy the ancestral territories rich in mineral resources. The Lumad people were forced to evacuate their homes by the Philippine National Police and transferred to sports fields and churches in nearby cities. The Philippine government has no plan to provide a permanent home for the Lumad people. Across the country, Indigenous territories are continuously being turned into dams, highways, and large-scale mining. The lack of legal protection for Indigenous peoples has meant people have fewer options to fight back.

I had the opportunity in the summer of 2016 to get a closer look at the situation of the displaced Lumad people in Mindanao, particularly in the Caraga Region, where I witnessed their daily challenges. The Philippine government allotted a sports complex in a nearby populated city as a temporary shelter. The space is crowded, has poor sanitation and little to no running water.

The Lumad people were able to create a community inside this confined space. They built makeshift houses, schools, and common spaces for gatherings. Although they attempt to live harmoniously in the city, the Lumad endures discrimination and harassment from the city locals and police. They are threatened by arson and sexual assault.

There is no question why the Lumad people continue to be fearless and relentless in their fight for justice and right to return to their land, rivers, and forests – their living conditions in the city are inhumane and unacceptable. There is no access to water and there are no trees to provide shade which makes the Lumad prone to heat strokes and dehydration. Also, there is no security, so their privacy, property, and well-being are prone to frequent threats and damages from the locals.

At first glance, Mindanao shows its beautiful tropical flora and the humble temperament of the people. I visited this island paradise frequently during my early childhood, my parents were born and raised there. I remember the fruit stands on the side of the highways, the glistening waters of the white sand beaches, and the sound of motorized tricycles racing along unpaved roads. My parents frequently warn me about areas in Mindanao they deemed dangerous, but I have always questioned where their worries come from. My visit in 2016 provided the answer.

Although Mindanao is a serene tourist destination, it is also a place of unrest and turmoil, and it has been for decades.

I call Mindanao the "Land of Warriors" because of its history of wars from Spanish colonization, Japanese invasion, religious conflicts that caused a number of massacres between the Christians and the Muslims, and the ongoing corporate plunders that are displacing Indigenous peoples. With the bountiful natural resources that sits in Mindanao, the constant war on the island is not surprising. Mindanao is dangerous but fragile, the people that should be taking care of it are treated less than human. The Lumad people, like many Indigenous peoples of the world, are tied to their land - as they are intrinsically connected in a symbiosis. They fight for self-determination to maintain the biological integrity of the land rather than ownership, because the land is to be lived in and not owned.

8.1 Recommendations

- Support environmental and climate activists in the Philippines in their demand to repeal the Anti-Terrorism Law of 2020 that targets eco defenders as terrorists (Republic of the Philippines, 2020). This "Anti-Terror" law will worsen the already existing human rights problem in the Philippines that greatly affects the lives of the Lumad people in Mindanao and the individuals working to defend Indigenous territories (United Nations Human Rights Office of the High Commissioner, 2020).
- Donate to the Save Our Schools Network in the Philippines and ALCADEV Lumad School to help the displaced Lumad children continue their education in evacuation centers during the COVID-19 pandemic and beyond. A simple search online will bring up ways to contact the organization leaders and find out how to remotely support the Lumad children. An easy and quick help is by amplifying their social media presence using the *#standwiththelumad* when sharing their sources.

Note

1 English Translation: "Join in the struggle of the people of Mindanao against militarization and plunder." This slogan can be found in many banners and posters held by protesters during rallies and marches all over the Philippines created for the event called *Manilakbayan ng Mindanao* that started in 2015 (Aryoso, 2015). Chants such as "Stop Lumad Killings!" and "Save Our Schools" often accompany this slogan, which are also used as hashtags (*#stoplumadkillings*, *#saveourschools*) in social media networks.

References

Aryoso, D. (2015, October 23). Retrieved from https://www.bulatlat.com/2015/10/23/manilakbayan-ng-mindanao-bringing-the-peoples-struggle-to-the-center/

Mining Technology. (2020). Tampakan Gold Copper Project. Retrieved from https://www.mining-technology.com/projects/tampakangoldcopperpr/

Philippines Mining Bureau. (1915). *Gold in the Philippines*. Philippines Bureau of Printing.

Republic of the Philippines. (1995, March 3). Republic Act No. 7942. Retrieved from http://www.mgb.gov.ph/images/stories/RA_7942.pdf

Republic of the Philippines. (2020, July 3). Republic Act No. 11479. Retrieved from Official Gazette https://www.officialgazette.gov.ph/2020/07/03/republic-act-no-11479/

United Nations Human Rights Office of the High Commissioner. (2020, June 4). Philippines: UN report details widespread human rights violations and persistent impunity. Retrieved from https://www.ohchr.org/EN/NewsEvents/Pages/DisplayNews.aspx?NewsID=25924&LangID=E

Interlude 2 Flood experiences in the United States

"I was flooded during Katrina and never moved back into that home. As a renter, it was hard to find any affordable housing after Katrina. Now, I avoid certain streets when it has been raining all day because they will be flooded." – Renter, New Orleans, LA

"During the hurricanes and flooding I have felt very afraid for my safety and that of my community. I have worried about the inability of the local government to rebuild (they don't have a great track record). And I have feared that my life could be blown or washed away. My home was not damaged; we were blessed THIS TIME. But what about the next storm? They keep getting worse and worse." – Homeowner, Port Arthur, TX

"I was a senior in high school when [Hurricane] Rita hit. The school had holes in the ceilings of many of the classes on the second floor. There were many of my classmates who had to relocate because their families could not or were too afraid to rebuild." – Student, Port Arthur, TX

"My house flooded with Matthew. And this past summer my business was closed from wind tide flooding two different times. And, we had the mandatory evacuation. It was an extremely hard year financially for the business. The combination of all the events cost us ten of thousands of dollars in lost revenue that we needed to save to get through the winter and reinvest into the business and our employees." – Business Owner, Virginia Beach, VA

Responses to survey question: What do you think of retreat as a strategy to deal with climate change and flooding?

- Yes, please! Sign me up! (Virginia Beach, VA)
- Don't like that land/property in the 9th Ward, formerly owned by African-Americans, was Sold Dirt Cheap! (New Orleans, LA)
- Will not support this because I been in my home for over 40 years (New Orleans, LA)
- WONDERFUL!!! (Pensacola, FL)

Flood experiences in the United States 105

Figure Int 2.1 "Swampy" the South Carolina, USA, flood activist.

Figure Int 2.2 United Flooded States of America.

Figure Int 2.3 Repetitive floods in a home.

Figure Int 2.4 Flood activist for managed retreat.

Flood experiences in the United States 107

Figure Int 2.5 Flood loss means more than loss of property.

Interlude 3 Rock That Fell to Earth: Storm Series

Martha Lerski

Rock That Fell to Earth: Storm Series is formed out of oak from a Northern California tree felled by an El Niño storm in 1998.

In this piece, gaping cracks and rough edges indicate impact. The piece was made using direct carving: an approach in which the final form is not fully planned but emerges.

The carving was inspired by my then-5-year-old daughter on visiting the American Museum of Natural History (AMNH) weeks after we had evacuated her school four blocks from the World Trade Center on 11 September 2001. She wondered whether a rock from outer space had felled the twin towers.

We had viewed fossil remains of dinosaurs earlier during our museum visit and talked about why dinosaurs no longer exist. Her response to seeing a meteorite alerted me that our local world had been altered in a cosmic way: Like the ensuing change in eras accompanying the demise of dinosaurs, we are in an epoch of great disruption.

While it can be difficult to imagine a trajectory in the midst of tragedy, there are paths of resilience and adaptation. The 5-year-old who inspired this sculpture – influenced and encouraged by trips to the natural history museum and extra time spent in the park while unable to return to school – eventually studied physics, working on computational climate models on dust emissions and also ocean flows, with plans to work in the realm of environmental policy. After first volunteering in the library of the elementary school which my daughter attended, I went to library school and became an academic librarian.

Pulling together after 9/11, New York eventually recovered its vitality and returned to its daily demonstration and celebration of diversity. In addition to the contentious formal World Trade memorial, alternative participatory, spontaneous, or community-managed "found space living memorials" were important contributions to healing (Svendsen & Campbell, 2014).

This piece, though partly inspired by a meteorite, was created out of wood: the material provided by climate circumstances at a time concurrent with my experiences. Direct carving integrates the existing natural attributes and inspirations of a given medium toward development of form or narrative, rather than imposing a preexisting idea on material. Sculptors employing the direct carving method combine improvisation as well as deliberate actions to create

DOI: 10.4324/9781003141457-103

new forms – using the materials, imperfections, philosophical and artistic approaches, technical skills, and tools present at a given time.

Figure Int 3.1 Rock That Fell to Earth: Storm Series, oak (sculpture by Lerski); photo by Javier Agostinelli.

Reference

Svendsen, E. S., & Campbell, L. K. (2014). Community-based memorials to September 11, 2001: Environmental stewardship as memory work. In K. G. Tidball & M.E. Krasny (Eds.), *Greening in the red Zone: Disaster, resilience and community greening* (pp. 339–355). Springer Science + Business Media. DOI 10.1 007/978-90-481-9947-1_25

Part III
Navigating transitions

9 Moving to higher ground: Planning for relocation as an adaptation strategy to climate change in the Fiji Islands

Beatrice Ruggieri

9.1 Introduction: Adapting to climate change in the Fiji Islands

The Fiji Islands are becoming increasingly vulnerable to the impacts of climate change. These include sea-level rise, changes in precipitation and wind patterns, ocean acidification and stronger cyclones (IPCC, 2014; The Government of Fiji, World Bank, and Global Facilities for Disaster Reduction and Recovery, 2017). When category 5 tropical cyclone Winston hit Fiji in February 2016, it became the most powerful tropical storm ever recorded in the Southern Hemisphere with over 60% of the population affected and losses estimated at USD 1.4 billion (The Government of Fiji, 2017a). Climate change is already causing significant disruptions to traditional customs and practices in Fiji and other Pacific Island societies (Warner & Van der Geest, 2013). Winston, for example, has had severe impacts on fisheries-dependent communities in Fiji by affecting their main livelihoods as shown by Thomas et al. (2019) and by damaging "raw materials necessary to produce costumes for rituals and prepare herbal medicine, totemic plants/trees, and crops and animals for rituals and ceremonies" (Takahashi & Nemani, 2016). However, as stated by several Island Studies scholars, it is fundamental to avoid associating *islandness* with vulnerability, as the societies have long-standing resilience attributes (Kelman & Kahn, 2013; Walshe et al., 2018). In Fiji, for example, traditional and Indigenous environmental knowledge can provide valuable support in strengthening disaster risk reduction and climate change adaptation.

In the recent decades, considerable attention has been placed on the concept of migration as adaptation and planned relocation has been reframed as a potential adaptation for vulnerable coastal communities (Bogardi & Warner, 2008; McLeman & Smith, 2006; UNHCR, Brookings Institute, & Georgetown University, 2015). While most of the literature on planned relocation in the Pacific has focused on long-distance and international resettlements by highlighting their negative effects on affected communities (Campbell, 2008; Weber, 2016), new studies have emphasized the helpful role of internal resettlement as climate change adaptation (Barnett & Webber, 2010; Ferris, 2015). Internal relocation will guarantee that no one from Fiji will be identified as

DOI: 10.4324/9781003141457-9

Figure 9.1 Mural in Natalei Eco Lodge, Fiji, with memorial to TC Winston.
Source: Photo by Ruggieri.

"climate refugee," a label that Pacific Islanders strongly refuse because of its implicit allusion to apocalyptic futures of sinking, powerlessness, and lack of agency (Farbotko & Lazrus, 2012; McNamara & Gibson, 2009). Instead, Pacific Indigenous people, political leaders, and climate activists are showing significant contribution to the shape of global and regional climate change policies as well as to the promotion of climate justice principles (Kirsch, 2020; McNamara & Farbotko, 2017). In this regard, the Republic of Fiji (2018) has released the Planned Relocation Guidelines, a document that aims to support internal relocation by ensuring the fundamental rights of affected communities and enhancing development opportunities.

This chapter examines the multifaceted complexities of relocation, understood as both a rational solution to climate-change displacement and an often-controversial process. I focus on a set number of case studies that highlights different relocation processes, impacts, and outcomes for affected communities. First, historical cases of relocation involving *iTaukei* communities will be presented. Then, based on primary and secondary data, I will focus on recent examples of internal planned relocation in Fiji, discussing both State-led and autonomous cases of resettlement (Figure 9.1).

9.2 Human mobility patterns in Fiji: Village relocation as a traditional coping strategy

The Fiji Islands, consisting of more than three hundred high and low-lying isles, are well known for the entrenched mobility patterns of its population. Inter-island travel and village shifting toward inland areas have always been an essential support to respond to environmental changes, food insecurity and conflicts, as highlighted by Nunn (2007, 2013, 2019) who observes that Fijian hilltops were once occupied by many fortified villages. Over several occasions,

Fijian Indigenous communities (*iTaukei*) have changed their site and the history of the Country is studded with examples of resettlement, both autonomously managed and institutionally forced. Historical sources show that, since the end of the 19th century, Christianization and colonization in Fiji led to the relocation of inland settlements toward the coast to enable trade contacts with *iTaukei*, facilitate evangelization and test new systems of societal control (Banivanua Mar, 2016; Bennett, 1974; Kaplan, 1989). These kind of "controlled resettlements" brought about several sociocultural changes, especially in terms of the land tenure systems and freedom of mobility so that "the right and capacity to move (or stay) for Indigenous and subject populations was colonized along with their lands" (Suliman et al., 2019, p. 7).

While acknowledging the substantial differences of sociopolitical context in which those resettlements took place, many scholars have pointed out that past relocations can offer valuable lessons for present and future planning, the understanding of associated challenges and how to overcome those (Connell, 2012; McAdam, 2014). Among the 86 cases of community relocation that took place in the Pacific region between 1920 and 2004, Campbell et al. (2005) noted that 37 of them were caused by environmental variability. Also, Janif et al. (2016) described several cases of village relocation in different locations of the Fiji archipelago, verifying that storm surges, flooding and subsidence have been influencing decisions to relocate for a long time. Fijian oral narratives, in particular, testify that inter-island mobilities and internal relocation have always been a significant response to climate and environmental stressors. Similarly, future shifting may be expected to adapt to changes in rainfall patterns and sea-level rise.

In Fiji, there is considerable resistance to abandoning ancestral home sites and/or struggle to give land to those relocating. Islander's attachment to place is a key aspect of Pacific cultural heritage (Farbotko et al., 2018) and customary land – *Vanua* in Fiji – has physical, sociocultural, and spiritual dimensions so that it is considered as an extension of the self (Ravuvu, 1988). Furthermore, as mobility patterns in island societies have been limited and controlled in the past, attachment to place as well as reluctance to move have increased and both represent common constraints to the implementation of relocation as an adaptation measure (Campbell & Bedford, 2014). Therefore, it is vital to recognize the potential of relocation as a practicable solution to keep sociocultural and economic viability of Fijian communities alive without causing additional disruptions. For this reason, adequate attention should be paid to guaranteeing climate and environmental justice in relocation processes and context by ensuring participation and transparency in decision-making, conducting appropriate risk assessments and avoiding further land exploitation (Schade, 2013).

9.3 Fiji National Planned Relocation Guidelines

Regarding Pacific Indigenous people, Suliman et al. (2019, p. 3) suggest that it would be better to talk about Pacific (im)mobilities, thus highlighting the complex holistic system and network of "mobilities and immobilities connecting people, ancestors, stars, canoes and other vessels, ocean, islands and continents." Others have observed that mobility patterns in Oceania are better seen in a perpetual oscillation between *roots and routes* (Bonnemaison, 1984; Lilomaiava-Doktor, 2009). This constant tension between movement and rootedness is a common feature in current decision-making processes of climate-induced migration: Indeed, despite being extremely mobile and having strong regional ties, decision to leave permanently their own land is particularly sensitive and often psychologically costly for Indigenous islanders. In addition, it should be considered that relocation projects usually imply considerable costs and that, in many cases, they are not successful for those who move (Connell & Lutkehaus, 2016; Edwards, 2013).

So far, in Fiji, three low-lying coastal communities have been fully relocated and a few others are in the process of moving while 45 communities have been listed for relocation within the next 5–10 years (Charan et al., 2017; The Government of Fiji, 2017b). To overcome potential disruptive outcomes and considering that relocation is becoming a more common adaptation response to environmental changes, the Fiji government developed its national Guidelines on Planned Relocation – in collaboration with the German Development Agency (GIZ) – after years of consultations between affected communities, government representatives and non-State actors. In addition, this document shows the clear commitment of Fiji in enhancing its national adaptation strategy. Planned relocation in this document is understood as a State-led and development-oriented measure in which a community is physically moved to another location and resettled permanently there (The Government of Fiji and GIZ, 2018). Built on international framework on planned relocation (UNHCR, Brookings Institute, & Georgetown University, 2015), and aligned with national adaptation plan and policies, the Fiji Guidelines reaffirm the relevance of three key principles of relocation:

1. the whole community has to give its consensus for the relocation to take place;
2. the request to move must come from the community; and
3. the recourse to relocation occurs only when all other adaptation options have been exhausted.

As stated by Prime Minister Bainimarama: "These guidelines provide us with a blueprint for engaging our communities in the process of relocation, ensuring proper coordination between our various agencies, sensitizing the process along the lines of gender, and taking into account how marginalized groups, such as children, the elder and those living with disabilities should be catered for" (Bainimarama, 2019). Since 2007, the Fijian government has been involved in financial and technical assistance of communities' relocation, together with the support of external development and aid agencies. Indeed, relocation is expensive and external funding is quite essential due to the low level of economic development of Fiji. To attract international financial resources, last September Bainimarama launched the world's first relocation fund after extensively publicizing its climate relocation policy in many international meetings such as COP23 in Bonn[1].

The "Guidelines" represent a useful tool in planning internal relocation in that they help to minimize potential tensions or prevent the emergence of new vulnerabilities. For instance, disputes over land rights[2] in Fiji are not rare and could even increase as a side effect of relocation if it cannot take place within the village customary land boundaries (Fonmanu et al., 2003; Gharbaoui & Blocher, 2017). In addition, relocation timeline can be subjected to unexpected delays due to several drawbacks: for example, long negotiations with host communities and uncertainties about the setting-up of new economic livelihoods (McNamara & Jacot Des Combes, 2015). In those cases, the Guidelines can be used to facilitate dialogue between stakeholders and better organize the relocation phases. However, it is crucial not to frame relocation as urgent and inevitable since that might hamper other present-day adaptation measures and prevent further investments in sites of climate vulnerability. As illustrated by McMichael et al. (2019, p. 303), in Tokou village (Fiji) the assumption of the "migration as adaptation" idea by local villagers made them "reluctant to invest in maintenance and development of their homes and villages given the threat of sea-level rise" (McMichael et al., 2019, p. 303).

The Fiji National Planned Relocation Guidelines are a first tentative to regulate multiple aspects of relocation, including "planning, coordination and organization, consultation across sectors, and, critically, community participation" (Nichols, 2019, p. 256). However, the document is not yet operative and has not been translated into *iTaukei* language.

9.4 Moving on our own terms: Examples of State-led and autonomous relocation

Tackling the complexity of relocation means acknowledging that there is usually a world of differences between the initial theoretical project and its implementation. In Fiji, internal relocation is being implemented through a cooperative approach between institutional stakeholders and local communities since this is perceived as the preferable way to ensure successful long-term relocation. Nonetheless, during relocation, which often requires years to be concluded, it is not rare to face a series of unexpected events: As stated by McAdam (2015, p. 32) resettlement is "rarely considered successful by those who move." Indeed, even the example offered by the relocation of Vunidogoloa village, which is presented as a good model of participation and cooperation, is far from unproblematic (Figure 9.2).

The village of Vunidogoloa is the first case of planned relocation that took place in Fiji (2014) because of coastal erosion and repeated flooding during high tides and heavy rains. The community asked several times for relocation and after 5 years of consultation and negotiation, a new site was identified almost

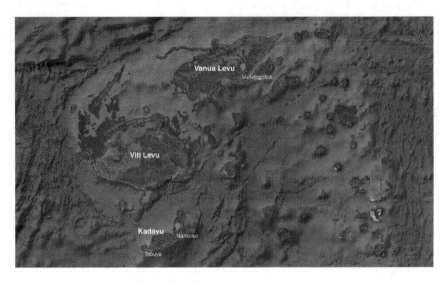

Figure 9.2 Aerial view of Fiji.

Source: Adapted from Earth Engine by Ruggieri.

2 km inland and uphill within village land boundaries, and thus avoiding major concerns regarding land property rights (Tronquet, 2015). However, many issues arose from the vulnerability assessment of the new location since the initial site was found to be prone to erosion and landslides (Tronquet, 2015). In addition, villagers expressed concern about housing structure and village layout. As observed by Bertana (2020), interviewers stated that the houses were smaller (no kitchen inside) and less comfortable than before[3] and that the village layout did not resemble the traditional Fijian village. Every relocation implies a price to be paid and it seems that, in accepting external assistance, "villagers lost agency over certain aspects of their village" (Bertana, 2020, p. 14). In the case of State-led processes, these often reveal asymmetric power relations between the community and external experts (Mortreux et al., 2018).

Similar obstacles emerged in the relocation of the coastal village of Narikoso (ongoing), located in the outer island of Ono in Kadavu Island chain. In 2012, Narikoso villagers sought assistance from the government to mitigate shoreline erosion and frequent flooding. The new site was a few hundred meters inland within the village land boundaries but too small to accommodate every household in addition to be landslide and soil erosion prone (Barnett & McMichael, 2018). Furthermore, as reported by Bertana (2019), the inadequate attention paid during transportation of required building equipment resulted in extensive environmental damages, thus contradicting the goals of strengthening community resilience in the long term and ensuring justice in climate change relocations. Low levels of planning and coordination among actors involved as well as difficulties in finding financial resources are part of the reasons why Narikoso is still waiting for its resettlement to be completed.

Since autonomous community coping strategies in developing countries change with peripherality, it is likely that most rural peripheral settlements in Fiji will develop and implement their own relocation strategies (Nunn et al., 2014; Nunn & Kumar, 2019). Here, in spite of significant alterations brought by colonial legacies and globalization, rural communities appear to be more self-reliant than it is usually reported in public discourses (Bayliss-Smith et al., 1988; Dumaru et al., 2011). This proves to be a critical attitude especially for remote communities experiencing the effects of severe climatic events such as cyclone Winston (Nakamura & Kanemasu, 2020).

The fieldwork I conducted in 2019 in the village of Tabuya revealed many different ways to conceive and manage community relocation without external intervention. I collected qualitative data in July 2019 through in-depth interviews and informal discussions with 10 residents of Tabuya, which is a small village of approximately 70 people located on the southern coast of Kadavu. Here, people live in a semi-subsistence regime by consuming self-grown vegetables and fish and by selling the remaining parts in the few nearest markets such as the one in Kavala. In addition, women are involved in the traditional art of weaving mats from *pandanus* leaves, which are both sold and used in local ceremonies and celebrations (Figure 9.3).

In the last decades, the tidal waves exacerbated by sea-level rise and stronger cyclones have repeatedly destroyed infrastructures and damaged homes in Tabuya, which is situated close to the ocean and between two creeks that are always flooded during intense rainfall. During the past decades, villagers have progressively shifted their homes a few meters away from the ocean to feel safer: "In the 1980s houses were around 20 m further to the ocean because it was a safe place," a woman observed. After rejecting the construction of a sea-wall, which would have been only a short-term solution, Tabuya villagers agreed to relocate in 2016. As another woman stated: "The only option for now is to move but it takes time. This village will stay but we will move inland. We will miss this view and this place but we don't have another choice." In this respect, it is worth noting that the possibility to relocate while being able to continue to access the original village is considered as an integral part of community resettlement in Fiji.

At the time of visiting, only a few houses had already been built and only two families were living in the new site, located approximately 500 meters inland compared to the lower village. The short-distance is a very important factor in relocation processes because the sense of community is not eroded in the long term. Now, during cyclones and potential flooding, these new homes are used as an evacuation and rescue center. However, the new village site does not provide a sufficient space to accommodate all the villagers and it is very likely that land negotiations will have to be conducted in the next few years and that social tensions will arise (Figure 9.4).

Figure 9.3 Village of Tabuya, Fiji, rebuilt uphill.
Source: Google Earth.

Figure 9.4 Homes in the Village of Tabuya, Fiji.
Source: Photo by Ruggieri.

When discussing relocation, participation and decision-making are challenging aspects to be underscored. Decision-making in rural communities in Fiji, for example, is often subordinate to a hierarchical and seniority system, which may exclude women and younger people from central decisions (Neef et al., 2018). However, in Tabuya, villagers meet regularly to discuss village issues and interviewees revealed an appropriate level of inclusivity and general participation in critical matters such as village relocation. Indeed, the suggestion to move was made by the Chief of the village but the decision has been taken after listening to villagers' opinions usually during periodic meetings and informal discussions (*Talanoa*). Both land-related negotiations and decisions, instead, continue to remain men-dominated issues because "they are the landowners," as frequently observed by interviewed women.

In Tabuya, results of community consultation indicate that relocation has not been identified as a mandatory measure and each household can freely decide whether to relocate or not and do so on their own terms. Nonetheless, it is worth noting that this kind of "staggered relocation" can be risky because the most vulnerable may not have the necessary resources to move without external financial and technical support. At the same time, locally managed decision-making on environmental matters should be made by well-informed people to avoid null adaptation and potential *maladaptation* (Barnett & O'Neill, 2011; Juhola et al., 2016). In Tabuya, for example, awareness about climate change effects is widespread but barriers due to age and/or religious beliefs can discourage long-term anticipatory adaptation responses.

9.5 Conclusion

The impacts of climate change are clearly visible in Fiji and demand the implementation of ad hoc adaptation strategies. Among those, planned relocation is being considered as a last resort option for many communities affected by sudden climate extremes and environmental degradation. As previously highlighted, short-distance and internal movements to higher and safer locations are increasingly identified as an adaptive measure instead of a failure to adapt. The Fiji Planned Relocation Guidelines are a first proactive attempt to manage the process of relocation in a coordinated, participative, and socioculturally sensitive manner (Barnett & McMichael, 2018) to minimize climate related loss and damages (McNamara et al., 2018).

However, it is acknowledged that planned relocation can lead to several challenges so that specific attention needs to be paid to the role of decision-making systems, the presence of asymmetrical power relations, and the availability of sufficient funding. In addition, adaptation policies need to take into consideration the sociocultural, health, spiritual and psychological issues of resettlement processes (Adger et al., 2011). For instance, in Fiji, social tensions around land and resistance to move are both potential drawbacks that, in some cases, can prevent the relocation. Moreover, peripherality can contribute to the increase of exposure and vulnerability of certain groups to climate change. In archipelagic countries like Fiji, it is likely that peripheral rural communities will take the lead in planning and implementing autonomous relocation as shown by Tabuya inhabitants who are expressing high levels of agency and resilience. However, risks associated with autonomous retreat include difficulties in land negotiations, potential exclusions from decision-making of certain groups of people within a community, perpetuation of inequalities, and underestimation of environmental threats (Neef et al., 2018).

To understand the multiple dimensions and forms that relocation can assume in island societies and enhance its adaptive role, more research is needed. First of all, the advance of knowledge in community relocation requires the adoption of new research methodologies that have to integrate context and cultural specificities, which may subtend different worldviews and responses. Indeed, too many adaptation projects continue to fail when their goals are designed with an insufficient involvement of local perspectives and cultural specificities (Piggott-McKellar et al., 2019). Second, it is fundamental to assume that external stakeholders and community's ideas about relocation may not coincide. This means that a principle such as the right to stay has to be assured. In addition, careful attention should be paid to the most vulnerable people across the three phases of relocation. Finally, it is essential to consider climate-induced internal relocation only as a complementary response that must not stop industrialized countries from taking more ambitious mitigation efforts.

9.6 Recommendations

- Academic research: Enhance research on planned relocation as adaptation by improving the adoption of culturally sensitive research methodologies so that its outcomes can be meaningful for those directly affected. For instance, it is essential to support the involvement of local researchers in that specific matter and ensure community inclusion and participation throughout the relocation process.
- Adaptation in practice: It is vital to advance knowledge on community-based adaptation strategies, which are complementary to institutional measures. As local communities' traditional knowledge and values are keys to increase national adaptation, there is a need to carefully incorporate that knowledge into adaptation plans.
- Policy: Policy makers should be aware that their ideas about managing relocation may not reflect those expressed by affected communities. In this respect, they should respect claims of voluntary immobility while ensuring that no one will become "trapped."
- Planned relocation in the Republic of Fiji: Improve communication and coordination between institutional stakeholders across different scales while ensuring clear and transparent communication with *iTaukei* communities in need of relocation.

Notes

1 For an in-depth and critical analysis of Bainimarama's work in UN negotiations and national adaptation policies, see Kirsch (2020).
2 In Fiji, *iTaukei* own almost 90% of the land communally through traditional land owning units, *mataqali* which in turn form the *yavusa*, a group of people "claiming descent usually in the male line from a common mythological ancestral spirit" (Parke, 2014, p. 57).

3 However, it has to be remarked that new houses were built as able to respond to a category five TC and equipped with solar panels and private toilets. The building of 30 identical houses was in accordance with the choice of the villagers to be treated equally (Tronquet, 2015).

References

Adger, W. N., Barnett, J., Chapin III, F. S., & Ellemor H. (2011). This must be the place: Underrepresentation of identity and meaning in climate change decision-making. *Global Environmental Politics*, 11(2), 1–25.

Bainimarama, F. (2019). We need to arm ourselves with the ability to act now. We can't wait for communities to be drowned out by the encroaching tides. https://cop23.com.fj/climate-relocation-and-displaced-peoples-trust-fund/

Banivanua Mar, T. (2016). *Decolonisation and the Pacific. Indigenous globalisation and the ends of empire*. Cambridge University Press.

Barnett, J., & McMichael, C. (2018). The effects of climate change on the geography and timing of human mobility. *Population and Environment*, 39, 339–356.

Barnett, J., & O'Neill, S. (2011). Maladaptation. *Global Environmental Change*, 20(2), 211–213.

Barnett, J., & Webber, M. (2010). Migration as adaptation: Opportunities and limits. In J. McAdam (Ed.), *Climate change and displacement: Multidisciplinary perspectives* (pp. 37–56). Hart Publishing.

Bayliss-Smith, T. P., Brookfield, M., Bedford, R., & Latham, M. (1988). *Islands, islanders and the world: The colonial and post-colonial experience of Eastern Fiji*. Cambridge University Press.

Bennett, J. A. (1974). *Cross-cultural influences on village relocation on the Wheather Coast of Guadalcanal, Solomon Islands, c-1870-1953*. University of Hawaii.

Bertana, A. (2019). Relocation as an adaptation to sea-level rise: Valuable lessons from the Narikoso Village Relocation Project in Fiji. *Case Studies in the Environment*, 3(1), 1–7.

Bertana, A. (2020). The role of power in community participation: Relocation as climate change adaptation in Fiji. *Politics and Space C*, 38(5), 902–919.

Bogardi, J., & Warner, K. (2008). Here comes the flood. *Nature Climate Change*, 2, 9–11.

Bonnemaison, J. (1984). The tree and the canoes: Roots and mobility in Vanuatu Societies. *Pacific Viewpoint*, 25(2), 117–151.

Campbell, J. (2008). *International relocation from Pacific Island Countries: Adaptation failure?* International Conference in Environment, Forced Migration & Social Vulnerability, Bonn, Germany.

Campbell, J., & Bedford, R. (2014). Migration and climate change in Oceania. In E. Piguet & F. Lackzo (Eds.), *People on the move in a changing climate: The regional impact of environmental change on migration* (pp. 177–204). Springer.

Campbell, J., Goldsmith, M., & Koshy, K. (2005). Community relocation as an option for adaptation to the effects of climate change and climate variability in Pacific Island Countries (PICs). Asia-Pacific Network for Global Change Research. Final Report.

Charan, D., Kaur, M., & Sing, P. (2017). Customary land and climate change induced relocation - A case study of Vunidogoloa Village, Vanua Levu, Fiji. In L. Filho (Ed.), *Climate change adaptation in Pacific Countries fostering resilience and improving the quality of life* (pp. 19–33). Springer.

Connell, J. (2012). Population resettlement in the Pacific: Lessons from a hazardous history? *Australian Geographer*, *43*(2), 127–142.

Connell, J., & Lutkehaus, N. (2016). *Another Manam? The forced migration of the population of Manam Island, Papua New Guinea, due to volcanic eruption 2004–2005*. IOM.

Dumaru, P., et al. (2011). *The adaptive capacity of three Fijian village communities: Bavu, Druadrua and Navukailagi*. University of the South Pacific.

Edwards, J. (2013). The logistics of climate-induced resettlement: Lessons from the Carteret Islands, Papua New Guinea. *Refugee Survey Quarterly*, *32*, 52–78.

Farbotko, C., & Lazrus, H. (2012). The first climate refugees? Contesting global narratives of climate change in Tuvalu. *Global Environmental Change*, *22*(2), 382–390.

Farbotko, C., McMichael, C., Dun, O., Ransan-Cooper, H., McNamara, K. E., & Thornton, F. (2018). Transformative mobilities in the Pacific: Promoting adaptation and development in a changing climate. *Asia & the Pacific Policy Studies*, *5*, 393–407.

Ferris, E. (2015). Climate-induced resettlement: Environmental change and the planned relocation of communities. *SAIS Review of International Affairs*, *35*(1), 109–117.

Fonmanu, K. R., Ting, L., & Williamson Ian, P. (2003). Dispute resolution for customary lands: Some lessons from Fiji. *Survey Review*, *37*(289), 177–189.

Gharbaoui, D., & Blocher, J. (2017). The reason land matters: Relocation as adaptation to climate change in Fiji Islands. In A. Milan (Ed.), *Migration, risk management and climate change: Evidence and policy responses* (pp. 149–173), Global Migration Issues 6. Springer.

IPCC. (2014). Small islands. In *Climate change 2014: Impacts, adaptation, and vulnerability. Part B: Regional aspects. Contribution of Working Group II to the Fifth Assessment Report of the Intergovernmental Panel on Climate Change*.

Janif, S. Z., Nunn, P. D., Geraghty, P., Aalbersberg, W., Thomas, F. R., & Camailakeba, M. (2016). Value of traditional oral narratives in building climate-change resilience. Insights from rural Communities in Fiji. *Ecology and Society*, *21*(2), Article 7.

Juhola, S., Glaas, E., Linnér, B., & Neset, T. S. (2016). Redefining maladaptation. *Environmental Science and Policy*, *55*(1), 135–140.

Kaplan, M. (1989). The "dangerous and disaffected native" in Fiji: British colonial constructions of the Tuka Movement. *Social Analysis: The International Journal of Anthropology*, *26*, 22–45.

Kelman, I., & Khan, S. (2013). Progressive Climate Change and disasters: Island perspectives. *Natural Hazards*, *69*(1), 1131–1136.

Kirsch, S. (2020). Why Pacific Islanders stopped worrying about the apocalypse and started fighting climate change. *American Anthropologist*. Online Version of Record before inclusion in an issue.

Lilomaiava-Doktor, S. (2009). Beyond "migration": Samoan Population Movement (Malaga) and the geography of social space (Vā). *The Contemporary Pacific*, *21*(1), 1–32.

McAdam, J. (2014). Historical cross-border relocations in the Pacific: Lessons for planned relocations in the context of climate change. *The Journal of Pacific History*, *49*(3), 301–327.

McAdam, J. (2015). Lessons from planned relocation and resettlement in the past. *Forced Migration Review*, *49*, 30–32.

McLeman, R., & Smith, B. (2006). Migration as an adaptation to climate change. *Climatic Change*, *76*(1), 31–53.

McMichael, C., Farbotko, C., & McNamara, K. E. (2019). Climate-migration responses in the Pacific Region. In C. Menjivar, M. Ruiz, & I. Ness (Eds.), *The Oxford handbook of migration crises* (pp. 297–314). Oxford University Press.

McNamara, K. E., Bronen, R., Fernando, N., & Klepp, S. (2018). The complex decision-making of climate-induced relocation: Adaptation and loss and damage. *Climate Policy*, 18, 111–117.

McNamara, K. E., & Gibson, C. R. (2009). "We do not want to leave our land": Pacific ambassadors at the United Nations resist the category of "climate refugees". *Geoforum*, 40(3), 475–483.

McNamara, K. E., & Jacot Des Combes, H. (2015). Planning for community relocations due to climate change in Fiji. *International Journal of Disaster Risk Science*, 6, 315–319.

McNamara, K. E., & Farbotko, C. (2017). Resisting a 'doomed' fate: An analysis of the Pacific Climate Warriors. *Australian Geographer*, 48(1), 17–26.

Mohanty, M. (2017). Fiji Kava: Production, trade, role and challenges. *The Journal of Pacific Studies*, 37(1), 2017.

Mortreux, C., de Campos, R. S., Adger, W. N., Ghosh, T., Das, S., Adams, H., & Hazra, S. (2018). Political economy of planned relocation: A model of action and inaction in government responses. *Global Environmental Change*, 50, 123–132.

Nakamura, N. & Kanemasu, Y. (2020). Traditional knowledge, social capital, and community response to a disaster: Resilience of remote communities in Fiji after a severe climatic event. *Regional Environmental Change*, 20(23), 1–14.

Neef, A., Benge, L., Boruff, B., Pauli, N., Weber, E., & Varea, R. (2018). Climate adaptation strategies in Fiji: The role of social norms and cultural values. *World Development*, 107, 125–137.

Nichols, A. (2019). Climate change, natural hazards, and relocation: Insights from Nabukadra and Navuniivi villages in Fiji. *Climatic Change*, 156, 255–271.

Nunn, P. (2007). Holocene sea-level change and human response in Pacific Islands. *Earth and Environmental Science Transactions of the Royal Society of Edinburgh*, 98(1), 117–125.

Nunn, P. (2013). The end of the Pacific? Effects of sea level rise on Pacific Island livelihoods. *Singapore Journal of Tropical Geography*, 34, 143–171.

Nunn, P. (2019). Forgotten citadels: Fiji's ancient hill forts and what we can learn from them. *The Conversation*. http://theconversation.com/forgotten-citadels-fijis-ancient-hill-forts-and-what-we-can-learn-from-them-121103.

Nunn, P., Aalbersberg, W., Lata, S., & Gwilliam, M. (2014). Beyond the core: Community governance for climate-change adaptation in peripheral parts of Pacific Island Countries. *Regional Environmental Change*, 14, 221–235.

Nunn, P., & Kumar, R. (2019). Measuring peripherality as a proxy for autonomous community coping capacity: A case study from Bua Province, Fiji Islands, for improving climate change adaptation. *Social Sciences*, 8(225), 1–26.

Parke, A. (2014). The Yavusa: The ideal and the reality. In M. Spriggs & D. Scarr (Eds.), *Degei's descendants: Spirits, place and people in pre-cession Fiji* (pp. 57–74). Terra Australis 41. ANU Press.

Piggott-McKellar, A. E., McNamara, K. E., Nunn, P. D., & Watson, J. E. M. (2019). What are the barriers to successful community-based climate change adaptation? A review of grey literature. *Local Environment*, 24(4), 374–390.

Ravuvu, A. (1988). *Development or dependence: The pattern of change in a Fijian village*. Institute of Pacific Studies and Fiji Extension Centre: The University of the South Pacific.

Schade, J., Faist, M., & Schade, J. (2013). Climate change and planned relocation: risks and a proposal for safeguards. In M. Faist & J. Schade (Eds.), *Disentangling migration and climate change. Methodologies, political discourses and human rights* (pp. 183–206). Springer.

Sofer, M. (2007). Yaqona and the Fijian periphery revisited. *Asia Pacific Viewpoint*, 48(2), 234–249.

Suliman, S., Farbotko, C., Ransan-Cooper, H., McNamara, K. E., Thornton, F., McMichael, C., & Kitara, T. (2019). Indigenous (im)mobilities in the Anthropocene. *Mobilities*, 14(3), 1–21.

Takahashi, A., & Nemani S. (2016). Integrating ICH in post-disaster needs assessments: A case study of Navala village. *ICH Courier*, 28, 8–11. https://ichcourier.unesco-ichcap.org/wp-content/uploads/ICH-Courier-V28-web.pdf

The Government of Fiji, World Bank, and Global Facilities for Disaster Reduction and Recovery. (2017). Fiji 2017: Climate vulnerability assessment-Making Fiji climate resilient. World Bank. Washington, DC. https://cop23.com.fj/wp-content/uploads/2018/02/Fiji-Climate-Vulnerability-Assessment-.pdf

The Government of Fiji. (2017a). Tropical cyclone Winston. Fiji Government and World Food Programme Joint Emergency Response. Lessons Learned Workshop Report. UN. https://reliefweb.int/sites/reliefweb.int/files/resources/fiji_lessons_learned_workshop_report_external.pdf

The Government of Fiji. (2017b). 5 year and 20 year national development plan. https://cop23.com.fj/wp-content/uploads/2018/03/5-Year-and-20-Year-National-Development-Plan.pdf

The Government of Fiji and GIZ. (2018). *Planned relocation guidelines*.

Thomas, A. S., Mangubhai, S., Vandervord, C., Fox, M., & Nand, Y. (2019). Impact of tropical cyclone Winston on women mud crab fishers in Fiji. *Climate and Development*, 11(8), 699–709.

Tronquet, C. (2015). From Vunidogoloa to Kenani: An insight into successful relocation. *The State of Environmental Migration*, 1–22.

UNHCR, Brookings Institute, & Georgetown University. (2015). Guidance on protecting people from disasters and environmental change through planned relocation. https://www.unhcr.org/protection/environment/planned-relocation-guidance-october-2015.html

Walshe, R. A., Seng, D. C., Bumpus, A., & Auffray, J. (2018). Perceptions of adaptation, resilience and climate knowledge in the Pacific: The cases of Samoa, Fiji and Vanuatu. *International Journal of Climate Strategies and Management*, 10(2), 303–322.

Warner, K., & Van der Geest, K. (2013). Loss and damage from climate change: Local-level evidence from nine vulnerable countries. *International Journal of Global Warming*, 5(4), 367–386.

Weber, E. (2016). Only a pawn in their games? Environmental (?) migration in Kiribati – Past, present and future. *DIE ERDE: Journal of the Geographical Society of Berlin*, 147(2), 153–164.

10 Voices of Arraigo: Redefining relocation for landslide-affected communities in the informal settlements of Bogota, Colombia

Duvan H. López Meneses, Arabella Fraser, and Sonia Hita Cañadas

10.1 Introduction: Informality, risk, and managed relocation in Bogota

Informal settlers have developed around 45% of Bogota's urban area (Torres, 2007). The geography of informal settlements coincides with precarity, exclusion, environmental stress, and the dynamics of political violence and drug trafficking in Colombia (Ordoñez et al., 2012; Zeiderman, 2016). In this context, the interaction with the biophysical dynamics of rainfall, runoff and erosion, provides the perfect scenario for producing vulnerability to climate risks and impacts (López & Hita, 2018) (see Figure 10.2). In the past 40 years, hydrometeorological events have caused almost 200 flood events and a similar number of landslides in Bogota, affecting 15,000 households, and causing around 100 fatalities (OSSO Corporation and EAFIT, 2011). Increasing heavy rainfall due to climate change, is likely to increase risk across Colombia's urban areas and in particular informal settlements (Lampis & Fraser, 2012).

Regulations in Bogota have included explicit procedures for managed relocation from 1998. Resettlement is defended as "a measure to address exposure as one of the components of vulnerability, resulting in the nullification of risk conditions by (…) changing the location of exposed elements, in this case the population" (Ramirez, 2011, p. 18). The resettlement of families is justified using the concept of "*nonmitigable risk*" (Ramírez & Rubiano, 2009, p. 37). Mitigability is defined by the state as "the condition when it is technically, economically, socially and politically feasible to intervene in a territory to reduce risk to a level that makes it possible to maintain population, infrastructure and economic activities, within a reasonable and socially acceptable range of security" (Ramírez & Rubiano, 2009, p. 36). In the case of landslides, the categories of "mitigability" and "nonmitigability" were introduced as soil properties to be delimited by technical studies (Decree 190, 2004, Art. 110, 138, 140, 146). The Colombian regulations impose strong land use restrictions on nonmitigable risk areas (Decree 190, 2004, Art. 140, 146, 387, 393), where inhabitants should be included in a "program of resettlement" (Decree 190, 2004, Art. 138, 158, 297, 301). There are no mechanisms for affected populations to question nonmitigable risk conditions, once defined (Agreement 26, 1996,

Decree 190, 2004, and Decree 255, 2013). National regulations allow public authorities to access properties under climate risk by voluntary or mandatory requests and evictions, operated by local authorities (Law 388, 1997, Art. 58).

The target established in 2004 was to relocate 10,000 households from non-mitigable risk areas before 2010 (Decree 190, 2004, Art. 158). In 2013, the municipal agency that once specialized in disaster prevention and response was transformed to take on the broader mandate of climate change adaptation (Agreement 546, 2013, Art. 8). In parallel, the target was expanded to include 12,000 additional households, with the objective of freeing up more than 20 km of "protection strips" around major water sources as a climate change adaptation measure (El Tiempo, 2013; Zeiderman, 2016). After more than 20 years of applying resettlement programs in Bogota, the results have fallen far short of the original aspirations of the program, and even further short of these more recent aspirations to clear corridors of the city to reduce climate change impacts (López, 2017). By 2012, 9,043 families had been included in the program, 43% of which had been resettled while the rest are waiting to be resettled (Campos et al., 2011). Despite major efforts, in the period 2013–2015, only 1,000 resettlements were carried out (Secretaría Distrital de Planeación, 2016). By contrast, informal settlements in risk conditions continued to grow (Alcaldía Mayor de Bogotá, 2016). In 2020, local media – quoting institutional sources – reported almost 26,000 "informal invasions" in the city (El Espectador, 2020). This gap between the results of institutional efforts to resettle and the ongoing dynamic of informal occupation reveals serious limitations in the program, but despite this, the applicability of resettlement policy in Bogota has never politically been called into question.

10.2 The concept and struggles of arraigo

A space for collaboration emerged in Bogota under the name ARRAIGO in response to risk-related resettlement policies. ARRAIGO is a platform of neighborhood organizations, social leaders, scholars, sponsors, volunteers and activists, which aims to (i) promote reflection and public discussion about managed relocation, (ii) protect the rights of citizens subject to resettlement measures, and (iii) seed ideas and innovative proposals for alternatives to what could be called "desarraigo" (or unrooting), instead looking to adaptation and the recovery of degraded urban spaces, from a rights perspective, putting human well-being, and dignity at the center. The ARRAIGO platform has more than 30 associated organizations, coordinating a common agenda and producing material, such as an audio–visual series "Voices of arraigo" (Hita, 2015a, 2015b), to allow people living in areas at risk to directly express their experiences.

The following section weaves together the voices of inhabitants affected by risk and resettlement who participate in the platform. Some testimonies were collected during fieldwork in Bogotá between June and November of 2015; others were collected via telephone in July 2020. People volunteered their testimony as a contribution to a project of making publically and politically visible the problems that affect them. However, in the most of the cases, real names were avoided and replaced to protect the identity of the participants.

Figure 10.1 A photo of a tree standing on the ruins of a resettled house to illustrate the arraigo and its persistence.

The expression "arraigo" in Spanish (literally "rooting") denotes an organic relationship between inhabitants and the space they inhabit (Figure 10.1). A community member of Cerro Norte (Usaquén), expresses it as follows:

> "That tree over there was planted 40 years ago and the roots of that tree are just like mine, 40 years old. I don't want to lose that at all. That's why I wouldn't leave this place." [Interview with SH: 12.10.2015]

Mrs. Celis, a resident of the San Cristóbal district in the south-east of Bogotá, gives an account of her "arraigo" in the following expression:

> "I have my roots here, my daughters and my grandchildren were born here too (…) I built this house here with sacrifice, penny to penny, brick to brick." [Interview with SH: 13.11.2015]

Giovanna, from the neighborhood of El Socorro, in the Rafael Uribe district, linked her "arraigo" to family tradition, and to a promise, which she maintains, to respect the will of her father:

> "My grandfather owned this land. My father bequeathed his part. I remain here because I was born and raised in this place. My dad told me to take care of this house, not to let it be lost, because it was the fruit of his effort and youth (…) I will carry out this promise until the last day of my life, I will take over my neighborhood and take care of it." [Interview with DL: 01.07.2020]

10.2.1 Struggles over rights, compensation and housing

Risk-related resettlement in Bogota has caused significant disruptions, affecting lives, communities and territories. Jair Clavijo's family came to Bogotá in 1979.

They received a piece of land in the Cordillera neighborhood, in Ciudad Bolívar district (Figures 10.1 and 10.2). He explains:

> "My family built a wooden ranch in Cordillera but, with time, the announcements came that there were going to be avalanches. My family received a relocation offer from the authorities: They would give us half the value of a relocation house and the other half would be facilitated by a loan, but my dad just turned it down.
>
> At that time, I was three years old. Bienestar Familiar (a national institution in charge of supporting vulnerable families, including providing shelter) announced that it was necessary to take the children who were in danger. So they took me away, and I grew up with an adoptive family until the age of 9 or 10 when I met my parents again, by accident. I could only return to them when I reached adult age, 18 years, after my father's death in 1995." [Interview with DL: 02.07.2020]

During the rainy season of 2011, an emergency occurred in the surroundings of Cordillera. That would trigger a massive relocation in the sector, and would provide another turn to Jair Clavijo's life.

> "I used to visit mom, wishing to improve the condition of her house. So, in November of 2011 I got enough savings and I demolished the wood house, keeping just the concrete slab there. I brought mama to my rented room in another district of the city and was ready to finish the construction, then that natural disaster occurred. So, the authorities prevented me from doing any building activity on the plot. As there was no housing at the time of the emergency, we were denied any right over our land. So, we were also denied relocation." [Interview with DL: 02.07.2020]

Figure 10.2 Panoramic views of Bogota from Cordillera Sur, an informal neighbourhood partially demolished after relocation.

At the beginning of 2002, in the neighborhood of Codito, District of Usaquen, Rodrigo Mendoza's father's house was declared at risk and they were evicted, by force. He was excluded from compensation payments due to a registration error in the resettlement process, which was recently compensated by a court.

> "... 80 riot police officers arrived, with judicial police, tankers and the local mayor. They started the evictions in the morning, taking things from the house down to the street and sealing it up, so you could no longer go back in.
>
> After we left here the change in life was drastic. My father, after having a 5-roomed house with 560 square meters of land, was reduced to small social housing chosen by the state institutions themselves, with a single room and bathrooms without ventilation. My brothers also got housing solutions but returned to Codito after dealing with delicate issues of insecurity and drug sales in their new neighborhoods.
>
> At the moment of the eviction I was not recognized by the authorities as entitled to resettlement. So I opened a legal action and now I am supposed to be a co-owner of that land with the institutions. What they denied to pay when they resettled us has now been restored to me. But they will never restore my father's life, who died really in sadness affected by what happened to his plot." [Interview with SH: 12.10.2015]

The testimonies from Jair Clavijo and Rodrigo Mendoza reveal the legacies of resettlement from one generation to the next. The disturbance of resettlement occurs from the very first moment, when plots are declared as "at risk." This operation implies a suspension in the regime of rights for those involved, who then become subjects of a transitory situation. Their access to a house depends on the practice of regulation among state institutions, in a time-consuming process without a defined deadline. Even when completed, resettlement implies a form of "victimization," whose effects endure after physical relocation has taken place.

Mrs. Celis reports her experience of the process, which started in 2007 and still continues:

> "This area was declared at imminent risk in 2007 when we were supposed to evacuate but, 13 years later, we are still here. Those who demolished their houses went to pay rent. A neighbor who demolished has been paying rent for four years and has not yet received his relocation house.

My house - as you may see - is 5 levels and includes an independent floor for each one of my daughters. So, we just couldn't live in an apartment like the one they are offering us. So, I always say: I will gladly accept the demolition if the authorities pay the real value of my house. I do not want to just go and pay rent." [Interview with SH: 13.11.2015]

Mariela Caro is facing the consequences of two successive resettlements: Since 1998, a first resettlement linked to the Water Company's protection program for ravine corridors. More recently, after acquiring a house outside the ravine corridor, she was again incorporated into a nonmitigable risk area established as the result of climate adaptation measures launched in 2013.

"I demolished my house in 2004 and I just recently received the new house, more than 15 years after being included in the resettlement program. Do you know why? Because for the housing respository option I initially chose, a second-hand house in this same neighborhood, the institutions took five months to go and evaluate that house, and because of that the vendors retracted. The institutions force you to choose only the house that they offer, which is really a matchbox. They say it is decent but it isn't.

I just received my new house and it is already affected, incurring serious damages due to a rupture of the sewage lines and flooding with total loss of my belongings. I was lucky to preserve the rented house, here in Quebrada Limas, we made a family effort to conclude a purchase agreement on it, and now it seems that, again, this other house is in the risk area of the ravine, now I am even more convinced that I want to stay here." [Interview with SH: 18.10.2015]

Mariela's story illustrates the multiple institutional weaknesses to provide effective relocation and beyond, to preserve emptied spaces against environmental degradation and socially produced risks. In contrast, it also exemplifies how community actors based on a strong sense of belonging, hereby interpreted as arraigo, are capable to even constitute physical and symbolic barriers against urban degradation and criminality (Hita, 2015a) (Figure 10.3).

Ferney Arenas, displaced from the rural regions of Colombia, leads a popular housing organization which promotes housing solutions for displaced victims of the Colombian armed conflict. Ferney and 12 families occupy a former risk resettlement area in the Paraíso neighborhood in the Ciudad Bolívar district (Figure 10.4). His testimony highlights the need for structural solutions to housing if risk reduction is to guaranteed:

Figure 10.3 Mariela's orchard. An ornamental and horticulture garden self-funded by a resettled woman in her former plot and surroundings, preventing environmental degradation.

Figure 10.4 Housing project promoted by Ferney Arenas.

> *"The government proposes that we make a programmed saving, on which they would make us a loan to complete the total, for a house that does not cover the family's needs.*

Where we currently live is a spacious and cozy space (it is 6 × 15 meters). A 32 square meter apartment would not be a deal, we prefer to be here fighting and hoping that one day the government will give us a realistic offer.

Some people were attracted by the offer, after they were deemed eligible for housing. But this only goes as far as giving them resources for 8 months of renting. Once that period was finished the subsidy was suspended but there was not still any solution, so many of them returned [to the risk area]." [Interview with SH: 01.11.2015]

10.2.2 Defining mitigable and nonmitigable risks

Whilst the concept of mitigability guides public decision-making about resettlement (Ramírez & Rubiano, 2009, p. 36), this is often at odds with peoples' preferences to stay or abandon their homes. A lack of transparency and agreement obscures what is deemed mitigable. The following testimonies illustrate people's experiences and perceptions of this process, illustrating how it is as much a social and political process, as a technical one.

Rodrigo Mendoza describes how his father's property was not in a risk zone, rejecting the technical authority of state institutions and warning about the inequality inherent in the operationalization of risk analysis:

"From my experience, I would suggest to those who have been given a classification of un-mitigable risk, to ask for another opinion. Authorities declare risk conditions without the existence of any risk, as happened to me, and then a court confirmed it.

The authorities said this land does not support water because it becomes mud. Meanwhile, a short distance from here, along the same mountain range, a Congressman was allowed to build two big houses attached to the foot of the slope; in front of such houses there are water spills, and a stream pass. Illogical things. Here is risk because the ground does not support humidity, although here there is no humidity. But there, they approved the construction although the humidity and the hill are much higher." [Interview with SH: 12.10.2015] (Figure 10.5)

Mrs. Celis and the people of her neighborhood also argue the same:

"The instability is due to our abandonment by the State, because there is no maintenance on a national road (the old highway to the Llano) that we have uphill. In my opinion, we are not at high risk, they just want to 'score a goal for us.'" [Interview with SH: 13.11.2015]

Another neighbor in claimed that "he will leave only from here to the cemetery," and added:

Figure 10.5 Uneven urban landscape and differential risk in coexisting real estate and urban informal projects over the hills of Usaquen in Bogota, Colombia.

> "Twenty days it took the responsible company to come. The root of the tree that was about to fall, turned and raised the pavement. The water, which has always spilled out from the rock, began to leak there. It is clean water, those who live upstream even cook with that water. We are rich in water, but poor in the eyes of the state." [Interview with SH: 13.11.2015]

Ferney Arenas illustrates perceptions of the uneven application of risk criteria (Figure 10.4).

> "In the countryside you live near a creek or a mountain and it is not declared as a risk zone, you do not see it as a risk. Here the State has declared mountain areas as high-risk areas, but it is not really so much for the real risk but for the interests that there are.
>
> The fact is that authorities do not have the resources for housing. But if we examine new projects of (formal) housing, they are placed on the hill." [Interview with SH: 01.11.2015]

Cerro Norte in Usaquen is next to the wealthiest locations of Bogota (Figure 10.5). Inhabitants there question the procedures used to rank their installations as "nonmitigable." Mercedes Santana explains:

> "We feel violated because the institutions want to have us removed. At the beginning, authorities had no will to invest and declared us in high risk, but we

Figure 10.6 Landscape of ruins 20 years after relocation in Altos de la Estancia.

fought, and we obtained, by this way, a very large amount of resources that allowed the realization of high-tech mitigation work. Despite this, they are still notifying us that we must leave due to the same risk condition that was mitigated by the works." [Interview with SH: 12.10.2015]

As the indicators of resettlement suggest and testimonies confirm, even when relocation takes place, people are neither resettled permanently and securely, nor are so-called "risk spaces" effectively controlled. In fact, areas declared at high risk are subject to multiple cycles of resettlement and reoccupation. The imposition of managed relocation thereby concerns those living around boundaries of nonmitigable risk zones as well as those within them. Jacinto Serrano has witnessed the historic evolution of Altos De Estancia, said to be the largest urban landslide zone in Latin America with 73 hectares declared under nonmitigable risk (Figure 10.6). He commented on the political and territorial dynamics of risk management policies:

"The landslide has resulted from the absence of the state, for not exercising its planning duties. The entities that have the competence to administer and guard this space are not visible. That allows reoccupation to occur. If the state had a more formal presence: daily, weekly, or monthly, it wouldn't happen. But they come and make an intervention and they leave after that, 20 days, one month, two months, or three months without doing anything, so others (occupants) come back gaining strength...When one government ends and the next one begins the process is also broken – the government that enters does not give continuity.

In this way, all the effort of the previous years is lost and the land is again vulnerable, due to the management." [Interview with DL: 01.07.2020]

In San Juan de Usme, a unique mountainous area in the city was, along with Altos de la Estancia, granted the title of Special Protection Park for Risk. The clearance of almost 6 hectares resulted from massive resettlement processes. However, in the words of Bernardo Romero, the authorities "failed to make the grade" in exercising good governance and incorporating the territory within the city:

"We have been established in this sector for 50 years and legalized since 1982. In 1999 it was declared high risk but mitigable. Some mitigation works were ordered but the government never accomplished it, sewerage systems were also ordered, it was done for wastewater but not for rainwater. However, what was done lacked quality.

There has also been mining here for more than 50 years, using dynamite, infiltrating water, and circulating with heavy traffic. The authorities calculated traffic limits and installed road signs but it was not respected. This influenced the rupture of the main tube that swept away houses in my neighborhood.

The rupture of the tube in December 2010 was the event responsible for the high-risk situation (…). The final word, at that time, by engineering diagnosis, was that a landslide had occurred. They declared high risk to whitewash the hands of the water company and the mining company.

They have been able to enter and evict our houses but not close the mining. Supposedly, 8 years ago, they should have closed but they are still active and have been renewing their permits. This is the responsibility of multiple entities. No one gives any answer regarding the environmental management of mining. It only matters that we must leave from here." [Interview with DL: 01.07.2020]

The declaration of "nonmitigable high risk" zones challenges the current territorial order and administrative capacities of governance against a backdrop of the complexity of the mountain landscape, its fragility, heterogeneity, marginality, and informality on the peripheries of Bogota (Carrizosa, 2014, p. 203). The need for a counter-proposition for the governance of urban space drives the purpose of the platform for people affected by risk and resettlement – Arraigo. As Stella Velandia commented:

"We are going to work with the Platform (ARRAIGO), to defend our rights and stop the narrative that the periphery is unworthy. We are not unworthy, we live in the periphery but we have a beautiful wealth. I like the periphery in the south, or the north. I am sorry for those wealthy people that don't like us. They removed our grandparents from the countryside, we arrived here and now, again, they want us out." [Interview with SH: 12.10.2015]

Those that make up Arraigo claim the need for risk mitigation, but according to conventional wisdom. The power of community resilience from this perspective is its creative capacity in relation to territory, and its function as a mechanism for risk management linked to cultural processes of re-appropriation and claiming the right to the city (López & Hita, 2018).

10.3 Conclusions

The testimonies presented earlier indicate the need to re-interrogate the basis of managed relocation policies, and to re-define their use in the context of re-shaping urban space in the interests of social justice and sustainability. In Bogota, a narrow focus on reducing exposure through relocation ignores the multiple and dynamic, social and spatial, factors at play in territories deemed "at risk." The failure to resettle all residents appropriately, permanently and securely, and the persistence of processes, in post-resettlement areas, such as urban degradation, risk expansion and risk re-configuration by reoccupation, lead to the reproduction of permanent states (and even the acceleration) of risk. The voices of Arraigo speak to attachments to place, the need for due recompense and respect for livelihoods, the importance of local perceptions (including accounts of the broader socioecological causes of risk) and the failure of inclusive urban governance in areas of poverty and informality. Such voices reveal the political bias of technocratic discourses promoting risk-related relocation, and how adaptation and resilience confer discretional authority that can surpass even basic rights, whilst obscuring the failure of public and private actors to take responsibility for equitable and just risk reduction. Such gaps lead us to echo calls for a refocusing of resilience and relocation policies towards the rights and entitlements of urban citizens, addressing the causes of risk beyond physical or ecological infrastructures (Ziervogel et al., 2017).Ziervogel 2017 Community platforms and networks such as Arraigo can play a vital role in making visible the human impacts of relocation, and thus informing the ways in which relocation policies can be implemented in partnership with communities, including finding new ways to define mitigability. However, technical, legal, and organizational support is fundamental to sustaining community engagement, and bringing about wider scale transformation, and not just resolutions to specific cases of injustice.

Recommendations

1. Managed relocation policies need to be implemented with due respect for the lives, livelihoods and well-being of affected populations, their vulnerability and resilience pre-, during, and post-resettlement. In urban areas, this may require understanding the needs and perspectives of highly heterogeneous groups (Fraser, 2016, for Bogota).
2. The analysis of risk dynamics in urban spaces on which relocation policies are based should use multiple approaches in order to avoid pre-determined definitions of risk. Instead of approaches that exclusively attribute risk

according to physical variables of a space, approaches that demonstrate how relations of justice configure spatial distributions of risks could be used to better inform the politics of mitigability, and how relocation options are distributed (see discussions by López, 2017).
3. Equitable and sustainable governance of relocation plans needs to move to a model of co-produced governance, in which the rights and voices of affected populations are meaningfully incorporated in policy and practice. This includes involving affected communities in processes of defining rationales for which populations and territories are considered for resettlement, such as risk mitigability.
4. Climate relocation policies should be accompanied by an independent monitoring and tracking process (a municipal "observatory" for resettlement) in which community organizations are involved in data collection, analysis, and use.

References

Agreement No. 26. Concejo de Bogotá D.C. (Bogota City Council). In Registro Distrital 1309 (1996, December 10th). Retrieved from https://www.alcaldiabogota.gov.co/sisjur/normas/Norma1.jsp?i=2027

Agreement No. 546. Concejo de Bogota D.C. (Bogota City Council). In Registro Distrital 5269 (2013, December 30th). Retrieved from https://www.alcaldiabogota.gov.co/sisjur/normas/Norma1.jsp?i=56152

Alcaldía Mayor de Bogotá. (2016). *Proyecto del Plan de Desarrollo Bogotá Mejor para Todos* [City Development Plan Project: Bogota best for all]. Retrieved from http://www.sdp.gov.co/gestion-a-la-inversion/planes-de-desarrollo-y-fortalecimiento-local/planes-de-desarrollo-distrital

Campos, A., Holm-Nielsen, N., Diaz, C., Rubiano, D., Costa, C., Ramirez, F., & Dickson, E. (2011). *Analysis of disaster risk management in Colombia: A contribution to the creation of public policies*. World Bank. Retrieved from https://openknowledge.worldbank.org/handle/10986/12308

Carrizosa, J. (2014). *Colombia compleja [Complex Colombia]*. Jardín Botánico de Bogotá José Celestino Mutis. Instituto de Investigación de Recursos Biológicos Alexander von Humboldt. Bogotá, D.C., Colombia.

Decree No. 190. Alcaldía Mayor de Bogotá D.C. [Mayoralty of Bogota]. In Registro Distrital 3122 (2004, June 22nd). Retrieved from https://www.alcaldiabogota.gov.co/sisjur/normas/Norma1.jsp?i=13935

Decree No. 255. Alcaldía Mayor de Bogotá D.C. [Mayoralty of Bogota]. In Registro Distrital 5137 (2013, June 12th). Retrieved from https://www.alcaldiabogota.gov.co/sisjur/normas/Norma1.jsp?i=53386

El Espectador. (2020, June 20th). *Desalojos en Ciudad Bolívar: las víctimas de una crisis humanitaria durante el COVID-19 [Evictions in Ciudad Bolívar: The victims of humanitarian crisis during COVID-19] (video file)*. YouTube. Retrieved from https://www.youtube.com/watch?v=mVnZjtie7-U

El Tiempo. (2013, June 17th). *Petro ordenó reubicar 12 mil familias que viven en zonas de riesgo [Petro ordered the relocation of 12,000 families living in high risk]*. Retrieved from https://www.eltiempo.com/archivo/documento/CMS-12876819

Fraser, A. (2016). The politics of knowledge and the production of vulnerability in informal, urban settlements: Learning from Bogota, Colombia. In M. Roy, D. Hulme, M. Hordijk, & S. Cawood (Eds.), *Urban poverty and climate change: Life in the slums of Asia, Africa and Latin America* (pp. 221–237). Routledge.

Hita, S. (2015a). *Testimonios de Arraigo. La huerta de Mariela, recuperando, habitando quebradas* [Voices of Arraigo: The allotment of Mariela. Recovering, inhabiting ravines] *(video file)*. YouTube. Retrieved from https://youtu.be/_wqwSrKy4uI

Hita, S. (2015b). *Testimonios de Arraigo. Las raíces del cerro* [Voices of Arraigo. The roots of the hill] *(video file)*. YouTube. Retrieved from https://youtu.be/4ZAHoXyAZAk

Lampis, A., & Fraser, A. (2012). *The impact of climate change on urban settlements in Colombia. UN Habitat Global Urban Economic Dialogue Series*. UN Habitat Nairobi, Kenya.

Law No. 388. Congreso de Colombia [Colombia National Congress]. In Diario Oficial No. 43.091. (1997, July 18th). Retrieved from https://www.alcaldiabogota.gov.co/sisjur/normas/Norma1.jsp?i=339

López, D. (2017). Attribution and distribution of climate risks: A critical analysis of policies for resettlement in Bogota, Colombia, to elucidate new practices of risk governance. In A. Rudolph-Cleff, C. Mendoza Arroyo, & B. Hekman (Eds.), *Proceedings: Towards Urban Resilience* (pp. 77–86). Towards Urban Resilience International Workshop. TU Darmstadt. Retrieved from http://tuprints.ulb.tu-darmstadt.de/6986/

López, D., & Hita, S. (2018). Altos de la Estancia: An applied project of risk governance in Colombia. In W. Filho & L. Esteves (Eds.), *Climate change adaptation in Latin America* (pp. 477–499). Springer.

Ordoñez, M., Ángel, A., & Lozano, D. (2012). *A través de la ventana: una apreciación paisajística de los efectos de la minería en los cerros de Bogotá* [Through the window: A landscape appreciation of the effects of mining on the hills of Bogotá]. In Toro, C., Fierro, J., Coronado, S., & Roa, T. (Eds.), *Minería, territorio y conflicto en Colombia* (pp. 381–395). Universidad Nacional de Colombia, Bogotá.

OSSO Corporation and EAFIT. (2011). *Base de datos de pérdidas históricas en Colombia (período 1970-2011)* [Database on historical losses in Colombia (Period 1970.211)]. Retrieved from http://online.desinventar.org

Ramírez, F., & Rubiano, D. (2009). *Incorporando la gestión del riesgo de desastres en la planificación del desarrollo. Lineamientos generales para la formulación de planes a nivel local* [Inserting disaster risk management into development planning. General guidelines for the formulation of plans at the local level]. Lima, Perú: Secretaría General de la CAN - PREDECAN.

Ramirez, F. (2011). Resettlement as a preventive measure in a comprehensive risk reduction framework. In E. Correa (Ed.), *Populations at risk of disaster: A resettlement guide* (pp. 15–27). Washington, DC: The World Bank: GFDRR, 142 pp.: xii.

Secretaría Distrital de Planeación. (2016, February 4th). *Informe de seguimiento a los compromisos del Plan de Desarrollo Bogotá Humana a 31/12/2015* [Monitoring commitments report on the development plan Bogota Humana at 31/12/2015]. Rretrieved from http://www.sdp.gov.co/gestion-a-la-inversion/programacion-y-seguimiento-a-la-inversion/seguimiento

Torres, A. (2007). *Identidad y política de la acción colectiva. Organizaciones populares y luchas urbanas en Bogotá. 1980-2000* [Identity and politics of collective action. Popular organizations and urban struggles in Bogotá. 1980-2000]. Bogotá: Colección en Ciencias Sociales. Universidad Pedagógica Nacional.

Zeiderman, A. (2016). Adaptive publics: Building climate constituencies in Bogotá. *Public Culture, 28*(2(79)), 389–413. doi: 10.1215/08992363-3427499

Ziervogel, G., Pelling, M., Cartwright, A., Chu, E., Deshpande, T., Harris, L., & Pasquini, L. (2017). Inserting rights and justice into urban resilience: A focus on everyday risk. *Environment and Urbanization, 29*(1), 123–138.

11 The climate crisis is a housing crisis: Without growth we cannot retreat

Deborah Helaine Morris MCP

Nicholas[1] treasures his freedom. He likes to come and go as he pleases and does not like to explain himself. Like most young adults, Nicholas wants to hang out with his friends and figure things out on his own schedule. Nicholas used to live with his grandmother Carol, but when their home, a rented bungalow in Edgemere, Queens, was damaged by Hurricane Sandy in 2012, Carol moved away, not wanting to live in a dank, moldy, flood-damaged space with a temperamental teenager. Nicholas was 18, and his friends lived nearby, so he stayed.

In 2018, 6 years after Hurricane Sandy, Nicholas received a notice that his landlord was selling the bungalow to the City of New York. The bungalow was located in a Coastal-A flood zone and was being acquired as a part of a broader community-led plan to manage and retreat from flood risk. The owner of Nicholas's home, his great-uncle William, lived on the same property, in another bungalow that had also been substantially damaged by Sandy. William and his wife, Josephine, were tired of living in a storm-damaged property and were moving to a new, elevated home a few blocks away. Their new house would be smaller, and the couple did not want the hassle of being landlords anymore. The buyout program would allow them to change their lives, hopefully for the better, by permanently reducing their risk. Nicholas needed to move.

Buyout programs are publicly managed property acquisition programs designed to permanently reduce the risks posed by environmental hazards by helping homeowners to move. After their property is purchased, the land is usually allowed to return to its natural state and then remains as open space. In Edgemere, the property would be used to expand a coastal wetland park. Buyout programs are a critical tool in the realm of managed retreat, a risk management approach whose advocates argue for the voluntary and permanent transition of residential communities away from areas facing significant environmental hazards. In the United States, I often hear buyouts described as cost-effective ways to reduce the long-term financial burden that flood-prone communities place on the rest of the country. There is a linear logic: Move people away from the physical risk and their vulnerability will disappear.

Buyout programs are often designed to address the financial stability and housing stability of the homeowner, but rarely are they designed to create financial or housing stability for displaced tenants. When I took over management of the City of New York's Buyout Program in 2016, I was surprised to learn of the existence of tenants, like Nicholas, within our program, because the prior program manager never mentioned them and described the program as an essential form of relief for homeowners. After learning about Nicholas, I thought I was seeing a disappointing failure of program administration: Tenants are legally required to receive assistance from the program in finding a new unit. A federal statute known as the Uniform Relocation Act of 1970 (URA, 1970) requires that federally funded programs, such as our buyout program, provide all rental tenants with moving expenses. If the tenant moves into a new unit that's more expensive than the old one, the program must pay the difference in rent for 42 months. The prior administrator of this program had neglected to implement this requirement. The complexity of assisting homeowners had been overwhelming, and he had just never gotten around to dealing with their tenants.

This lapse in program management represents more than the compliance failure of a single managed retreat program. Buyout programs are the primary implementation tool for climate migration in the United States, yet I believe my program's early carelessness toward tenant support is symptomatic of an over-emphasis on land use policy in climate adaptation. The challenges facing tenants are more systemic and cannot be easily solved by a simple change in address.

In the United States, homeownership and housing stability are closely linked to wealth: Housing is not only a sheltering service, but also it is often the owner's largest financial asset. Access to this asset is uneven: Decades of government policies were explicitly designed to promote racial segregation, frequently limiting housing choice to areas with environmental risk, poor environmental quality, or a lack of environmental infrastructure. These policies perpetuated inequality in access to opportunity across generations (Rothstein, 2017). Thus, the root causes of physical vulnerability are not meteorological: There are structural deficiencies in the provision of basic welfare and human needs.

As I worked to set up a retroactive tenant assistance program, I began to see the challenge of climate vulnerability within a constellation of social challenges, and to see a set of solutions in the broad provision of stable affordable housing.

11.1 Competing crises: Housing and climate risk

New York is a city of renters: As of 2018, two-thirds of New York residents rented their homes (Furman Center, 2019). Renters cross the economic spectrum from very high to very low income. Yet, finding housing for everyone in New York is challenging: The city's vacancy rate is 3.5% and housing is expensive (NYC Housing Survey, 2017).

This combination of high cost and limited availability is a daily disaster. Over 60,000 people seek housing in an emergency shelter every night (Coalition for the Homeless, 2020). Despite an aggressive affordable housing construction program, New York does not have housing sufficient for all who want to live within it. Yet, locally and regionally, construction of housing is suppressed through zoning and environmental regulations. In suburban parts of the region, strict zoning restricts the size and location of housing, substantially increasing housing prices (Albouy & Ehrlich, 2018). Often, to protect property values, homeowners will use environmental regulations to oppose affordable housing development programs, citing issues such as stormwater management and overburdened infrastructure to prevent housing construction. In a region as dense and developed as New York, this combination of restrictions and opposition makes it virtually impossible to develop new affordable housing.

The population of the New York region has grown by over a million residents over the past two decades, yet the supply of housing has stagnated (U.S. Census). Over one million people live within the city's 100-year floodplain: A number that will increase with sea-level rise (Regional Plan Association, 2017). When Sandy made landfall in October of 2012, it made the regional housing crisis significantly worse. Over 70,000 homes within New York were significantly damaged (NYC, 2013). All 521 miles of New York's coastline and every possible type of residence were affected – from luxury towers to public housing to single-family beach bungalows. However, the distribution of damage was not even across the city's housing stock: Older, one and two story, wood-frame buildings suffered the most severe structural damage – representing just 18% of the buildings in the areas inundated but three-quarters of all buildings tagged as significantly damaged (NYC, 2013).

New York's few remaining communities of modestly priced small homes are generally far from Manhattan, located at the periphery of Queens, Brooklyn, and Staten Island (NYC, 2013). These houses tend to be older, certainly predating the official demarcation of New York City's floodplain in 1983, and often their construction predated the major revision of the city's building code in 1938. Sandy affected white communities in Broad Channel, Oakwood Beach, Midland Beach, and Sheepshead Bay, as well as predominantly African-American communities of Edgemere, Arverne, Hammels, and Canarsie. These were originally seasonal communities with summer homes providing easy access to the ocean and its cooling breezes. Over time, they were adapted for year-round habitation. All of these communities have extremely long commutes to the city center and most lack sufficient access to public transportation. As older buildings, far away from the economic center, and in fair-to-poor condition prior to Hurricane Sandy, these homes were amongst the cheapest to own, and cheapest to rent, in the city.

Because so much of the damage was to small homes, the victims of the hurricane were generally thought to be owner-occupants. As a result, the City of New York used Federal Community-Development Block Grant for Disaster Recovery (CDBG-DR) funding to develop a housing recovery program, known as Build It Back, that focused on providing financial and construction assistance to homeowners. The buyout program I managed, under the Build It Back umbrella, was conceived as a cost-effective way to help storm victims move on while transitioning storm-damaged property into more appropriate land uses.

Long-term buyouts were viewed by myself and colleagues as a responsible way to manage the city's growing risk from sea-level rise and coastal storms.

In my 6 years of working on New York City's hurricane recovery team, I participated in hundreds of conversations with city planning staff regarding storm-damaged neighborhoods and areas that face severe future flood risk. In discussions of policy tools and potential solutions for these environmental threats, the major constituent was consistently framed as a homeowner: Living oblivious to or in defiance of physical risk. Rental tenants do not fit easily into this narrative. The well-being and stability of rental tenants has little to do with access to information about flood risk. Instead, it is the historic patterns of inequity and the weakness of the American safety net that create choice-limiting constraints. Buyout programs may be both a logical and fiscally responsible solution to changing land uses; however, without a significant investment in the development of comparably priced housing in communities without flood risk, reducing the available housing inventory displaces the problem and does not reduce overall human vulnerability.

11.2 Housing of last resort

Housing access is often limited by policies that encourage segregation along lines of race and income. Nicholas grew up in Edgemere, a community located on Jamaica Bay, in the eastern Rockaways. Edgemere is in a forgotten corner of New York. Although it is almost entirely developed, in Edgemere, you can see the environmental evolution of the Rockaways as a barrier island. Jamaica Bay is a tidal wetland and an important stop on the Atlantic flyway, so during migration season, the wetland edges of Norton Basin are filled with a variety of terns, gulls, and plover. Sometimes peregrine falcons soar through. The subway is only a 5-minute walk away, but it takes over an hour to get to Manhattan.

Sandy inflicted damage across the Rockaway peninsula, but things were especially bad in Edgemere. The housing stock is primarily made up of one- or two-family wood-frame homes that were built as summer bungalows at the turn of the 20th century. When the Rockaways were abandoned as a beach destination in the 1930s, many of the bungalows fell into disuse. Over time, African-American families moved to the Eastern Rockaways as they were pushed out of other housing elsewhere in New York, and dilapidated, unwinterized bungalows became a source of low-cost housing. Thus, a barrier island and summer retreat became a year-round community, not because it was a place people wanted to live but because there were too few options elsewhere.

Redlining practices by the Federal Housing Administration led to the 1935 maps created by Home Owners' Loan Corporation, which labeled Edgemere as "Definitely Declining" due to its growing Black population (Nelson et al., n.d.). This loan rating made it challenging for people in certain areas to access traditional mortgage financing, making it impossible to get a mortgage or finance improvements for decades. The 1977 Community Reinvestment Act outlawed these discriminatory lending practices, but the effects of nearly five decades without public and private investment are still visible in Edgemere. As the quality of the housing stock declined, the city government made no meaningful capital improvements infrastructure until 1996.

Five years after Sandy, the house where Nicholas lived was unrepaired, without electricity, and with weak plumbing. The roof leaked, there was significant mold and very few furnishings. In addition to stable, safe housing, Nicholas needed a long-term social worker to help him with mental health issues and to find employment. While the owner of Nicholas's house, William, had chosen to participate in a managed retreat program which was part of a broader neighborhood initiative called Resilient Edgemere, Nicholas had not. As a renter, Nicholas did not have a say: If the homeowner participated in the program, Nicholas had to move. William was retired and had few financial resources. Although he was able to move, he could not financially help his tenant, even though Nicholas was his grandnephew.

The URA benefit allowed our agency to provide Nicholas with a housing choice voucher. Housing choice vouchers help make housing affordable for low-income individuals: Voucher holders pay 30% of their income toward rent and utilities, while the government pays the remainder. Nicholas struggled to find a landlord who was willing to accept his voucher as payment. Although racial discrimination is illegal, source of income discrimination, which is a new vehicle for racial discrimination, is common (Cunningham et al., 2018). The demand for the type of housing and services Nicholas needs exceeds the supply. When the waitlists for affordable housing and supportive housing are thousands of people long, the fact that Nicholas happened to be a "climate refugee" does not give him priority treatment. When I last heard from Nicholas, he had moved in with a friend and was still looking for a landlord who would accept his housing voucher.

Proponents of managed retreat often replicate the narrative of moral crisis that is used to blame vulnerable communities for the social ills they endure. Arguments for managed retreat have been used to justify and normalize the removal and exclusion of low-income people, people of color, and immigrants. To be both effective and equitable, managed retreat programs must recognize the need for more substantial solutions than just moving. The more tenants I met, the more I realized that most tenants, like Nicholas, had relatively few problems rooted in climate risk.

11.3 Risks beyond a flood

Rachel, a renter, lived in Howard Beach, Queens. It took over 90 minutes to commute to her job, taking multiple buses to the train. Her apartment was not in great shape even prior to Sandy. As soon as Rachel heard that the owner was selling, she moved out. She hated that apartment and hoped that URA would be her ticket to someplace better. Rachel first moved in with her mother, while she waited for her benefits to be calculated, thinking she could save money for new furniture. As anyone who has shared a small space with a relative might understand, the situation became untenable quickly. Rachel found a new apartment in the same neighborhood and outside of the floodplain, but it was significantly more expensive than her former home.

Rachel understood URA and that the program would cover this differential, so Rachel took the plunge on the more expensive unit so she could leave the floodplain. Because the URA program lagged behind in being administered, Rachel did not receive her first payment until 6 months after her move. Rachel made it work because her employer advanced paychecks and relatives lent her money, but the precarity of her new situation became clear. When Rachel's 42 months of rental assistance ended, it would be difficult for her to stay in the new apartment unless her income changed significantly. While the owner of her former residence was able to make a permanent move, Rachel's move did nothing to resolve her underlying precarity: She still lived one paycheck away from an empty bank account. After moving into her new unit, Rachel called our office frequently, asking if there was a way for our program to find her another apartment that had the same rent as her prior, storm-damaged unit, but that was not located in the floodplain, so she could afford the housing long term.

Rachel knew that she did not want to experience another natural disaster and that she did not have the financial resources to navigate another flood. But was climate risk the cause of Rachel's precarity? Rachel worked full time, but her low wages demonstrate how tenuously earnings are related to the actual cost of living. Moving out of the flood zone, when the home is more expensive, does not address the poverty which is at the root of her vulnerability.

Another tenant, Adam, rented a one-bedroom home with his wife and three children in Midland Beach, Staten Island. Although they would have liked more space, they enjoyed having a small backyard and being close to the neighborhood's large parks. Adam's wife stayed home with their children. She thought about working full time, but once you factored in the cost of childcare, her contributions would not increase their household's income enough to make it possible to rent a larger home.

Our program could not find any affordable or even comparably priced housing in Adam's neighborhood. Midland Beach is part of Staten Island's Lower Density Growth Management Area, a designation that limits new residential construction to primarily large-lot single-family homes (Staten Island Growth Management Task Force, 2003). A portion of the area does overlap with a significant freshwater wetland, and intensive development could impair the geomorphology of this sensitive natural system.

The result of these zoning and environmental conditions is that it is extraordinarily challenging to construct new affordable housing (Vitullo-Martin, 2003). Building small houses and apartments is illegal in the Growth Management Area. Instead, new homes must meet minimum lot size and minimum open space requirements. Building a new home the same size as Adam's would require a property with three times as much land. By prohibiting the concentrated development of apartments, growth management regulations put all new construction out of reach for families like Adam's. The majority of smaller, rental units remaining in Midland Beach are in older houses built before the zoning restrictions went into effect in 2003. These units are also in the floodplain, just like the unit Adam left. The home Adam and his family moved into, outside of the floodplain, was substantially more expensive than his prior home. After several months, they broke the lease and moved back into a cheaper and smaller home within the floodplain.

When faced with a choice between financial or environmental security, Adam's family chose the financial stability of long-term affordability. Our buyout programs did not have the tools to create a different outcome, which would have required providing substantially more than moving expenses and time-limited rent differential. Adam and his family loved Midland Beach, but the neighborhood's zoning it made it challenging to find housing for families like theirs. Adam's family cannot afford to live in a home physically large enough to legally house its five members. Without significant changes to the zoning restrictions, housing opportunities for working families like Adam's will continue to be sparse. Further, Adam's wife wanted to work, but the high cost of childcare made it impossible. Without access to essential family support services, like, universal low-cost childcare, Adam's family could not make a physical move work out economically.

Ronald was a tenant in a home on the interior of a court of small bungalows in Sheepshead Bay, Brooklyn. The interior buildings on these courts lacked street access and a legal connection to the public sewer system. The homes were built prior to the construction of the city's sewer and stormwater system, and the bungalows could not easily be brought up to the current building code because they lack street frontage. Legislation passed in 1962 prevents the legal winterization of the bungalows, with the futile goal of preventing their year-round use. However, as cheap housing, they existed in a record-keeping no-man's-land. It never made sense to invest public funds in connecting structures that do not legally exist to a public sewer. The city grew up around them, creating a virtual bathtub that floods with every rainstorm.

Ronald and his girlfriend shared the 1200-square-foot bungalow with their clowder of cats. Ronald was retired and had a limited income. When Sandy hit, Ronald's landlord decided he was ready to sell. What made it even more appealing was that his property would be used to help manage rainwater in the rest of the bungalow courts.

When Ronald learned about his URA benefits, he considered the potential of a lump sum benefit payment. It took perseverance, adjusted expectations, and totally uprooting his life away from his community and connections, but Ronald was able to buy a home in coastal New Jersey and use the URA assistance to make a down payment. It was not financially possible for Ronald to stay nearby, or even within New York City, and his new home is also in the floodplain.

As a retiree living on only Social Security benefits, Ronald's economic choices were very constrained. Our program allowed him to develop the potential financial security of homeownership, even though it came at the expense of proximity to his lifelong social networks, and even though it did not reduce his overall flood risk. Although transitioning a renter into a homeowner would normally be hailed as a program success, Ronald's position is now much more tenuous; by purchasing a home within a floodplain, Ronald will have the financial risk of the unpredictable cost of flood insurance as well as the physical risk of future flood events, while living further from him his friends and family. But in purchasing a home, Ronald hopes to have climbed a rung of the economic ladder.

11.4 Retreat to where?

What Adam, Rachel, Ronald, Nicholas, and the three dozen other tenants relocated by my program shared were limited choices. Very few felt significantly attached to their neighborhood – only Adam's family expressed a strong desire to remain in the same area. Most worried about surviving another flood, yet only Rachel moved out of the floodplain. Everyone shared significant worries about the instability created by the precariously high cost of their new housing.

The climate migration for all of these tenants was not fully or appropriately "managed." Undoubtably, they would have all been better served with more timely provision of benefits.

The URA does create an assistance floor capable of providing minimum costs of a physical move from one location to another, but it does not push managed retreat programs, like mine, to imagine assistance in any other form than a remittance. Rather than address the disinvestment and exclusion that have placed certain populations in communities of risk, our program placed the burden on these vulnerable households.

It is very easy to say that people should not live in areas of high risk without specifying which communities must welcome displaced people. Approaching this adaptation challenge from a negative perspective – focusing on the physical spaces where people should not live – is unproductive. By focusing on imposition of new limitations, rather than on the creation of new opportunities, managed retreat programs perpetuate the structural system that has created precarious patterns of inhabitation.

The source of climate vulnerability is not only climate risk; it is the sociopolitical system that constrains social mobility, including equal access to safe and affordable housing. By working full time, Rachel should not have to struggle to find safe affordable housing. Nicholas should be able to find landlords willing to accept his housing choice-voucher. Freshwater wetlands, such as those in Adam's neighborhood, are not an appropriate environment for development. But for Adam to have a place to live, affordable housing must be abundantly constructed in areas without quagmire, along with access to affordable childcare and family support services. Ronald needed housing he could afford as a retiree and would have been better served through finding housing near his family and social networks. The changes required to solve these challenges are more than a physical move from one place to another and require more than purely financial compensation.

The collective responsibility for the segregation and unaffordable housing that have systematically pushed people to live in landscapes of risk extends far beyond the most vulnerable neighborhoods. As global populations continue to increase and a changing climate significantly limits the habitability of many parts of the planet, the need for a comprehensive physical and social redesign can only grow. This redesign requires communities with lower risk to invest in building the housing and providing the support needed to serve Nicholas, Adam, Rachel, Ronald, and the millions of future climate migrants who will need safe housing and social support. Managed retreat alone is not the solution. The climate crisis is also a housing crisis. Programs that continue to operate as if they are solely solving a climate challenge will continue to fail people who have been failed in the past. Ameliorating climate risk and solving the housing crisis will require recognition of the extent of the challenge and intentional, intensive, and inclusive new growth.

11.5 Recommendations

1 Buyout programs must address permanent housing stability for displaced owners and tenants: This requires rethinking these assistance programs as more than real estate transactions and instead as vehicles for creating permanent housing stability.
2 Managed retreat programs can only be effective if they are accompanied with significant development programs that facilitate the creation of safe and affordable housing in communities with fewer environmental stressors. It is not productive to reduce housing supply in one place if it is not accompanied by an increase in housing someplace else.
3 Managed retreat programs must recognize the history and policy systems that have placed certain populations into geographies of risk. If the retreat programs do not address the broader and systemic sources of risk, vulnerabilities will simply be relocated to a new location.
4 In the context of the United States, the URA is the backbone of the benefit and pricing structures of all federally funded relocation programs, for owners and tenants. Advocates of managed retreat should advocate for significant congressional reforms to the URA to create more flexible and equitable rehousing solutions.

Note

1 Names have been changed to protect the privacy of program participants.

References

Albouy, D., & Ehrlich, G. (2018). Housing productivity and the social cost of land-use restrictions. *Journal of Urban Economics, 107*, 101–120.

City of New York. (2013). *A stronger more resilient New York report.* Accessed online: https://www1.nyc.gov/site/sirr/report/report.page

Coalition for the Homeless. (2020). *State of the homeless 2020.* Accessed online: https://www.coalitionforthehomeless.org/wp-content/uploads/2020/03/StateofTheHomeless2020.pdf

Cunningham, M., Galvez, M., Aranda, C. L., Santos, R., Wissoker, D., Oneto, A., Pitingolo, R., & Crawford, J. (2018). A pilot study of landlord acceptance of housing choice vouchers report prepared for U.S. Department of Housing and Urban Development. *Urban Institute.* Accessed online: https://www.huduser.gov/portal//portal/sites/default/files/pdf/Landlord-Acceptance-of-Housing-Choice-Vouchers.pdf

Furman Center. (2019). *State of New York city's housing and neighborhoods in 2018* (p. 21). New York University.

Nelson, R. K., Winling, L., Marciano, R., Connolly, N., et al. Mapping inequality. In R. K. Nelson & E. L. Ayers (Ed.), *American Panorama.* Accessed online: https://dsl.richmond.edu/panorama/redlining/#loc=10/40.716/-74.108&city=queens-ny&area=C105 (map C105)

Regional Plan Association. (2017). Fourth Regional Plan 2017. Accessed online: http://fourthplan.org/action/protect-dense-communities

Rothstein, R. (2017). *The color of law: A forgotten history of how our government segregated America* (1st ed., Democracy and urban landscapes). Liveright Publishing Corporation.

Staten Island Growth Management Task Force. (2003). *Final report - Recommendations to Mayor Michael R. Bloomberg.* Accessed online: http://home2.nyc.gov/html/gmtf/pdf/si_final_report_nov_26.pdf

Uniform Relocation Assistance and Real Property Acquisition Policies Act of 1970, 42 USC Ch. 61, subchapter II, §4624.

U.S. Census, 1990 and 2000 decennial census data. Accessed online: https://www1.nyc.gov/site/planning/planning-level/nyc-population/nyc-population.page

U.S. Census, *New York City Housing and Vacancy Survey: 2017 microdata.* Accessed online: https://www.census.gov/data/datasets/2017/demo/nychvs/microdata.html

Vitullo-Martin, J. (2003). Rethinking Staten Island. Manhattan Institute. Accessed online: https://web.archive.org/web/20031222081801/; https://www.manhattan-institute.org/email/crd_newsletter10-03.html

12 Voices of Ghoramara Island, India: The case for planned relocation

Oana Stefancu

The communities of Ghoramara Island in the Ganges Brahmaputra Delta, India, have been marginalised for many decades, but in the face of climate change, their only means of survival may be planned relocations. Will the government relocate them? And will the communities want to move?

In 2019, the monsoon season in India saw the highest amount of rainfall in over 25 years and lasted longer than any monsoon in India's recorded history. This led to severe floods and mudslides, causing hundreds of thousands of people to flee their homes. The monsoon was followed by cyclone Bulbul, which struck the coasts of India and Bangladesh, causing storm surges, heavy rains, and flash floods. In the region of West Bengal, approximately 3.5 million people were directly affected. Cyclone Bulbul damaged or destroyed over 500,000 homes and 1.5 million hectares of crops. The Indian government struggled to respond. When asked about the impact of Bulbul, Anjana Bhuiya[1], a resident of the Ganges Brahmaputra Delta, says *"My house was wrecked due to this storm, but nothing has been given."* Another resident, Pradeep Manna, says *"Our crops are ruined and our roof fell off, and we only got one sheet of plastic [from the authorities], nothing more"* (Figure 12.1).

Located on the south west edge of the Ganges Brahmaputra Delta, Sagar Block is a group of islands exposed to coastal flooding, storm surges, and cyclones, causing high rates of coastal erosion and salinity intrusion (Ghosh et al., 2014). One of these deltaic islands, Ghoramara, saw 70% of its landmass lost to coastal erosion since the 1920s. Currently, it measures only 1.8 square miles (Ghosh et al., 2014). Once home to 40,000 people, Ghoramara's population has decreased to roughly 3,500 people today, and five of the nine main villages of the island are now submerged (Ghosh et al., 2003; Jana et al., 2012).

12.1 Life on Ghoramara Island

Ghoramara Island was once a lush and beautiful ecosystem. Twenty-two years ago, Ruhana Bibi was a little girl that used to visit Ghoramara with her father: *"We used to come to the Moharram [religious festival]. It felt so good! I liked it so*

DOI: 10.4324/9781003141457-12

Figure 12.1 A tree sinks in Ghoramara Island, India, where due to coastal erosion, five of the nine main villages of the island are now submerged. Photo by author

much! So my father said, 'Let's arrange your marriage here.'" But only three years after her marriage, the river swept away her land, house, chili garden, and betel leaf farm.

Pradeep Manna and his family are the human face of today's reality on Ghoramara. When Pradeep was young, his grandfather had 230 acres of land. His father used to cultivate the land and farm fish; his mother used to be in charge of the household work, whilst Pradeep and his five siblings all attended the local school. But during his lifetime, circumstances have changed drastically for the worse. "*Between 1968 and 1980 we lost everything. This is not our original house. This is our seventh house. I had to move seven times. This is the seventh house and we had to build it on other's [people] land. Can you understand?*" asks Pradeep "*I am living on other people's land! I am landless!*"

Pradeep's story is common on Ghoramara. Most of the islanders were historically landowners and farmers. They recall growing up with plenty of land and food. "*Then we had agricultural lands we used to cultivate. The rice paddy cultivation could sustain us for a year,*" says Rahim Ali. The inhabitants of the coastal areas of Ghoramara, in particular, have lost their farmland and hence their year-round food supply.

With few opportunities to diversify their livelihoods, residents on Ghoramara Island have limited capacity to adapt without external support. However, external support is difficult to find. The broader Ganges Delta region is a marginalized rural area that receives little investment due to its severe exposure to cyclones and flooding (Mukhopadhyay, 2009). Critical infrastructure and material well-being are low, with 59% of the population not having access to drinking water, 47% living with some food shortage, and 34% living below the poverty line (Centre for Sciences and Environment, 2016). Hence, the capacity of many residents on Ghoramara to invest in major adaptations to their deteriorating environment is limited.

Now their only means of survival is to rely on insecure daily income to support themselves and their families. With almost no work opportunities left on the island, many had to migrate to find work elsewhere and send remittances home. Ruhana Bibi describes her family situation: *"We have no more land left. Our condition is very bad. We have only one male and our family has six members. My husband is in Kerala for work. We don't have anything except Kerala. He's there and has to see all the good and bad, my poor husband."* Sampa Das, an 80-year-old widow, is in a similar position: *"I have one son, and he has to go away for work. Wherever he can, for whatever labor works he finds."*

Daily labor in fishing or the construction industry does not provide sufficient income to maintain an adequate living standard. Whilst many residents report that they were able to receive basic education, they are now unable to provide the same opportunities to their children. *"I have these two girls, and we had to stop their education,"* Ruhana Bibi says tearfully. *"Their father said 'You both need to make fish nets to help.' What else could we do? We cannot provide everything our daughters need. They have expenses. One is saying that she doesn't have a pant, the other needs a saree, so they cannot sit idle while we are in this situation. They also have to work, earn their own money. We spend our poor lives like this."*

Due to the financial hardships brought by the environmental changes, many women had to abandon their traditional role of taking care of the household and children and enter the workforce. Ruju Das says of his wife: *"Twenty years ago, she used to do household work. During those days, men and women, especially women, did not go out [of the house] much. Now women go farming and do other work too. We somehow have to find ways to sustain ourselves"* (Figure 12.2).

The inhabitants of Ghoramara find themselves in a very precarious situation. They have lost their lands, and therefore the ability to sustain themselves through farming, and are left to rely on income from daily work that only allows them to live day-by-day. Many residents have left Ghoramara Island and relocated to neighboring Sagar Island or to mainland India, but they have been unable to buy land in these new locations. Others, knowing that they cannot buy land elsewhere, remain on Ghoramara and witness the island slowly being lost to the sea. Many, when reflecting on their circumstances, say they feel "trapped," as the environmental threats increase, and their ability to escape decreases (see Black et al., 2013).

Rahim Ali, a 74-year-old man living on the mud embankment, says *"That for us [the erosion] means we are in much tension due to this. Can you understand? We do not*

Figure 12.2 Support for a traditional division of gender roles has declined as women joined the workforce to help sustain their household. Photo by author

have any such resource or money saved up or any such business finance with which we can move out and directly buy any land, to do that we do not have the power." Similarly, Ruhana Bibi says "While we sleep at night, we have to be very careful. The tide water often enters [the house] and we face great difficulty at night. We cannot sleep properly at night. Like sho... sho... sound is coming, then the sea is pulling away the land and taking it with itself. But we have nowhere to go. What can we do?"

12.2 The plea of the people of Ghoramara for relocation

The lives of the people of Ghoramara are severely affected by climate change and they want to relocate, but their voices are largely unheard in academic and political arenas.

Asima Bibi, a young mother living on the eastern coast of Ghoramara says "*Shortly, our house will be lost. After the next monsoon, our house will be lost. We do not have any land in the inner part [of the island] and at the rate that erosion is happening, the probability that we will buy land and be able to live the rest of our lives in the inner part, is little. Such a good sign cannot be seen. So we will have to move out. There is no option left other than moving out [off the island].*" Whilst residents say the only solution is to "move out," many do not have the means to do so by themselves. Uma Jana says "*The government needs to move us. Otherwise where will we live? If the river erodes so fast, then where will we live? We have to live somewhere, anywhere. We are drifting away here.*" Similarly, Ansar Gosh says "*The island is being washed away. What can we do? If the Government can arrange for our houses to be in a different location, then we can survive, otherwise we will have to die of hunger*" (Figure 12.3).

Across the world, the reality of climate change is bringing the option of potentially assisting acutely vulnerable populations to move into sharp relief (Hino et al., 2017; McAdam & Ferris, 2015). These so-called "planned relocations" are governmental interventions employed to move communities and individuals away from environmental risks. They involve buying up land, compensating owners, and providing new places to settle. Whilst some distinguish between relocation as the process of physically moving people and resettlement as the process of assisting people in the new location, others argue that for relocation to be an effective adaptation strategy, governments cannot relocate communities without providing for their resettlement (Ferris, 2015).

Planned relocations are controversial. They are frequently portrayed as the only viable or rational option, thereby overriding or negating adaptation strategies that are desired by those directly affected (Bravo, 2009; Hulme, 2009; O'Brien & Wolf, 2010). However, when individuals want to relocate, then dismissing their needs or their pleas for relocation assistance, marginalizes their voices.

The concept of planned relocations is well known by the inhabitants of Ghoramara. In 1977 the government of West Bengal declared Ghoramara Island, and the neighboring, now sunken, Lohachara Island, to be a "no man's land" (Mortreux et al., 2018). As a consequence, the government withdrew funding for support and services because of the high rates of erosion and the lack of long-term viability of the islands. That same year, a change of government took place, with the Communist Party coming into power. The newly elected party introduced a long-term adaptation strategy for coastal communities: Relocating some of the residents of Ghoramara and Lohachara to Sagar Island. Sagar Island was chosen due to its proximity and availability of land for relocations.

The relocation policy came at a time in which the new Communist Party, through the newly elected local self-government of Sagar Panchayat Samity in 1978, was establishing its legitimacy and popularity through land reform commitments (Löfgren, 2016). It was in the interests of the new Communist government to prove its commitment to land redistribution and social welfare in its

Figure 12.3 A house on Ghoramara Island, soon to be lost to coastal erosion. Photo by author

early years of power (Mortreux et al., 2018). Relocation programmes helped to establish the new government's legitimacy, but as time wore on this incentive diminished, as did available land for relocation and relocation assistance.

People who relocated during the 1970s and 1980s have now been able to rebuild their lives. Many report that they have been able to provide for their families and bond with the new place and people around them. Those left behind are left only with the hope that in the future they will also be relocated. However, there have been no relocation programmes offered in Ghoramara since 1990. The population of Ghoramara is highly marginalized and has limited capacity to protest government inaction. The marginalized and vulnerable nature of the population means that the community does not have the resources to put pressure on the government for additional support. The result is a reality where social services and infrastructure have broken down, and those who still live there lack the ability to move, becoming a self-described "trapped population" (Black & Collyer, 2014).

So far, the only governmental intervention seen in Ghoramara in the past years is the annual rebuild of the mud embankment. There are no plans to build a permanent embankment. Each monsoon season washes the mud embankment away and further erodes the coasts of Ghoramara. *"The loss of the embankment*

Figure 12.4 The embankment, made of wood and sacks of mud, gets washed away every monsoon season. Photo by author

causes many damages to people. The Panchayat then repairs it but then again gets washed away. Damages occur to people every year. Because the embankment is made of mud, it washed away every year," says Uma Jana (Figures 12.4 and 12.5).

The constant rebuild of the embankment also means that every year it is reconstructed further inland. Sometimes it is even built behind houses that are still standing. *"When three years ago the embankment was constructed, then it was lost in one night! Again a new embankment was constructed. Nothing has remained of it, everything got destroyed. Now the embankment is behind our house,"* says Sampa Das, *"Oh my god, when the high tide will enter, how will we live? And we do not have anywhere [to go]. Where will we go?"*

Despite the dire situation the people of Ghoramara are in, and their desire to leave the island, they still exhibit strong emotional bonds to the place, mainly due to the existing social attachment. *"The [social] environment in Ghoramara is*

Figure 12.5 Some houses are no longer protected by the embankment, risking to be washed away any day. Photo by author

very good. You have to agree upon this!" says Rahim Ali, "There is no animosity, no fighting in here. Here we don't have any violence, no discrimination among people. People are one, who is Hindu, who is Muslim, everyone lives together as brothers. That's why this place has always been peaceful. But due to the erosion, the land is not safe, so why live here?" Similarly, Ruhana Bibi says "The land is eroding, and we are scared, but we are also very happy. It [Ghoramara] is better than other places. We don't have any fighting, no arguments. But this beautiful place cannot be sustained. The sea is pulling away the land gradually, but people are still loving this place."

However, these attachments do not seem to prohibit the people of Ghoramara from wanting to move to a safer place: "We are all like brothers, we stay together, we help each other" says Putul Das, "but in the future, Ghoramara won't be here; no hope can be seen." Nupur Manna also says, "If we could live here, *then it would have been good, [because] it's my birth land. But will you live here in your birth land despite the erosion just because it is your birth land?"*

12.3 What are feasible and just ways forward for trapped populations?

People living on Ghoramara Island have a deep love for the island and its people, and this is juxtaposed with fear for the future and a desperate desire to be given the opportunity to move elsewhere. Governmentally led planned relocations are seen as the only means of survival.

When governments decide to not intervene and relocate populations, many are left feeling trapped and hopeless. Environmental degradation can push communities into deprivation, while also limiting their opportunities to escape from poverty. Those who wish to move but have limited access to the resources and networks to escape deteriorating environmental conditions are however unable to migrate, reinforcing conditions of vulnerability (Black et al., 2013; Black & Collyer, 2014; Logan et al., 2016; Milan & Ruano, 2014).

Relocating communities of Ghoramara to minimize environmental risks may be a viable option, and it is what residents want, but it is a momentous change

Figure 12.6 People of Ghoramara waiting for the daily boat to go to the market on the mainland. Fresh produce, clothing, and medicines cannot be bought on the island. Photo by author

and is not a panacea for all the challenges residents face. Just ways forward, do not single-mindedly refer to the relocation of people as a last resort strategy but a deeper engagement with investment, economic support, and consideration of the other steps that could have been taken to preserve life on Ghoramara Island before the situation becomes dire. Just ways forward give marginalized people a voice and hear their concerns. Just ways forward empower vulnerable communities and give them opportunities to survive, adapt, and thrive.

12.4 Recommendations

- To avoid vulnerable and marginalized people becoming "trapped" or displaced, governments should facilitate, plan, and manage migration and relocation programmes with the participation of the communities.
- To successfully facilitate, plan, and manage migration and relocation programmes, those affected should be given a voice to express their concerns and priorities.
- People's concerns and priorities are based in culture and values, and governments should not impose externally formed ideals surrounding mobility onto vulnerable populations (Figure 12.6).

Note

1 For privacy reasons, the names of the people referred to in text are pseudonyms; the people represented in the photographs are not the same people quoted in text; the identity of all participants in this research has been anonymised. All pictures are of Ghoramara Island and its people.

References

Black, R., Arnell, N. W., Adger, W. N., Thomas, D., & Geddes, A. (2013). Migration, immobility and displacement outcomes following extreme events. *Environmental Science and Policy, 27*, S32–S43.

Black, R., & Collyer, M. (2014). Populations trapped at times of crisis. *Forced Migration Review, 45*, 52–56.

Bravo, M. T. (2009). Voices from the sea ice: The reception of climate impact narratives. *Journal of Historical Geography, 35*, 256–278.

Centre for Sciences and Environment (CSE). (2016). Living with changing climate: Indian Sundarbans. CSE, Delhi Website. Accessed: 22 September 2020. Available at: http://www.cseindia.org/userfiles/adaptation_paradigm.pdf.

Ferris, E. (2015). Climate-induced resettlement: Environmental change and the planned relocation of Communities. *SAIS Review of International Affairs, 35*(1), 109–117.

Ghosh, T., Bhandari, G., & Hazra, S. (2003). Application of a 'bio-engineering' technique to protect Ghoramara Island (Bay of Bengal) from severe erosion. *Journal of Coastal Conservation, 9*, 171–178.

Ghosh, T., Hajra, R., & Mukhopadhyay, A. (2014). Island erosion and afflicted population: Crisis and policies to handle climate change. In W. L. Filho (Ed.), *International perspectives on climate change* (pp. 217–225). Springer.

Hino, M., Field, C. B., & Mach, K. J. (2017). Managed retreat as a response to natural hazard risk. *Nature Climate Change, 7*, 364–370.

Hulme, M. (2009). *Why we disagree about climate change.* Cambridge University Press.

Jana, A., Sheena, S., & Biswas, A. (2012). Morphological change study of Ghoramara Island, Eastern India using multi temporal satellite data. *Research Journal of Recent Sciences, 1*(10), 72–81.

Löfgren, H. (2016). The communist party of India (Marxist) and the left government in West Bengal, 1977–2011: Strains of governance and socialist imagination. *Studies in Indian Politics, 4*, 102–115.

Logan, J. R., Issar, S., & Xu, Z. (2016). Trapped in place? Segmented resilience to hurricanes in the Gulf Coast, 1970–2005. *Demography, 53*, 1511–1534.

McAdam, J., & Ferris, E. (2015). Planned relocations in the context of climate change: Unpacking the legal and conceptual issues. *Cambridge Journal of International and Comparative Law, 4*(1), 137–166.

Milan, A., & Ruano, S. (2014). Rainfall variability, food insecurity and migration in Cabricán, Guatemala. *Climate and Development, 6*, 61–68.

Mortreux, C., Safra de Campos, R., Adger, W. N., Ghosh, T., Das, S., Adams, H., Hazra, S., et al. (2018). Political economy of planned relocation: A model of action and inaction in government responses. *Global Environmental Change, 50*, 123–132.

Mukhopadhyay, A. (2009). On the wrong side of the fence: Embankment people and social justice in the Sundarbans. In P. K. Bose & S. K. Das (Eds.), *Social justice and enlightenment: West Bengal.* Sage.

O'Brien, K., & Wolf, J. (2010). A values-based approach to vulnerability and adaptation to climate change. *WIREs Climate Change, 1*, 232–242.

13 The need for a resettlement pathway for Guyana's vulnerable coastal communities

Dina Khadija Benn

13.1 Introduction

Guyana sits on the northern coast of South America, where 90% of the population lives on 5% of the nation's total land area. As climate change threatens coastal dwellers, the country must consider how to assist those most at risk – particularly indigenous persons and informal settlers who live near the shore. Coastal vulnerability and resettlement issues are presented here from the perspective of three communities. Based on interviews with impacted residents and government officials, site visits, and review of policy documents, this chapter examines the experiences of vulnerable communities, the broader challenges of managing settlement in Guyana's low-lying coastal zone, and the need for early and appropriate resettlement actions to safeguard vulnerable communities.

13.2 Conservation and community at Almond Beach

Almond Beach is a remote stretch of land situated near the apex of Guyana, close to the Venezuela border. This sensitive ecosystem is characterized by recent shell deposits along the shore (Daniel, 2001), and the beach is a vital nesting ground for four species of endangered marine turtles: The Leatherback, Hawksbill, Olive Ridley, and Green Turtle (Pritchard, 1969; Protected Areas Commission, 2014). The beach forms part of the Shell Beach Protected Area[1], a 90-mile stretch of mostly undisturbed coastland that also comprises the nation's largest stretch of contiguous mangrove forests (Protected Areas Commission, 2014). It can take up to 8 hours to arrive at Almond Beach from the capital city Georgetown (Kandaswamy, 2015); in fact, the area is so isolated that in the mid-20th century its sole visitors were turtle poachers (Pritchard, 1969). Visitors now travel there during the February to June annual nesting season for research and tourism activities (R. John, personal communication, September 2020) (Figure 13.1).

Nestled in this pristine tropical setting is the Almond Beach community, a small Arawak settlement that emerged in the late 1980s. Audley James, a key informant who is a former resident interviewed as part of this research, was the first person to settle at Almond Beach in 1988. Originally from Moruca,[2] he

DOI: 10.4324/9781003141457-13

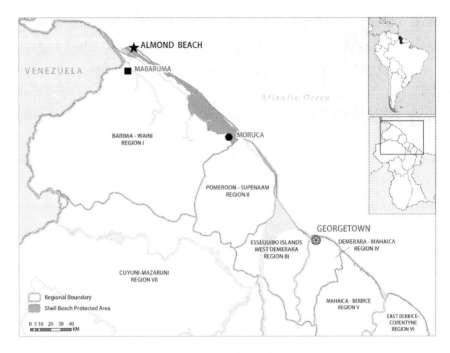

Figure 13.1 Map of the Guyana coast. Figure by author.

spent much of his youth working as a gold and diamond diver across Guyana. Audley first visited Almond Beach in 1949, when he canoed there at spring tide with other hunters to collect turtle meat and eggs to feed his family (Danny, 2013; Pritchard, 1969). Audley recounted that he later observed a decline of the nesting turtles, which seemed to be linked to their turtle poaching activities in the area. Eventually, he met the British zoologist Peter Pritchard, whose research on marine turtle habitats frequently brought him to Guyana. After learning about the importance of these animals, Audley settled at Almond Beach and worked with Dr. Pritchard to establish a conservation project that would protect the nesting turtles (Danny, 2013). Over three decades, Audley and his son, Romeo, managed turtle monitoring activities that included measuring and tagging turtles, patrolling beaches, and maintaining seasonal records of turtle populations.

The settlement grew organically as other Amerindians settled there, and at its peak it became home to about 40 families. The Almond Beach settlement today forms about 100 acres of land that mostly stretches along the coast, although the community structures cover less than 5% of that area. The main land use activities have been for agricultural, residential, and conservation purposes, and all are situated within 600 feet from the shoreline. Traditional wooden, low-stilted houses feature troolie palm roofing and detached communal kitchens. A small health post and primary school were built as the population grew with more women and children. The community relied on natural resources for their livelihoods, and women in particular would partake in agricultural activities such as fishing, crabbing, and growing crops such as cassava and beans. The community eventually established a thriving coconut farm, and they sold the harvested nuts at places such as Mabaruma, the capital city Georgetown, and Trinidad and Tobago (Guyana Times International, 2011). Members of the Almond Beach community also actively participated in turtle conservation and beach protection duties. Eventually, the turtle populations improved, and a guesthouse and observation post were constructed to cater for scientists and turtle watchers who would visit during the nesting season. Audley proudly recalls that in the year 2000, they recorded more than 4,000 leatherback turtles – a triumphant number that reflected the community's efforts. Turtle poaching became less prevalent through public outreach efforts and the support of various conservation bodies (Danny, 2013; EPA, 2019; Spitzer, 2014).

13.3 Disruptive weather and few relocation prospects

Almond Beach residents recalled that in the past, the beach was wider and flood events were rare. But over time, they noticed a worrying trend: There seemed to be less annual rainfall, but thunderstorms seemed to be more frequent and severe. They described how wave action from the Atlantic Ocean ingressed deeper into the beach until the shoreline had noticeably receded (Figure 13.2).

Figure 13.2 Felled coconut palms on the Almond Beach. Photo by author.

The Shell Beach Protected Area Management Plan reported that satellite data from 1987 to 2007 showed regression of some parts of the coast by up to 1,100 meters (Protected Areas Commission, 2014). Several Almond Beach families who lived closer to the shore were forced to shift their homes and farms inland to avoid erosion and flood damage. Floodwaters would remain on the land for several days, each time bringing mosquito swarms and crop losses. Erosion also forced the relocation of community structures inland, including the primary school, guesthouse, and the Protected Areas Commission's turtle monitoring campsite buildings (Benn et al., 2020; Department of Public Information, 2017; Guyana Marine Turtle Conservation Society, 2018). As erosion worsened, Almond Beach families eventually moved away (Solomon, 2017); Audley says that at least 11 families gradually relocated to Mabaruma and Moruca.

Despite these events, Audley's family and others explained that they remained at Almond Beach simply because it was their home. They wanted to continue working with the turtles, and they felt they could manage the impacts of floods and erosion on their own by occasionally retreating several feet inland. But their views on this shifted in January 2017 after what they described as the most severe weather they experienced there. Abnormal high tides brought intense storm surges that lashed the shore for hours. The aftermath of the storm was pronounced as it had further eroded the beach and flooded the community for weeks (Stabroek News, 2017a). Residents described how the floods reached as much as four feet in some sections of the community. Families with flooded homes temporarily relocated to less affected neighbors, and mobility was limited in the settlement for weeks. Children could not attend school for some time and much of the subsistence crops perished. The community also sustained major agricultural losses as hundreds of coconut trees were felled by the storm (Solomon, 2017).

> *"We knew it was just a matter of time before we'd have to leave."*
> *– Audley James*

It was this event that finally caused the James family to consider relocation. The government quickly deployed an emergency team to Almond Beach to assess flood impacts (Stabroek News, 2017b), and the community received relief supplies from both government and nongovernment organizations. Given the heightened tidal and erosional activities in the area, the regional government prioritized the evacuation of teachers from the community (Y. Greene, 2020) and the Protected Areas Commission suspended tourist activities at Shell Beach for 6 months (Department of Public Information, 2017; Stabroek News, 2017c).

Recognizing the community's inability to cope with the impacts of these climate events, regional officials encouraged the Almond Beach families to relocate (Solomon, 2017). However, residents indicated that they were uncertain about the form of relocation assistance that would be provided. Several months later, the government identified Khan Hill, a vacant area at Mabaruma, as resettlement land for Almond Beach residents. But Audley and Romeo found that the land would be too small for their needs, farming would be difficult due to rock outcrops, and a motor vehicle would be required to access the main settlement. This was the only resettlement land identified by authorities, and most Almond Beach families decided against moving there. It soon became evident that any other relocation effort would need to be largely self-directed. Following another bout of intense weather in 2019, the James family arrived at the same decision as some of those prior: It was time to retreat to safer land.

13.4 Community losses and uncertainty

Reflecting on his family's relocation experience, Audley indicated that the move meant that they lost their home, farmland, and sources of income; as well as their close-knit community and the turtles that they had grown fond of after more than thirty years of conservation work at Almond Beach. The loss of hundreds of mature coconut trees was a major setback for the village. He emphasized that the beach was considered among the community's losses from the 2017 storm: "Besides losing the [coconut] crops, having less beach area meant there would be less nesting area for the turtles."

Audley's wife, Violet, explained the careful planning that was required over several weeks to organize the family's resettlement. This included deciding what items to leave behind and preparing enough food and supplies for the journey. The month-long move required a small crew, three motorboats, and five half-day trips. The James family has since resettled at Moruca, where they constructed their new home on fertile family land. Audley attributed their successful move to the support received from family and friends, as well as the barrels of fuel provided by regional officials. But he explained that they required additional in-kind supplies: "We needed food, we needed more assistance with transportation. We needed nails and lumber … this would have helped us to rebuild." He shared that assistance was also needed after the move as they needed new sources of income.

> *"We now have to start all over again."*
>
> – Almond Beach resident

Back at Almond Beach, the six remaining families weigh their future. The community has dwindled and life there remains uncertain. The receded shoreline continues to impact agriculture and conservation activities, which provide important sources of income. Some families were forced to shift their houses further inland from erosion, but several structures remain exposed. One resident with nine children said he is unsure about how to begin moving. Others expressed concern about their livelihoods, whether they will have enough resources to resettle, and how the move might impact the children – and the turtles. Regardless of how things unfold, Audley stands ready to support the remaining families. He expects at least two families to eventually settle near his

Moruca home, and plans to establish a large playfield for all the children he expects to move there soon.

Elsewhere in the region, flood-hit villages have presented health issues or caused entire small riverine communities to relocate (Drakes & Benn, 2017). Regional government officials who were interviewed indicated that despite significant financial and capacity limitations, the region offers what little support it can to climate-affected communities. For instance, persons affected by severe drought often receive direct assistance in the form of fuel or borrow boats so that they can collect fresh water from inland sources. Officials expressed concern about how climate change has impacted people living in rural areas and acknowledged that more assistance is needed for these communities. But education, health, and infrastructural needs usually dominate the regional budget allocations, with little funds available for this type of social assistance (Y. Greene, personal communication, October 2020). Government has mainly engaged climate-affected communities by encouraging them to move to higher ground and providing relief items. Limited land is available for housing at the Mabaruma Township, and other lands are yet to be identified for relocation purposes. Further, the region has had to respond to these scenarios in the absence of a clear policy or plan for rendering assistance to climate-affected communities.

13.5 A perpetual battle with the Atlantic Ocean

At a broader level, national development strategies by different governments have committed the country to combating climate change through readiness, adaptation, and protection of at-risk communities (Government of Guyana, 2010, 2019). These policy considerations are critical given that the flood-prone, inhabited coast sits 6 feet below mean sea level (UNEP, 2009). The mobile shoreline is continually reshaped by erosional-depositional cycles that naturally occur roughly every 30 years (NEDECO, 1972; Pritchard, 1969). Short-term coastal dynamics are also driven by the westward movement of mud shoals that slowly shift with wave action along the continental shelf. The coastal biome typically remains stable along the length of the mud shoal, but an erosional front exists between two shoals that can destroy mangrove stands and shorelines (Daniel, 2001). This may eventually cause intrusion of the ocean when wave action enables a breach of the natural embankment. Research has shown that sea level rise and severe precipitation events can intensify erosional activity on the coast (Bruun, 1962; Schwartz, 1967; Zhang et al., 2004). This can be disruptive for inhabitants and sensitive ecosystems; for instance, at Almond Beach, the Protected Areas Commission reported that beach loss due to erosion is the main challenge affecting marine turtle nesting (Benn et al., 2020).

The economically important and populous coastal zone is protected from inundation by a combination of artificial and natural sea defenses (Government of Guyana, 2012; NEDECO, 1972). The lowlands feature remnants of colonial Dutch flood engineering comprising seawalls, groynes, kokers (sluices), and

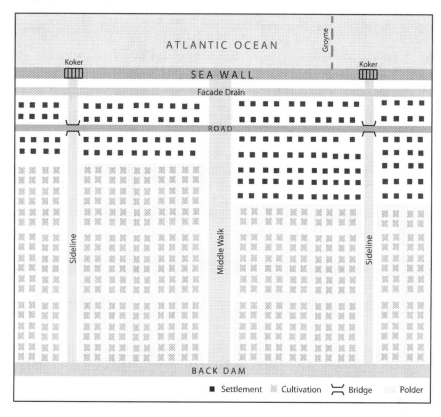

Figure 13.3 Schematic of irrigation and drainage on the Guyana coast. Figure by author.

dams (Figure 13.3). *Groynes* break wave action and slow sediment movement, while *kokers* regulate water levels in irrigation canals (Daniel, 2001). The *sea dam* prevents tidal flooding of homes and farms, while the *back dam* regulates water from marshy peat swamps and prevents inundation of agricultural lands from biannual seasonal precipitation (Bernard, 1999). This aging, discontinuous system is supported by fragile mangrove stands that have long functioned to stabilize sediment movement caused by longshore drift (Dalrymple & Pulwarty, 2006; Daniel, 2001). Mangroves that were disturbed by natural and human activity have been re-established through ongoing restoration programs (Conservation International, 2018). Authorities regularly reinforce the hard sea defenses by installing riprap, extending or raising the wave wall, and improving pumping capacities and internal drainage (Lees et al., 2009). But the capital city and many other coastal settlements remain vulnerable to sea level rise, and the sea defense system can fail if poorly maintained or overwhelmed during the wet season (ECLAC, 2005; Pelling, 1997; UNEP, 2009; Williams & Johnson-Bhola, 2009).

13.6 Plight of vulnerable coastal settlements

One of the nation's worst natural disasters occurred in 2005, when unusually heavy rainfall brought devastating floods on the Guyana coast (ECLAC, 2005; GFDRR, 2005). Here, families typically occupy detached concrete flats or stilted houses with enclosed lower levels in flood-prone areas with poor drainage and infrastructure (Pelling, 2003; Williams & Johnson-Bhola, 2009). With as much as 5 feet of water in some places, the floods submerged coastal communities for weeks. The scale of this disaster overwhelmed response institutions. Thousands of persons were displaced or trapped and many required evacuation to emergency shelters. Other impacts included disease outbreak, property damage, and disruption of economic activities and social services (ECLAC, 2005; Williams & Johnson-Bhola, 2009).

Coastal flooding continues to occur almost every wet season (Government of Guyana, n.d.); for instance, tall waves during spring tide have severely overtopped the sea defenses and flooded homes and farms (Johnson, 2018). Guyana's Second National Communication to the United Nations Framework on Convention on Climate Change (Government of Guyana, 2012) projected that increased storm surges and intense rainfall will continue to be experienced along the coastal plain. Vulnerability is further expected to increase given sea level rise projections of up to 51 cm by 2071, while storm surge projection modeling indicates that more than 100,000 hectares of coastland may be inundated by the 2070s (Government of Guyana, 2012; Office of Climate Change, 2019).

Low-income groups are especially vulnerable and may struggle the most to withstand climate change impacts (Office of Climate Change, 2019). Among the most critically exposed communities in the urban zone are small settlements at places known as "zero tolerance areas" – a term that connotes self-settlement in places deemed unfit for human occupation (V. Roberts, personal communication, 2013). Plastic City in Region III is one such settlement that exists on the seaward side of the defenses. Small informal wooden and zinc dwellings are pocketed among the mangroves and scattered along sea defense structures. Many women and young children occupy these precarious buildings, often in overcrowded conditions. Drivers of settlement at this area include the inability to afford adequate housing, access to jobs, and proximity to social services including markets, schools, and places of worship (Bynoe et al., 2018). Drakes (2016) reported that the community is frequently exposed to floods, gales, and poor sanitation. Settlement adjacent to the sea defenses can also impair structural maintenance and increase coastal vulnerability to extreme weather events (Drakes, 2016; Ragobeer, 2019). While foreshore squatters at other areas accepted government houselots to relocate from zero tolerance areas, Plastic City residents said they did not move mainly due to the greater distances of the new locations from the urban center and their sources of income (Bynoe et al., 2018; Ragobeer, 2019).

> "For weeks, the rice lands and the ocean appeared as one."
> – Mahaica (Region V) resident

The Civil Defence Commission is the nation's disaster management agency, which has worked for decades to protect communities and respond to the needs of affected persons. The agency has steadily improved its capacities to go beyond emergency response, and has worked to implement strategies and systems for disaster mitigation and preparedness. Among its efforts are development of comprehensive risk assessments, scalable disaster management plans, and early-warning systems (B. Henry, personal communication, 2020; Velasco, 2014). But unusual climate events across the country can challenge the success of these initiatives (Figure 13.4). An episode in late 2019 illustrates these difficulties, when two successive spring tide events coincided with natural mangrove depletion and other breaches to the sea dam. The Civil Defence Commission issued early-warning bulletins about above normal tides estimated at up to 11 feet, and placed low-lying coastal communities in five administrative regions (across 160 miles of coastland) on high alert (Guyana Chronicle, 2019; Kaieteur News, 2019).

Mahaica residents in Region V, many of whom are rice and livestock farmers, prepared for the floods by blocking entryways with sandbags, raising furniture, and moving livestock to higher ground. Yet, the floods overwhelmed the farming communities and caused millions of Guyana dollars in property losses (Bourne, 2019; Meusa, 2019a). Although residents tried to better prepare for the second spring tide event that occurred 3 weeks later, they did not expect that flood levels would be higher than before (Meusa, 2019b). The floods caused school closures, temporary evacuation, and health risks from flooded outhouses and mosquito infestations (Ramsay, 2019). The Civil Defence Commission distributed sanitation hampers to dozens of affected families and assisted some persons with evacuation (R. Harding, personal communication, 2019). One resident who was interviewed reflected that for weeks, it was hard to distinguish where the rice lands ended and the Atlantic Ocean began. The land was eventually drained through emergency works to fortify breached stretches of revetment and earthen embankment; but more than a year later, families still struggled to grow cash crops from the saltwater deposits.

Climate-impacted communities have historically contributed little to the global climate problem, yet often suffer disproportionately from its effects. National and regional assessments indicate that coastal and indigenous communities are among the most hazard exposed in Guyana (Drakes, 2016; Drakes & Benn, 2017; Velasco, 2014), with women and children from urban poor and indigenous communities holding lower coping and adaptive capacities (Office of Climate Change, 2019; UNICEF, 2016, 2017, 2018). A combination of poor socioeconomic conditions, limited access to health and education, and, in the case of indigenous communities, greater distances from urban centers, can exacerbate vulnerability (Pelling, 1997; UNICEF, 2016, 2018; Velasco, 2014). Although indigenous villages in Guyana have reported perceived changes in weather patterns, many lack the social programs needed to boost their resilience to climate change (UNICEF, 2017).

Figure 13.4 Spring tide waves from the Atlantic Ocean crash over a seawall defending Anna Catherina, Guyana. Photo by author.

13.7 Pursuing relocation as an adaptation strategy

The accounts in this chapter indicate preliminary relocation challenges and needs of different communities and social groups. Poorly executed resettlement can result in impoverishment, food insecurity, and loss of livelihoods (EBRD, n.d.). Community buy-in can also be difficult if external assistance appears fragmented or unstructured (Bertana, 2020; Lujala et al., 2014). As with Almond Beach and Plastic City, government-identified lands for relocation purposes may be considered as inadequate or unviable for their needs. Impractical relocation options and insufficient disaster assistance may place communities at greater risk, derail resettlement efforts, and weaken adaptive capacities (Lujala et al., 2014).

Resettlement has been put forward as an adaptation strategy in the 2001 Climate Change Action Plan, which proposed encouraging gradual retreat to higher ground through decentralization of coastal settlement and services (Government of Guyana, 2001). This Plan suggested that retreat could be promoted through affordable interior housing programs, new industries, and loans to stimulate small enterprises in hinterland areas. A study of the 2005 flood disaster (Williams & Johnson-Bhola, 2009) concluded that although relocation to higher areas was suggested by affected persons, such an initiative could be challenged by the required costs and levels of coordination. Community involvement and restitution packages have been recommended in the National Multi-Hazard Disaster Preparedness and Response Plan (Civil Defence Commission, 2013) as possible resettlement actions proposed as part of an early recovery framework for Guyana, which would aim to reduce future disaster impacts on the most poor and vulnerable groups. The Climate Resilient Strategy and Action Plan (Government of Guyana, 2015) proposed that the government should examine the potential for gradual retreat through feasibility studies and reiterated that policies be introduced to encourage resettlement. Coastal development has largely remained steady and will likely intensify given Guyana's new oil and gas sector, but like other developing states, Guyana has worked to implement responsive strategies for climate change mitigation and adaptation, and is developing sectoral capacity through climate readiness projects (Government of Guyana, n.d.; Green Climate Fund, 2015; Office of Climate Change, 2019). The government also appealed for greater climate financing to respond to communities affected by localized weather events, and is pursuing multilateral action to achieve national climate adaptation and meet the 2030 Agenda (Guyana Chronicle, 2020).

Specific action is now needed in the event that more climate-impacted communities require disaster and relocation assistance (UNDRR, 2015). Through this work, priority resettlement needs that were identified include the following: Access to adequate and cultivable land, access to building materials, proximity to social services, opportunities for income generation, and the ability to maintain community. As the government works to develop a national disaster recovery policy (Stabroek News, 2021), an ensuing resettlement action plan can be developed to support the resettlement of vulnerable communities (EBRD, n.d.; IFD, n.d.; Stapleton et al., 2017). This adaptation strategy can benefit from the recommendations of prior national climate action and adaptation plans (Government of Guyana, n.d., 2001, 2015; Office of Climate Change, 2019). The resettlement plan should receive direct input from coastal communities through impact surveys and needs assessments; ensure multisectoral involvement to successfully activate systems; and implement global best practices to help reduce climate risk.

Table 13.1 outlines a number of resettlement strategies that Guyana could incorporate. Resettlement planning should also adequately respond to the different needs of communities. Special consideration may be required for locale-specific scenarios like Almond Beach, whereby the indigenous residents' sense of place attachment is linked with their role as environmental guardians. Doherty and Clayton (2011) indicated that uncertainty can delay community mobilization; in this case, uncertainty about how the community's relocation might impact marine turtle protection. In this context, the government has clearly valued and respected indigenous contributions, given the close consultations and decision-making with stakeholder community members during the Protected Area's establishment (Kalamadeen, 2011; Protected Areas Commission, 2020). Similarly, should the entire community need to relocate, residents' concerns about future turtle conservation can be assuaged if they are meaningfully engaged by the authorities.

Participatory decision-making at all stages of resettlement is especially critical for securing more equitable outcomes for historically marginalized groups and underserved communities. From global experiences, displaced communities that are actively involved in state-led relocation efforts tend to achieve more positive outcomes (UNHCR, 2014). Guyana has emphasized the need for social equity

Table 13.1 Potential resettlement strategies at the pre- and post-resettlement stages

Pre-resettlement strategies	Post-resettlement strategies
Providing viable resettlement land options in respect of location, size, productivity, and proximity to services (IFD, n.d.).	Providing adequate social amenities as a fundamental right for women, children, elderly, and differently abled persons (Doherty & Clayton, 2011; Gromilova, 2014).
Implementing an assistance program such as a cash-land aid package to support resettlement needs and encourage timely mobilization (EBRD, n.d.).	Strengthening the coping capacities of women affected by the resettlement process (Gromilova, 2014).
Leveraging public-private partnerships to construct homes and community buildings at resettlement sites (UNHCR, 2014).	Providing vocational training to help resettled persons (especially youth) pursue alternative livelihoods (UNHCR, 2014).
Improving agricultural opportunities at resettlement sites as an incentive to relocate communities with high agricultural dependency (Karanth et al., 2018).	Continuing engagement to strengthen the social capital of affected groups (Stapleton et al., 2017).

of vulnerable groups in national climate change processes as a policy objective (Office of Climate Change, 2019), stating that the rights and protections of women, youth, elderly, and indigenous persons should be protected. The proposed resettlement action plan should therefore advance appropriate solutions that cater to the needs of these cohorts, along with the broader needs of indigenous communities and informal settlements like Almond Beach and Plastic City, respectively. This may become more complex should larger settlements eventually need to relocate, but just outcomes for all communities are more likely to be achieved through an early and inclusive adaptation planning process.

13.8 Conclusion

Climate affected communities should never feel abandoned in the effort to relocate. While the Almond Beach, Plastic City, and Mahaica examples comprise small-scale vulnerability scenarios, they present useful preliminary insight on the critical need for a resettlement pathway to assist vulnerable coastal communities. Guyana can work to bolster its disaster preparedness efforts and secure just solutions in the resettlement process. Fundamentally, a resettlement plan can strengthen national disaster preparedness and risk reduction efforts by informing coordinated actions at the national, regional, and community levels. Should large-scale retreat from the coast become necessary, Guyana would be better prepared to equitably implement relocation actions, render disaster assistance, and activate postrelocation social and economic recovery programs.

13.9 Recommendations

An early opportunity exists for Guyana to examine the impacts of climate induced displacement of vulnerable communities, with a view to implement relocation based on community needs and best practices garnered from the experiences of other nations. Three key recommendations are put forward:

1. The Guyana government should ensure equal access to the decision-making process so as to derive feasible and just outcomes for affected groups. Communities should have access to viable prospects for employment, agriculture, health, education, and other social services. Participatory and inclusive resettlement should facilitate timely transition to resettlement lands and pursue the physical, social, and cultural protections of vulnerable communities.
2. Geospatial scenario modeling can be used to identify the most vulnerable populations, determine alternative management strategies, and inform evidence-based decisions on planned retreat. Institutional strengthening programs should be used to improve capacities and resources so that agencies are ready and able to respond appropriately to climate resettlement issues.
3. A multisectoral approach should be adopted to develop a resettlement policy framework and action plan, with a view to integrate adaptation planning into existing policies, plans, and programs. This should coordinate the work of stakeholder agencies and set the framework for the restoration of livelihoods through appropriate resettlement lands, assistance packages, and continued community engagement.

Notes

1 Shell Beach was declared a Protected Area (IUCN Category VI – Managed Resource Protected Area) in 2011, and has since been managed by the Protected Areas Commission.
2 An Arawak settlement situated about 80 miles southeast of Almond Beach (Figure 13.1).

References

Benn, S., Bumbery, R., & Edghill, S. (2020, June 16). *World Sea Turtle Day: Sea turtle conservation efforts in Guyana* [Webinar]. WWF-Guianas.

Bernard, D. M. (1999). *A new geography of Guyana*. Macmillan Caribbean.

Bertana, A. (2020). The role of power in community participation: Relocation as climate change adaptation in Fiji. *Environment and Planning C: Politics and Space*, 38(5), 902–919. 10.1177/2399654420909394

Bourne, S. (2019, October 27). Spring tide hits Mahaicony again. Department of Public Information. https://dpi.gov.gy/spring-tide-hits-mahaicony-again/#gsc.tab=0

Bruun, P. (1962). Sea-level rise as a cause of coastal erosion. *Journal of the Waterways and Harbors Division*, 88(1), 117–132.
Bynoe, P., Edinboro, E., & Benn, D. K. (2018, June 27–29). *Confronting the challenges of managing informal settlements: A critical review of the case of 'Plastic City', Guyana* [Conference Presentation]. Caribbean Urban Forum (CUF2018), Kingston, Jamaica.
Civil Defence Commission. (2013). *National Multi-Hazard Disaster Preparedness and Response Plan – Guyana*.
Conservation International. (2018). *State of mangroves in Guyana: An analysis of research gaps, and recommendations*.
Dalrymple, O. K., & Pulwarty, R. S. (2006, June 21–23). *Sea level rise implications for the coast of Guyana: Sea walls and muddy coasts* [Conference presentation]. Fourth LACCEI International Latin American and Caribbean Conference for Engineering and Technology, Mayagüez, Puerto Rico.
Daniel, R. K. (2001). *Geomorphology of Guyana: An integrated study of the natural environment* (2nd ed.). Occasional paper No. 6, D. M. Bernard (Ed.). Department of Geography, University of Guyana.
Danny, K. (2013, June 30). Endangered marine turtle populations increasing in Guyana. *Kaieteur News*. https://www.kaieteurnewsonline.com/2013/06/30/endangered-marine-turtle-populations-increasing-in-guyana/
Department of Public Information. (2017, June 20). Shell Beach remains closed until mid-July. Department of Public Information. https://dpi.gov.gy/shell-beach-remains-closed-until-mid-july/
Doherty, T. J., & Clayton, S. (2011). The psychological impacts of global climate change. *American Psychologist*, 66(4), 265–276.
Drakes, O. (2016). *Hazard and vulnerability assessment of Essequibo Islands – West Demerara (Region Three)*. Civil Defence Commission: Georgetown.
Drakes, O., & Benn, D. K. (2017). *Hazard and vulnerability assessment of Barima-Waini (Region One)*. Civil Defence Commission: Georgetown.
EBRD. (n.d.). *Resettlement guidance and good practice*. European Bank for Reconstruction and Development.
ECLAC. (2005). *Guyana flood reports*. Economic Commission for Latin American and the Caribbean.
EPA. (2019). World Turtle Day 2019: Preparing for the impending ban on single-use plastics (Part 4). Environmental Protection Agency-Guyana. http://www.epaguyana.org/epa/news/155-world-turtle-day-2019-preparing-for-the-impending-ban-on-single-use-plastics-part-4
GFDRR. (2005). *Guyana: Preliminary damage and needs Assessment following the intense flooding of January 2005*. Global Facility for Disaster Reduction and Recovery. https://www.gfdrr.org/sites/default/files/publication/pda-2005-guyana.pdf
Green Climate Fund. (2015, December 9). *Guyana signs agreement with GCF in Paris* [Press release]. https://www.greenclimate.fund/news/guyana-signs-grant-agreement-with-gcf-in-paris
Gromilova, M. (2014). Revisiting planned relocation as a climate change adaptation strategy: The added value of a human rights-based approach. *Utrecht Law Review*, 10(1), 76–95.
Government of Guyana. (n.d.). *National Adaptation Plan Guyana: Inception Report*.
Government of Guyana. (2001). *Guyana Climate Change Action Plan in response to its commitments to the UNFCCC*.
Government of Guyana. (2010). *A low carbon development strategy: Transforming Guyana's economy while combating climate change*. Office of the President: Georgetown.

Government of Guyana. (2012). *Guyana's second national communication to the United Nations Framework Convention on Climate Change*. Ministry of Agriculture: Georgetown.

Government of Guyana. (2015). *Climate resilience strategy and action plan for Guyana*. Ministry of the Presidency: Georgetown.

Government of Guyana. (2019). *Green state development strategy: Vision 2040*. Ministry of the Presidency: Georgetown.

Guyana Chronicle. (2019, October 29). CDC Mobilised. *Guyana Chronicle*. https://guyanachronicle.com/2019/10/29/cdc-mobilised-2/

Guyana Chronicle. (2020, October 30). COVID-19 threatens fight against climate change. *Guyana Chronicle*. https://guyanachronicle.com/2020/10/30/447166/

Guyana Marine Turtle Conservation Society. (2018, June 24). *Almond Beach Primary School before it was relocated board by board* [Attached image] [Status update]. https://www.facebook.com/317183365285264/photos/almond-beach-primary-school-before-it-was-relocated-board-by-board-due-to-the-en/644988799171384/

Guyana Times International. (2011, May 19). Almond Beach community exploring prospect of bottling coconut water. *Guyana Times International*. https://www.guyanatimesinternational.com/almond-beach-community-exploring-prospect-of-bottling-coconut-water/

IFD. (n.d.). *Handbook for preparing a resettlement action plan*. International Finance Corporation: Environment and Social Development Department.

Johnson, D. (2018, March 3). Spring tides smash West Dem sea defence, houses. *Stabroek News*. https://www.stabroeknews.com/2018/03/03/news/guyana/spring-tides-smash-west-dem-sea-defence-houses/

Kaieteur News. (2019, October 24). Spring tide advisory – CDC urges residents in low-lying areas to take precautions. *Kaieteur News*. https://www.kaieteurnewsonline.com/2019/10/24/spring-tide-advisory-%E2%80%95cdc-urges-residents-in-low-lying-areas-to-take-precautions/

Kalamadeen, M. (2011, February 1). *The case of Shell Beach*. SWOT Report 6. State of the World's Sea Turtles. https://www.seaturtlestatus.org/articles/2011/1/the-case-of-shell-beach

Kandaswamy, S. V. (2015, June 21). Trekking Almond Beach. *Guyana Times Sunday Magazine*, pp. 3, 5.

Karanth, K. K., Kudalkar, S., & Jain, S. (2018). Re-building communities: Voluntary resettlement from protected areas in India. *Frontiers in Ecology and Evolution*, 6, 183. 10.3389/fevo.2018.00183

Lees, D., Wernerus, F. M., & Simpson, M. C. (2009). *Strategic environmental assessment of the Sea Defences sector policy in Guyana*.

Lujala, P., Lein, H., & Rød, J. K. (2014). Climate change, natural hazards, and risk perception: The role of proximity and personal experience. *Local Environment*, 20, 1–21. 10.1080/13549839.2014.887666.

Meusa, S. (2019a, October 1). Mahaicony farmers counting millions in losses after flooding. *Stabroek News*. https://www.stabroeknews.com/2019/10/01/news/guyana/mahaicony-farmers-counting-millions-in-losses-after-flooding/

Meusa, S. (2019b, October 29). Mahaicony residents in deeper flooding. *Stabroek News*. https://www.stabroeknews.com/2019/10/29/news/guyana/mahaicony-residents-in-deeper-flooding/

NEDECO. (1972). *Report on sea defence studies*. Netherlands Engineering Consultants. The Hague, Holland. Government of Guyana: Ministry of Works, Hydraulics and Supply.

Office of Climate Change. (2019). *National climate change policy and action plan 2020-2030 (Draft 2.0)*. Office of Climate Change, Ministry of the Presidency: Georgetown.

Pelling, M. (1997). What determines vulnerability to floods; a case study in Georgetown, Guyana. *Environment and Urbanization*, 9(1), 203–226. 10.1177/095624789700900116

Pelling, M. (2003). *The vulnerability of cities: Natural disasters and social resilience*. Earthscan.

Pritchard, P. C. H. (1969). Sea turtles of the Guianas. *Bulletin of the Florida State Museum*, 13, 85–140.

Protected Areas Commission. (2014). *Shell Beach Protected Area management plan 2015-2019*. Protected Areas Commission: Georgetown.

Protected Areas Commission. (2020, July 29). Shell Beach Protected Area (SBPA). *Guyana Chronicle*, p. 10.

Ragobeer, V. (2019, April 23). Gov't directs focus on 'squatting'. *Guyana Chronicle*. http://guyanachronicle.com/2019/04/23/govt-directs-focus-on-squatting

Ramsay, R. (2019, September 30). High tides invade coastal Guyana. *Kaieteur News*. https://www.kaieteurnewsonline.com/2019/09/30/high-tide-invades-coastal-guyana/

Schwartz, M. (1967). The Bruun theory of sea-level rise as a cause of shore erosion. *The Journal of Geology*, 75(1), 76–92.

Solomon, A. (2017, January 18). Government mulls relocating Shell, Almond Beach residents – following recent floods. *Guyana Chronicle*. http://guyanachronicle.com/2017/01/18/govt-mulls-relocating-shell-almond-beach-residents-following-recent-floods

Spitzer, R. (2014). *Human dimensions of community based conservation of endangered sea turtles on Shell Beach, Guyana [Project Report]*. Department of Wildlife, Fish and Environmental Studies, Swedish University of Agricultural Sciences – Umeå.

Stabroek News. (2017a, January 16). Almond Beach in north west battered by high waves. *Stabroek News*. https://www.stabroeknews.com/2017/01/16/news/guyana/almond-beach-north-west-battered-high-waves/

Stabroek News. (2017b, January 18). Team dispatched to assess Almond Beach after flooding, erosion. *Stabroek News*. https://www.stabroeknews.com/2017/01/18/news/guyana/team-dispatched-assess-almond-beach-flooding-erosion/

Stabroek News. (2017c, April 23). Authorities order Shell Beach closed over floods, erosion. *Stabroek News*. https://www.kaieteurnewsonline.com/2017/04/23/authorities-order-shell-beach-closed-over-floods-erosion/

Stabroek News. (2021, January 17). National disaster recovery policy to be developed – CDC Director General. *Stabroek News*. https://www.stabroeknews.com/2021/01/17/news/guyana/national-disaster-recovery-policy-to-be-developed-cdc-director-general/

Stapleton, S. O., Nadin, R., Watson, C., & Kellet, J. (2017). *Climate change, migration and displacement: The need for a risk-informed and coherent approach*. Overseas Development Institute and United Nations Development Programme.

UNDRR. (2015). *Sendai framework for disaster risk reduction 2015-2030*. United Nations Office for Disaster Risk Reduction: Geneva.

UNEP. (2009). *Urban environment outlook: An integrated environmental assessment of Georgetown*. United Nations Environmental Programme: Geneva.

UNHCR. (2014). *Planned relocation, disasters and climate change: Consolidating good practices and preparing for the future* [Project Report]. UNHCR – The UN Refugee Agency.

UNICEF. (2016). *Analysis on children and women in Guyana*. UNICEF Guyana: Georgetown.

UNICEF. (2017). *Study on indigenous women and children in Guyana*. UNICEF Guyana: Georgetown.

UNICEF. (2018). *Climate landscape analysis for children: An assessment of the impact of climate, energy and environment on children in Guyana*. UNICEF Guyana: Georgetown.

Velasco, M. A. (2014). *Progress and challenges in disaster risk management in Guyana*. Civil Defence Commission: Georgetown.

Williams, P. E., & Johnson-Bhola, L. P. (2009). Causes and consequences of coastal flooding in Guyana. In D. McGregor, D. Dodman, & D. Barker (Eds.), *Global change and Caribbean vulnerability: Environment, economy and society at risk* (pp. 74–99). UWI Press.

Zhang, K., Douglas, B. C., & Leatherman, S. P. (2004). Global warming and coastal erosion. *Climatic Change, 64*, 41–58. 10.1023/B:CLIM.0000024690.32682.48

14 Mobile livelihoods and adaptive social protection: Can migrant workers foster resilience to climate change?

Haorui Wu and Catherine Bryan

14.1 Introduction

Typhoon Haiyan (2013), also known as Super Typhoon Yolanda, was one of the deadliest Philippine typhoons on record (BBC News, 2013). A category 5 storm, it destroyed 90% of the island of Leyte, with immediate and protracted consequences of 14 million people across 44 provinces (National Disaster Risk Reduction and Management Council, 2013). As a result, the livelihoods of 5.9 million workers were disrupted; 4.1 million people were displaced; more than 6,300 were killed; and 7 years later, nearly 2,000 people remain unaccounted for (World Vision, 2020). In the aftermath of Super Typhoon Yolanda, the Government of the Philippines adapted existing social protection measures to provide direct cash and in-kind support for individuals and families who lost property and assets. An expansion of the "Pantawid Pamilyang Pilipino Program" – a conditional cash transfer program, the state implemented several new lending programs intended to mitigate severe deprivation. These included house repair loans, salary loans, and pension advances (International Federation of Red Cross and Red Crescent Societies, 2014).

Running in tandem to these efforts, international aid organizations offered different kinds of cash transfer and in-kind transfer to address food consumption needs, housing repair, and livelihood and education-related expenses. Regarded a success by the Filipino government and various relief-providing agencies (Bailey & Harvey, 2015; Reyes et al., 2018; Smith, 2015), activists and critical scholars are less optimistic. Here, critique has largely focused on the short-term nature of official responses to Yolanda, and the extent to which these failed to address long-term development issues, and more precisely, to bolster residents' long-term resilience to future climate-induced events. Additionally, there has been concern that cash transfers obscured extent to which post-Haiyan rehabilitation and reconstruction favored big, internationally financed companies to the determinant of local business and labor (Yee, 2018).

Since the early 2000s, scholars and practitioners have considered the connection between social protection and migration (Crawley & Skleparis, 2018). At the intersection of disaster management scholarship and migration studies, this chapter contributes to this ongoing analysis by considering the relationship

DOI: 10.4324/9781003141457-14

between state development projects focused on labor export and those intended to off-set the social and economic consequence of climate catastrophe. It does so with an emphasis on the dual-role of migrants who are explicitly and implicitly centered in each approach. A near constant absent-presence, migrants' grassroots efforts subsidize the often-nominal contributions of the state vis-à-vis the material well-being of its citizenry, both preenvironmental and postenvironmental crisis. Drawing on in-depth fieldwork with Filipino temporary foreign workers in Manitoba and their nonmigrant kin in the Philippines, this chapter explores the strategies of migrant families when faced with social and economic hardship following environmental disaster, while highlighting the ambiguity generated by postdisaster responses in national contexts informed by state-driven migration-as-development paradigms. Revealed in three case studies, these strategies entailed redirecting remittances toward postdisaster rebuilding, and were often accompanied by a relocation. Of importance, these new mobilities, despite their grounding in long-standing familial projects of migration, varied in scale, direction, and anticipated outcome. As we illustrate in the preceding sections, labor migration and postdisaster relocation represent potentially intersecting and mutually constitutive forms of mobility that map onto existing inequalities that simultaneously compel labor migration and generate climate vulnerabilities.

Migration-as-development has long been a strategy of the Philippine state (Bagasao, 2005; Battistella, 2012; Calzado, 2007). Originating in the late 1960s, state-led labor export initiatives were formalized in the early 1970s. Under martial law (1972–1986), the Marcos regime consolidated all development and economic planning. Aided by rising world commodity prices, debt rescheduling and IMF-sponsored stabilization programs, the 1970s saw some economic growth in the Philippines (Bello et al., 2005). However, those who had formulated the regime's development strategy sought to retain, rather than disrupt, the distribution of power and resources within the country, using high-levels of poverty to justify their pursuit of top–down development. In this context, while per capital income in the Philippine grew, real wages fell, and the poor became poorer. Following a series of failed land reform initiatives, faced with growing internal tensions, and eager to take advantage of global employment opportunities, the regime firmly and aggressively committed itself to a strategy of labor export (Tyner, 2010). Institutionalized in the 1974 Labour Code, Filipino labor export has since been consistently, across administrations, deployed, as a means of utilizing the international economy to bolster domestic economic growth. As a result, over 50 years later, labor migration has become a ubiquitous feature of Filipino social, familial, cultural, and economic life (Aguilar, 2014; Asis et al., 2004; Barber, 1997).

The data for this analysis come from ethnographic fieldwork conducted by Bryan over a 5-year period of time in the Philippines and Canada with migrants and their nonmigrant kin. An amalgamation of in-depth life history interviews and participant observation, these data have been read through the substantive and policy expertise of Wu. As such, our analysis occupies a unique methodological and conceptual space, one simultaneously informed by anthropology's ethnographic method and long-standing engagement with migration studies, and hazards and disaster research. From these diverse perspectives, the chapter argues that the efforts of migrants vis-à-vis their nonmigrant kin in climate-vulnerable locations represents a form of adaptive social protection, and unlike institutionalized social protection strategies, embedded in mainstream relief paradigms, this adaptive instrument importantly provides ongoing support that may enhance a family's long-term resilience to disaster. In these way, migrant contributions serve as additional support to the short-term efforts of states and NGOs, which concurrently fail to generate long-term capacity vis-à-vis future potential climate catastrophe. That said, mirroring the faulty logics of migration-as-development in both the sending- and receiving-state – in this instance, the Philippines and Canada – this resilience remains tied to tumultuous and precarious labor markets that are supported by governments primarily invested in economy. As a result, the capacity of migrant families to withstand environment degradation and extreme climate events remains tenuous, contingent, and bound to premigration, predisaster class status.

14.2 Social protection in disaster efforts

Vulnerable and marginalized populations are often hardest hit by extreme events (Wu & Karabanow, 2020). This is due to their precarious geographic locations, material realities, and other related societal issues reflective of inequality (Wu & Drolet, 2016). These vulnerability factors reinforce one another, weakening individual and collective coping capacities and resulting in significantly slower recovery processes relative to other groups. This limited resilience capacity, in turn, worsens their social, economic, cultural, and physical status, exposing them to multiple adjacent risks and further stalling community-based social and economic development, particularly in developing countries (Cutter, 2018; Liu & Huang, 2014). Despite the policy and programming efforts of international NGOs, intragovernmental agencies like the United Nations (UN), and individual states, climate change- and/or disaster-related vulnerabilities remain major barriers for developing countries located in the disaster-prone regions.

The social protection model offers promise for poverty reduction and economic development (Devereux et al., 2016), and is regarded as an effective instrument for the support of marginalized groups faced with increasing levels of risk and vulnerability (Tenzing, 2019). As a set of interventions, social protection ideally combines different types of social services and income/in-kind transfer to prevent and/or reduce the risk of poverty (UNRISD, 2010). The model is also regarded as having considerable potential to increase the capacity of marginalized groups to contend with the consequences of extreme climate events (Stern, 2008). Institutionalized social protection strategy in the global context of climate change adaptation and disaster reduction always follows a top–down trajectory, concentrating on economic development by reducing poverty and protecting local inhabitant's rights (Devereux et al., 2016). Despite the potential variability of social protection interventions, in such contexts, cash and in-kind transfers are most often used to meet disaster survivors' basic living requirements: For example, obtaining food, water, and daily necessities, and access to social, medical, and other related services (Medair, 2013). Such an approach is increasingly favored by many sponsoring organizations because it offers immediate visible outcomes that are relatively easily achieved and measured.

In response to some of the perceived limitations of current models of post-disaster social protection that prioritize short-term needs, Devereux and Sabates-Wheeler (2004) argue that a transformative framework for social protection responsive to mandates of social justice must encapsulate rights, basic short- and long-term demands, and empowerment. Offering the term "adaptive social protection," Davies and his colleagues (2008) expand this position explaining that social protection should enhance, in the long-term, disaster survivors' preparedness, response, adaptation, and recovery capacity in the face of extreme events, namely, resilience capacity. Put differently, social protection must hold in balance both short- and long-term needs and objectives, and it should prompt an equitable redistribution of resources, such that people – regardless of their predisaster condition – are better equipped to withstand future events (Wu, 2021). Concurrently, these scholars argue that these efforts should, at once, be institutionalized and cultivated at the grassroots level. In many ways, we might understand the contributions of migrants in the aftermath of climate catastrophe to represent such a grassroots social protection effort. To illustrate this potential, in the following section, we explore three case studies centered on the responses to Typhoon Haiyan by Filipino migrants working in Canada. However, as we elaborate in the fourth section, tethered to the imperatives of state-driven migration-as-development strategies, such grassroots efforts fail to meet the laudable goals set out by Devereux, Sabates-Wheelers, and Davies in their collective work. Moreover, and of particular importance for practitioners invested in social justice for the communities they serve (migrant and non-migrant alike), these efforts might inadvertently reinforce the very systems and structures that set the stage for both out-migration and climate crisis vulnerability.

14.3 Migration and climate catastrophe

Much of the literature on migration and climate change emphasizes the extent to which environment degradation and extreme climate events prompt mobility (Black et al., 2011). Much as economic development and civil conflicts have been positioned as the major drivers stimulating international migration (International Organization for Migration [IOM], 2020), climate change, environmental degradation, and disaster progressively add to the complexity of international migration (Bassetti, 2019; Lonesco, 2019). The Internal Displacement Monitoring Center (2019), for example, reported over 17.2 million new climate-induced, environmentally triggered, and/or disaster-related displacements across 148 countries and territories in 2018. Some of this poses new challenges for states and nongovernmental agencies supporting migrants. Less explored, however, is the role of migrants in supporting nonmigrant kin affected by environmental disaster, and the extent to which this support enhances long-term resilience to future climate events. Our case studies focus on the well-worn migration pathway between Canada and the Philippines. The Philippines has over 10% of its population working and/or residing across almost all the countries and territories on the earth (Asis, 2006). In Canada, Filipinos consistently rank amongst the top three categories of newcomers, arriving through a range of permanent and temporary immigration programs (Canadian Filipino Net, 2020). Although Filipino Canadians generally have lower incomes than the national average and other immigrant groups (Statistics Canada, 2007), they send more money back to their families abroad than any other immigrant group (Statistics Canada, 2019). Given the extent of climate vulnerability in the Philippines, these remittances are often used to support families to recover from extreme events (Dunham, 2019).

14.4 Migrant worker responses to Typhoon Haiyan

Between 2009 and 2014, a small hotel in rural Manitoba hired 71 migrant workers. These workers arrived from across the Philippines, with ten from the province of Leyte, which would be hit in 2013 by Typhoon Haiyan. Most were recruited through Canada's Temporary Foreign Worker Program (TFWP). The TFWP facilitates and manages the employment of migrant workers by Canadian employers on a temporary basis. By tying residency to employment, the TFWP establishes the conditions by which workers are dependent on their employers, rendering them vulnerable to a number of exploitative practices (Nakache & Kinoshita, 2010). Critically, however, all but one worker of these 71, would transition to permanent residency. Since the late 1990s, immigration policy in Canada has been increasingly decentralized. Empowered by a series of bilateral agreements with the federal government, the provinces develop and manage their own immigration schemes to meet unique demographic and economic needs (Carter et al., 2008). Although the criteria for permanency under the Manitoba Provincial Nominee Program (MPNP) have shifted over time, between 2009 and 2014, most TFWs in Manitoba could anticipate a transition to permanent residency status through the MPNP, if they had full-time permanent jobs. Of the ten migrant workers at the Hotel affected by Super Typhoon Haiyan, our case study focuses in three.

14.5 Migrant case studies

14.5.1 Joseph: Relocation and return

Joseph arrived in Manitoba in 2010. By early 2013, Joseph had secured permanent residency and had moved to Victoria, British Columbia in search of a higher salary and, as he explained, greater opportunity. He enjoyed his life and work in Manitoba, but the town was small and there were limited possibilities for personal and professional growth. He craved a larger city and moreover, with new earnings, planned to build his parents a new house in Leyte, somewhere further inland. He would also begin to put aside for his niece's education. There was also the issue of his father's deteriorating health, and his growing need for a greater number of daily medications. Joseph's family lived in Tacloban, on a busy commercial and residential street that traced the coastline. Their house, like others on the street, was a small, two-story wooden structure. Their proximity to the bay had earlier enabled the family's livelihood, but it had been several years since his father had fished. When Typhoon Haiyan arrived, their neighborhood was directly in the storm's path. In conversation a year after the event, Joseph recalled: "You feel like you're going to burst. You can't contact your family members. And everything you see on TV – dead people, the ravages of the typhoon." Joseph continued, "from the photos on Facebook and the clips from the news, you expect the worst, and it's hard to accept; it's hard to be strong."

Unable to confirm the location of his parents, Joseph returned to the Philippines. Denied his request to take a leave of absence from his job in Victoria, he was forced to resign his position. Securing an expensive, last-minute ticket, he eventually found his parents living in a shelter amongst other survivors.

Their home destroyed, Joseph began the work of finding a new residence further inland. While this corresponded broadly to his initial pre-Haiyan plan, the family's relocation within Leyte did not offer the hoped-for opportunity for upgrade beyond enhancing the distance between their home and the coast. Depleting his savings from years of overseas employment, Joseph began the work of rebuilding his family's life, relocating them to a new, but not more robust home, replacing all of their material possessions, and supplying them with several months-worth of nonperishable provisions. After a few weeks, Joseph returned to Canada. However, rather than returning to Victoria where he no longer has employment, he arranged to return to Manitoba, where he would resume his work at the Hotel. There, he anticipated greater support from his employer, the comfort of an extended social network and familiar Filipino community, and an improved capacity to send money home due to a significantly lower cost of living. Saving was, at this point, not an option. All of the money he earned was spent on his own immediate needs or those of his parents in the Philippines. Years of work, time spent away from kin, and careful planning for the future – the sacrifices of overseas employment, undermined.

14.5.2 Ester: Resuming the responsibility of migrant-breadwinner

Ester arrived in Manitoba in 2009. When we met in 2012, she was a permanent resident and was living with her 9-month-old son born in Canada. At the time, she was waiting for her husband's residency visa to be issued so he could join them. Like Joseph, Ester's adult life had mostly been lived away from the Philippines where her family – her husband, two brothers, their wives, and children, her sister, several cousins, and her parents – lived in Leyte. First from Taiwan and then Manitoba, Ester ensured their survival and little-by-little sought to improve their fortunes. Like Joseph, Ester's plan was to build a more secure and spacious home for them, and so, as she was able, she replaced the nipa (palm) and corrugated metal with concrete. Ester had long accepted her role as her family's primary provider, but was relieved when her parents and brothers were able to generate additional income: Her parents, through a small church they ran, and her brothers through employment with a local construction company. Their earnings meant a redistribution of responsibility for the family as they were increasingly able to cover the family's immediate and daily needs. In turn, Ester redirected some of her income toward her own ambitions, saving to attend a nursing program and opening an educational account for her son.

Though their home in Leyte was sturdier than those of their neighbors, when Haiyan landed, nearly a decade of work and investment was undermined as the structure and everything it contained was washed away. Similarly, her parents' church – located near the harbor – was destroyed, and the sector that had employed her brothers, obliterated. Unable to connect with her family for days, Ester eventually learned of their fate through a photo posted by the Red Cross of her brother at a rice distribution center. "That's how I knew," she said in 2014, "that's how I knew he was alive." The photo depicted her family's profound and sudden need: Her brother surrounded by a swell of people, his hands held out toward the camera, hoping to secure a small pail of rice. The image provoked great anxiety, and yet in the absence of other news, Ester found comfort in it. With the knowledge they were safe, she set herself to the task of caring for her family from Manitoba. The burden of her family's survival firmly once more on her shoulders, she emptied her bank account and began sending the majority of her earnings to the Philippines. Her own prospects in Canada regarding upward mobility curtailed, she began rebuilding their lives and their family home in Leyte.

14.5.3 Eli: New opportunities through relocation

Eli's family was also impacted by the disaster. Eli had arrived in Manitoba in 2012, after having lived and worked in the United States, and by 2014 was living and working in Victoria at a high-end hotel. The third of four siblings, Eli's sisters had secured lucrative employment in Singapore and Dubai. The recipients of their remittances and already relatively well-off (his father had had a career in local politics), Eli's parents lived in a solid concrete home in a coveted neighborhood at some distance from Tacloban. In preparing for his own overseas employment, Eli had pursued postsecondary education in Hotel and Restaurant management. While this was similar to both Ester and Joseph, Eli had been able to supplement this education with culinary training. With his sisters' remittances and then his own, his family had opened a small bakery near their home. When the Typhoon hit, their house was badly damaged, but it was not destroyed; the solid, concrete walls offering some protection to the inhabitants and their possessions. Importantly, and in no small part a reflection of their relative class privilege, after Haiyan, Eli's family fared better than many in the region. For them, rebuilding did not require the liquidation of the family's shared savings; nor did they have to supplement Eli and his sister's remittances with cash transfers from the state. Instead, they were able to pool their resources to expand their pre-hurricane livelihood (the bakery), and even more significantly, they were able to relocate to a neighboring province, which was less vulnerable to hurricanes and typhoons.

14.6 Discussion: Migration and adaptive social production

When describing why they sought employment abroad, the Hotel's migrant workers consistently reflected on social instability, economic precarity, and environmental degradation endemic in the Philippines – these risks are compounded by the state's limited investment in infrastructure and the unevenness that characterizes resource distribution in the country. Migration is a well-documented response to the inequalities that determine the social and material fate of many Filipinos. As such, it emerges as a possible source of grassroots social protection, running adjacent to state efforts. And yet, determined by the logic of migration-as-development, the social justice potential of this strategy is undermined, as resources come to be distributed according to pre-existing social hierarchies and inequalities.

Demonstrated in our three case studies, the ability of Eli, Ester, and Joseph's families to recover from Typhoon Haiyan was significantly based on the contributions – both past, present, and future – of their respective migrant-member, which served preventative and protective functions that alleviated immediate deprivation. For Ester's family, this meant circumventing the reality of post-disaster relocation faced by many, and facilitated by Ester's employment in Manitoba and willingness to resume the role of migrant breadwinner, remaining and rebuilding in-place. For Joseph, relocation was a goal that predated the Typhoon. Reflective of the social, financial, and physical insecurity prompting his migration, this long-standing objective had been set in anticipation of the very real potential of extreme climate events and environmental disaster. As such, while the conditions and outcomes of their relocation were not as hoped for, their move was otherwise welcome. Importantly, in Eli's case, recovery also provided an opportunity for expansion and upward class-mobility, and as such safeguarded his family from future environmental disaster. Ester, Eli, and Joseph were similarly positioned when they arrived in Manitoba; yet, the experiences of Ester and Joseph, whose families' class status did not permit specialized training or significant savings, differed significantly. If Ester and Joseph had to start from scratch as their family's only migrant breadwinner, Eli – one of four sibling-migrant earners – was able to invest in his family's shared business enterprise. By the time the Typhoon hit, and in its aftermath, Eli's relative privilege had proven particularly advantageous. For him, migration emerged as a mechanism of social protection that safeguarded life and livelihood, *and* increased his family's capacity to withstand future environmental disaster by facilitating their relocation within the Philippines and providing them with expanded livelihood opportunities. In contrast, given their social and economic locations in the Philippines and in Canada, Joseph and Ester's ability to redress long-term risk was limited despite their long-standing integration in global labor markets.

And yet, further to this, a number of characteristics distinguish Ester, Eli, and Joseph from many other Filipino Overseas Workers – notably, that despite their initial temporary status, by the time the Typhoon struck, each had become a permanent resident of Canada. As such, relative to other migrants, they had a certain amount of security. Indeed, though problematized in a small but growing literature (Bonifacio, 2015; Bryan, 2019; Polanco, 2016), the transition to permanency redresses many of the exploitative tendencies of the Temporary Foreign Worker Program by undercutting worker dependency. A second critical distinction is the support they received from the Hotel and from the community in rural Manitoba. The local paper, for example, profiled the migrants affected by the disaster with sensitivity and care to generate awareness. In addition to facilitating time-off and accommodating short-term travel to the Philippines, the Hotel, in turn, sponsored a fundraiser that successfully generated over $10,000 all of which was distributed to the affected families. Joseph's case offers unique insight into these dimensions of this particular cohorts' experience. In addition to the relative financial security engendered by permanent residency that enabled him to relocate and continue supporting his family, permanency facilitated his own secondary relocation and return to Manitoba. Here, Joseph's mobility, though not localized in the Philippines, was – all the same – prompted by the material needs of his parents in the aftermath of the Typhoon. Still, despite their permanency in Canada and the support of their employer, their experiences of the Typhoon reflect the vulnerabilities, contingencies, and uncertainties endemic in so many global migration scenarios. Simultaneously, while each case offers insight into the potential of migration and mobile livelihoods as a form of social protection to natural disaster, they also reveal the tensions inherent in doing so.

These tensions follow from the social inequalities and economic conditions that compel workers to seek out employment in global labor markets, and that concurrently reduce local capacity to withstand climate events. That Ester, Joseph, Eli, and their families fared relatively well cannot be taken as exemplary. Rather, their status in Canada, coupled with their employment security read against the challenges faced by their kin is the illustrative feature of their experiences. Relying on migrants to bolster inadequate state-based and NGO-delivered social protection merely meets short-term needs. Moreover, in the absence of safe working conditions, equitable legal status, and independence vis-à-vis employers, the adaptive social protection they offer to nonmigrant kin comes at an enormous cost. In a historic context where migration for most is a risky attempt to remedy poverty, unemployment, and discrimination, reliance on it as a social protective measure is fraught at the best of times. States benefit from it because they are absolved of the responsibility of caring for their citizens. And in a manner similar to cash transfers, and other short-term social protection measures, monetary injections from individual migrants fail to redress systemic forms of inequality and structural violence, that accelerate the effects of disaster. The grassroots adaptive social protection effort via ongoing cash transfer addresses nonmigrant kin's immediate challenges, and it may fulfill their long-term development requirements, ultimately building their families' resilience capacity; however, mirroring the mandate of migration-as-development initiatives, it does so in an ad hoc, contingent, and inconsistent manner. The consequence of which is an uneven redistribution of resources, and the exacerbation of existing inequalities.

14.7 Conclusion

In the global context of climate change and disaster, academic scholarship and policy-based research has considered the relationship between migration and social protection since the early 2000s. Importantly, in this literature, Sabates-Wheeler and Waite (2003) distinguish between migration as a strategy of social protection, and migration as a phenomenon that generates a need for social protection. Indeed, given the exploitative potential of guest-worker programs globally, the extent to which worker mobility from the Philippines is part of a larger state-led project of development, and the precarity that often characterizes migrant labor, migration also demands social protection for those most immediately implicated, the migrant workers themselves, who are subject to a range of abusive and exclusionary practices and policies. In the context of climate-disaster and crisis, we see migrant workers left behind by the states and local populations who, otherwise, depend on their labor. For many, in the context of COVID-19, for example, this has been exacerbated, reflected in involuntary confinement (Devet, 2020), increased rate of infection, and death (CBC, 2020). In turn, this chapter demonstrates how the migrants may promote the traditional organization-led cash transfer approach to develop adaptive social protective measures for their nonmigrant kin, while illustrating the variability and contingency of such measures.

Among our participants, each sought overseas employment to support nonmigrant kin. For those with family affected by Haiyan, the Typhoon dramatically reoriented the objectives of the migration project. Revealed in their narratives are a series of relational and structural dynamics that migrants navigated as they redirected their goals in the service of short- and long-term disaster relief. Their efforts, we argue, represent a form of adaptive social protection. However, unlike institutionalized social protection strategies, embedded in mainstream relief paradigms, this adaptive instrument provides long-term support. Consequently, migration may enhance an extended family's long-term resilience to disaster. However, and as we outline, this dynamic also reinforces some of the most problematic features of international labor migration. Indeed, migration as a source of disaster relief fails to attend to the structural inequalities and pervasive injustices that increasingly set the stage for both out-migration and climate crisis vulnerability.

14.8 Recommendations

- Social protection intervention policy of cash/in-kind transfer should merge with the organization-led and Grassroots-led approaches, to not only fulfill the disaster survivors short-term basic living requirements, but also provides ongoing support addressing the survivor's challenges for long-term development.
- Migrant grassroots adaptive social protection strategies should be encouraged in community-based climate-induced resettlement practice because these strategies target their nonmigrants' rooted vulnerabilities and contribute to building their nonmigrant families' resilience.
- Although remittance sending as an adaptive social protection strategy inevitably maps onto existing social and economic inequalities were demonstrated as a partial measure in this chapter, future studies should comprehensively examine other benefits of this grassroots approach and/or explore other approaches to advance the traditional social protection intervention policies for climate-induced resettlement.

References

Aguilar Jr, F. V. (2014). *Migration revolution: Philippine nationhood and class relations in a globalized age*. Nus Press.

Asis, M. M. B. (2006, January 1). The Philippines' culture of migration. Migration Policy Institute. Retrieved from https://www.migrationpolicy.org/article/philippines-culture-migration

Asis, M. M. B., Huang, S., & Yeoh, B. S. (2004). When the light of the home is abroad: Unskilled female migration and the Filipino family. *Singapore Journal of Tropical Geography*, 25(2), 198–215.

Bagasao, I. F. (2005). Migration and development: The Philippine experience. *Remittances: Development impact and future prospects*, 133–142.

Bailey, S., & Harvey, P. (2015). *State of evidence on humanitarian cash transfers: Background note for the high level panel on humanitarian cash transfers*. Retrieved from https://www.odi.org/sites/odi.org.uk/files/odi-assets/publications-opinion-files/9591.pdf

Barber, P. G. (1997). Transnationalism and the politics of "home" for Philippine domestic workers. *Anthropologica*, 39, 39–52.

Bassetti, F. (2019, May 22). Environmental migrants: Up to 1 billion by 2050. *Foresight*. Retrieved from https://www.climateforesight.eu/migrations/environmental-migrants-up-to-1-billion-by-2050/

Battistella, G. (2012). Multi-level policy approach in the governance of labour migration: Considerations from the Philippine experience. *Asian Journal of Social Science*, 40(4), 419–446.

BBC News. (2013, November 22). Typhoon Haiyan death toll rises over 5,000. *BBC News*. Retrieved from https://web.archive.org/web/20131122124347/http://www.bbc.co.uk/news/world-asia-25051606

Bello, W., Bello, W. G., De Guzman, M., Malig, M. L., & Docena, H. (2005). *The anti-development state: The political economy of permanent crisis in the Philippines*. Zed Books.

Black, R., Adger, W. N., Arnell, N. W., Dercon, S., Geddes, A., & Thomas, D. S. G. (2011). The effect of environmental change on human migration. *Global Environmental Change - Human and Policy Dimensions*, 21(SI), S3–S11. doi: 10.1016/j.gloenvcha.2011.10.001

Bonifacio, G. T. (2015). Live-in caregivers in Canada: Servitude for promissory citizenship and family rights. In *Migrant domestic workers and family life* (pp. 145–161). Palgrave Macmillan.

Bryan, C. (2019). Labour, population, and precarity: Temporary foreign workers transition to permanent residency in rural Manitoba. *Studies in Political Economy*, 100(3), 252–269.

Calzado, J. R. (2007). Labour migration and development goals: The Philippines' experience (International Dialogue on Migration). WMO Conference Center, Geneva.

Canadian Filipino Net. (2020). Filipinos in Canada double in numbers in last decade. Retrieved from https://www.canadianfilipino.net/news/filipinos-in-canada-double-in-numbers-in-last-decade

Carter, T., Morrish, M., & Amoyaw, B. (2008). Attracting immigrants to smaller urban and rural communities: Lessons learned from the Manitoba Provincial Nominee Program. *Journal of International Migration and Integration/Revue de l'integration et de la migration internationale*, 9(2), 161–183.

CBC News. (2020, June 1). Temporary foreign worker dies due to COVID-19 as disease hits southerwestern Ontario farms hard. *CBC News*. Retrieved from https://www.cbc.ca/news/canada/windsor/southwestern-ontario-farms-covid19-migrant-worker-dies-1.5593046

Crawley, H., & Skleparis, D. (2018). Refugees, migrants, neither, both: Categorical fetishism and the politics of bounding in Europe's "migration crisis". *Journal of the Ethnic and Migration Studies*, 44(1), 48–64. doi: 10.1080/1369183X.2017.1348224

Cutter, S. L. (2018). Compound, cascading, or complex disasters: What's in a name? *Environment: Science and Policy for Sustainable Envelopment*, 60, 16–25. doi: 10.1080/00139157.2018.1517518

Davies, M., Oswald, K., Mitchell, T., & Tanner, T. (2008). Climate change adaptation, disaster risk reduction and social protection briefing note. Retrieved from https://www.ids.ac.uk/download.php?file=files/IDS_Adaptive_Social_Protection_Briefing_Note_11_December_2008.pdf

Devereux, S., & Sabates-Wheeler, R. (2004). Transformative social protection. IDS Working Paper 232. Retrieved from https://opendocs.ids.ac.uk/opendocs/bitstream/handle/20.500.12413/4071/Wp232.pdf?sequence=1

Devereux, S., Roelen, K., & Ulrichs, M. (2016). Where next for social protection? *IDS Bulletin*, 47(4), 103–118. doi: 10.19088/1968-2016.158

Devet, R. (2020). Thousands of seafarers stuck on board with no way to get repatriated and no shore leave. *The Nova Scotia Advocate*. Retrieved from https://nsadvocate.org/2

020/06/09/thousands-of-seafarers-stuck-on-board-with-no-way-to-get-repatriated-and-no-shore-leave/
Dunham, J. (2019). Filipinos in Canada sent more money abroad than any other group in 2017: Study. *CTV News*. Retrieved from https://www.ctvnews.ca/business/filipinos-in-canada-sent-more-money-abroad-than-any-other-group-in-2017-study-1.4384006
Internal Displacement Monitoring Center. (2019). Global report on internal displacement. Retreived from https://www.internal-displacement.org/sites/default/files/publications/documents/2019-IDMC-GRID.pdf
International Federation of Red Cross and Red Crescent Societies. (2014). Case study: Unconditional cash transfers response to Typhoon Haiyan (Yolanda). Retrieved from https://www.calpnetwork.org/wp-content/uploads/2020/01/philippines-ctp-case-study-en.pdf
International Organization for Migration [IOM]. (2020). World migration report 2020. Retrieved from https://www.un.org/sites/un2.un.org/files/wmr_2020.pdf
Liu, M., & Huang, M. C. (2014). *Compound disasters and compounding processes: Implications for disaster risk management*. United Nations Office for Disaster Risk Reduction, National Graduate Institute for Policy Studies, Asian Development Bank Institute. Retrieved from https://www.preventionweb.net/english/hyogo/gar/2015/en/bgdocs/inputs/Liu%20and%20Huang,%202014.%20Compound%20disasters%20and%20compounding%20processes%20Implications%20for%20Disaster%20Risk%20Management.pdf
Lonesco, D. (2019). Let's talk about climate migrants, not climate refugees. Retrieved from https://www.un.org/sustainabledevelopment/blog/2019/06/lets-talk-about-climate-migrants-not-climate-refugees/
Medair. (2013). Medair use of cash transfer through disaster risk reduction activities. Retrieved from https://www.humanitarianresponse.info/en/operations/afghanistan/document/success-storiesmedair-use-cash-transfer-through-disaster-risk
Nakache, D., & Kinoshita, P. J. (2010). The Canadian temporary foreign worker program: Do short-term economic needs prevail over human rights concerns? *IRPP Study*, 5. Retrieved from https://ssrn.com/abstract=1617255
National Disaster Risk Reduction and Management Council. (2013). *Final report re effects of Typhoon" Yolanda" (Haiyan)*. Retrieved from http://ndrrmc.gov.ph/attachments/article/1329/FINAL_REPORT_re_Effects_of_Typhoon_YOLANDA_HAIYAN_06-09NOV2013.pdf
Polanco, G. (2016). Consent behind the counter: Aspiring citizens and labour control under precarious (im)migration schemes. *Third World Quarterly*, 37(8), 1332–1350.
Reyes, C. M., Albert, J. R. G., & Reyes, C. C. M. (2018). Lessons on providing cash transfers to disaster victims: A case study of UNICEF's unconditional cash transfer program for Super Typhoon Yolanda victims. Retrieved from https://think-asia.org/handle/11540/8171
Sabates-Wheeler, R., & Waite, M. (2003). Migration and social protection: A concept paper. GSDRC. Retrieved from https://gsdrc.org/document-library/migration-and-social-protection-a-concept-paper/
Smith, G. (2015). Cash coordination in the Philippines: A review of lessons learned during the response to Super Typhoon Haiyan. Retreived from https://www.calpnetwork.org/publication/cash-coordination-in-the-philippines-a-review-of-lessons-learned-during-the-response-to-super-typhoon-haiyan/

Statistics Canada. (2007). *The Filipino community in Canada*. Retrieved from https://www150.statcan.gc.ca/n1/pub/89-621-x/89-621-x2007005-eng.htm

Statistics Canada. (2019). *Study on international money transfers, 2017*. Retrieved from https://www150.statcan.gc.ca/n1/daily-quotidien/190417/dq190417c-eng.htm?CMP=mstatcan

Stern, N. (2008). *Key elements of a global deal on climate change*. London School of Economics and Political Science, London, UK.

Tenzing, J. D. (2019). Integrating social protection and climate change adaptation: A review. *WIREs Climate Change, 11*(2), 1–16. doi: 10.1002/wcc.626

Tyner, J. A. (2010). *The Philippines: Mobilities, identities, globalization*. Routledge.

United Nations Research Institute for Social Development [UNRISD]. (2010). *Combating poverty and inequality: Structural change*. Retrieved from https://www.unrisd.org/unrisd/website/document.nsf/(httpPublications)/BBA20D83E347DBAFC125778200440AA7?OpenDocument

World Vision. (2020, April 6). Hitting "Send" on digital emergency aid during COVID-19. *World Vision*. Retrieved from https://www.worldvision.ca/stories/health/hitting-send-on-digital-emergency-aid-covid19

Wu, H. (2021). Bottom-up adaptive social protection: A case study of grassroots self-reconstruction efforts in post-Wenchuan earthquake rural reconstruction and recovery. *International Journal of Mass Emergencies and Disasters, 39*(1), 65–86.

Wu, H., & Drolet, J. (2016). Adaptive social protection: Climate change adaptation and disaster risk reduction. In J. Drolet (Ed.), *Social development and social work perspectives on social protection* (pp. 96–119). Routledge.

Wu, H., & Karabanow, J. (2020). COVID-19 and beyond: Social work interventions for supporting homeless population. *International Social Work, 63*, 1–4. doi: 10.1177/0020872820949625

Yee, D. K. P. (2018). Violence and disaster capitalism in post-Haiyan Philippines. *Peace Review, 30*(2), 160–167. doi: 10.1080/10402659.2018.1458943

15 Identity and power: How cultural values inform decision-making in climate-based relocation

Rachel Isacoff

15.1 Introduction

The projected impacts of climate change-related sea level rise and extreme weather have forced policy makers, practitioners, and communities to acknowledge the grim reality that tough choices need to be made among competing interests. The deleterious impacts of climate change on cultures will include tangible effects on landscapes and structures (e.g., flood damage) and intangible social and economic consequences that jeopardize entire cultures (e.g., uprooting communities and their associated livelihoods) (Veerkamp, 2009), especially those involved in climate-based relocation. Government agencies and local communities frequently have competing values, contingent on perceptions and attitudes to risk (e.g., when to relocate) as well as cultural preferences (e.g., who decides, who moves, and where to), which shape relocation discourse, decisions, plans, and implementation.

If there is a conflict in views and subjective priorities, whose values count? Using both rural and urban examples in the United States, this chapter explores the role of cultural values and power dynamics involved in relocation processes both at the community level and for policy makers and practitioners. I question the role of policy makers and practitioners in determining outcomes for relocation processes and call on them to examine their own privilege and worldviews. Ultimately, equity, justice, and the intangible costs to cultures must be balanced with other desirable agendas to meet community needs.

15.2 Culture and values

Culture and climate are inextricably linked through the emotional bonds between people and place. Culture is the beliefs, rituals, art, and stories that create collective outlooks and behaviors. It captures a group's identity, shapes values, and develops community cohesion. What people think, do, and produce develops culture, which in turn becomes "shared, learned, symbolic, crossgenerational, adaptive, and integrated" (Adger et al., 2011). Communities are built through shared experiences, belonging, cultural practices, memories, and

DOI: 10.4324/9781003141457-15

social networks – which themselves can be negotiated and change over time (Adger et al., 2012; Williams, 2002).

Culture is central to understanding decision-making processes for relocation – the identification of risks, decisions about response, and means of implementation are all mediated by culture (Adger et al., 2012). Decisions guiding climate-based relocation are informed by cultures ranging from urban areas to rural settlements, as well as marginal locations. Places are valued by those with cultural associations to them – everyone from descendants of farmers, to Indigenous people, to the longtime patrons of the local family pizzeria. Culture and ethnocentric ideologies then inform the actions of governments, academia, and businesses. As culture and politics interact with existing beliefs and behaviors, tension may emerge between stakeholders when determining what information and process is deemed legitimate.

As core elements of a culture, values serve as standards and guide action (or inaction), choice and rationalization. Values also provide the framework for how societies and institutions manage risk and allocate scarce financial resources. Relocation is a multiscalar process, requiring acknowledgment, cooperation, and investment from multiple levels of government and a variety of individuals and stakeholders (Adger et al., 2009; Koslov, 2016). Some stakeholders, like governments and planners, prioritize strategies that give precedence to rational, financial analyses while others, like frontline communities, are aligned with group identity – integrating local knowledge and prioritizing established livelihoods, equity, and self-determination (O'Brien, 2009). The peril of the involuntary loss of places, identities, and perceived individual rights, coupled with the urgency for financially cost-effective solutions, are deep-seated concerns for communities facing climate-related relocation. The clash of views between and within governments and frontline communities imposes a critical impediment to fostering just relocation solutions.

Since relocation is complicated by problems of control and influence, more powerful actors favor an approach most aligned with their interests, often framed as rational economic imperative (Ajibade, 2019). This dynamic is further complicated by political cycles in which elected officials in US cities and states hold office for 2- to 4-year terms and focus on short-term wins that can support them or their affiliated party in re-election campaigns. Similarly, government planners are typically committed to a program or project for the term of a grant or their employment period. In some locations, policy makers are reluctant to embrace relocation for fear of upsetting their constituents, especially those that envision adapting in-place (Jessee, 2020), or because they see relocation as failure and defeat but value waterfront development as growth and progress (Koslov, 2016). Another concern is the administrative burdens of a relocation program and resulting loss of a tax base (Grannis, 2011), although preventing future flood damage can result in extensive savings in the long term. Government agency budgets are created with this lens, often demonstrating the prioritization of short-term, cost-efficient projects, rather than long-term plans and projects that support relocating communities.

15.3 Justice and power

The values that are pursued versus those that are ignored can easily become tangled in the politics of relocation. Government authorities will only support community relocation and other risk reduction measures if these measures are perceived to be a priority in their political agendas. For example, in 2015, during my time at the White House Council of Environmental Quality, President Obama traveled to Alaska in an attempt to understand and respond to community needs. As part of that effort, we announced public, philanthropic, and private commitments to support climate resilience in Alaska through coastal erosion strategies and voluntary relocation coordination efforts (The White House, 2015). In the lead-up to the trip, I worked with the Office of Management and Budget to secure funds for an access road in Kivalina, an Alaska Native community. Kivalina had previously received $15.5 million from federal and state agencies between 2006 and 2009 to protect the community with seawalls that had a lifespan of 10–15 years (Bronen, 2015). Nearly 10 years later, the aging seawalls were increasingly ineffective, and the community had decided that the relocation of their entire community was the only strategy to protect them from advancing coastal erosion and repetitive flooding (Bronen, 2015). Advocates for the community pressed the Administration to support Kivalina's transition with a safe and reliable evacuation route.

The evacuation road was a shovel-ready project with a clear price tag. But when the administration could not secure funding in time to include it in the press release for the President's trip, the urgency to support Kivalina waned and the project was buried under a mound of paperwork. All of the White House offices involved moved on to prepare for the next press release or the next priority project. My team shifted focus to initiate the development of equitable principles for relocation – an important effort to infuse equity practices in federal processes, but not a project that the Kivalina community would benefit from before their next storm. And, unsurprisingly, the development of those principles halted with the change of federal administrations (Mandel, 2017).

As with the example above, power is most commonly understood through the allocation (or withholding) of resources – money and the influence attached to those who control it. The capacity of a community to relocate is shaped not only by government agencies with more power to control relocation processes, but also by the extent of economic opportunity, degree of urbanization, access to insurance, and existing planning regulations at national and local levels (IPCC, 2012). Cost, feasibility, and unequal distribution of benefits can be significant obstacles for relocation (Thornton & Comberti, 2013). Inadequate governance mechanisms or budgets to support relocating communities can intensify community impoverishment and loss of culture (Maldonado et al., 2013). Since relocation is often limited to those with more economic resources and human capital, lack of action may lead to immobility (Adger et al., 2012). Kivalina and many other Alaska Native communities have for decades voted repeatedly to relocate but have not received resources to do so, despite multiple government studies demonstrating imminent danger (Bronen, 2015; Marino, 2015).

In the United States, systems of oppression, racism, and exploitation have created long-term vulnerabilities for marginalized communities with resource-dependent livelihoods and/or lack of political influence and financial resources. These systemic inequities engender deep-seated distrust in government-led relocation processes. In parallel, governments and funders distrust the coordinating capacity of communities, and community-led relocations are confronted by uneven power relations. Culture in this way is linked to the dynamics of power and governments' imposition of a specific set of values and expectations, preventing community groups from defining and achieving their unique goals. To attain their goals, impacted communities must be active participants in relocation processes and negotiations (Adger et al., 2009). If the White House had led an inclusive process and invited leaders from Kivalina to the table during those negotiations, perhaps the outcome would have been different.

Beyond money and authority, however, it is important for communities and practitioners to consider other formal and informal modes of power. Power can take the form of time and knowledge, each of which can hinder or help communities. For example, governments may set up grant funding periods that are too short to allow meaningful community engagement during relocation processes. Also, to keep project costs down, consultants often conduct desktop research and leverage existing project templates rather than go on site visits and establish processes based on community needs Alternatively, many subsistence and Indigenous societies retain traditional ecological knowledge (TEK) of their environments, enabling them to monitor, observe, and manage environmental change – making them the experts on the impacts of climate change on their communities. Through consideration of social hierarchies and perceived influence, respect, and expertise, we can identify alternative positions of authority.

Given that conflicts arise when community and government cultures clash, as well as the role of perceived power dynamics within these conflicts, government agencies, practitioners, and communities must question how each of their worldviews, including their values and biases, shape their own influence in decision-making processes.

15.4 Failures and successes of community-driven relocation based on identity and power

The two US-based examples below demonstrate how various priorities, values, and ideologies can interrupt or sway relocation processes.

15.4.1 Disempowerment in Isle de Jean Charles

In 2014, the State of Louisiana applied to the federal National Disaster Resilience Competition (NDRC). The NDRC was a $1 billion funding opportunity for jurisdictions with recent disaster declarations. The goal of the NDRC was to incentivize integrated planning and stakeholder engagement in the development of climate adaptation projects. Louisiana's proposal included a request for funding to relocate the Isle de Jean Charles community,[1] which was awarded $48 million in 2016.

The barrier island located in the Gulf of Mexico is home to the Isle de Jean Charles Biloxi-Chitimacha-Choctaw Indians of Louisiana (IDJC tribe). Though in 1955 the island was 22,400 acres, today – due to coastal erosion, subsidence, oil drilling, and recurrent flooding due to extreme weather – less than 2% of the original land mass of the island remains. Between 2002 and 2012, the population of the island declined from approximately 325 people to only 70 – causing a rift in the way the IDJC tribe transfers knowledge, history, and culture (Comardelle, 2020; Maldonado, 2014). The IDJC Tribal Council had been crafting a plan for over a decade to reunite their displaced people and restore their traditional ways of life together on safer ground (Jessee, 2020), which the State leveraged for the national competition.

The awarded relocation program promised whole-community relocation about 40 miles inland, emphasizing housing, access to employment, and support for cultural and social networks. The press release from the State described "minimum disruption to tribal livelihoods and lifestyles" and added that "tribal cultural traditions will stay alive as the tribe will now be able to come together to live in one community, not scattered as it is now" (State of Louisiana, 2016).

However, after 4 years of well-intentioned but misaligned efforts, the relocation stalled. The IDJC community frequently expressed their uncertainty about the State's plan, recalling a long history of being excluded from the decisions that affect their lives. In the months that followed the NDRC award, the State switched from referring to the IDJC Tribal Council as "partners" to "stakeholders" and imposed a top–down, rather than community-driven, planning process – which made Tribal leaders question the State's commitment to their community efforts (Jessee, 2020). The IDJC tribe voiced concern about the lack of involvement in planning, even learning about the State's purchase of land for the new site by reading a press release (Comardelle, 2020). Conflicts over tribal sovereignty were also complicated by the lack of federal recognition of the IDJC tribe. Even though this relocation process was the first to receive an outsize grant from a federal agency, the problems in execution were so formidable that the US Government Accountability Office issued a report in July 2020 recommending that Congress establish a federally led pilot program to help communities relocate (GAO, 2020).

As part of the State's consultant team, I witnessed a bifurcated process of consultants tasked with analysis for the new site and those who focused on community engagement with the IDJC community. I was on the team that conducted a market analysis of the new resettlement site and made recommendations for housing, commercial, and retail schemes that reflected the "highest and best use" of the site. This technocratic approach to planning did not mirror the economic characteristics of the existing community and did not directly consider how well the proposed uses responded to the needs of or supported the workforce training and livelihoods of the IDJC tribe. In fact, because of the way our team (and all consultants) valued and capped time through billable hours and leaned heavily on experience from conducting past economic development analyses, no one from my firm met with the tribe during the process. For this task, we valued efficiency and enhanced economic activity over participatory processes for the new site that lifted up community goals, despite what the NDRC application promised.

The planning field tends to assume that analyzing information logically and systematically is equivalent to doing so "without interest or bias," deprioritizing nonquantitative value systems (Baum, 2015). This presumed neutrality can be seen in other elements of relocation processes. The risk of irreversible loss is hard to reconcile with economic metrics of value used in decision-making about climate change (Adger et al., 2011), including relocation. Reducing the problem of impacts on and losses to places, cultures, and environments into such decision-making frameworks is to fit "philosophically incomparable values into inappropriately technical procedures" (Adger et al., 2011). Conflicts emerge as various outlooks on what is worth preserving or achieving through relocation processes – that is, what governments and practitioners' value – are prioritized through techno-centric rather than more socially grounded means (Corner et al., 2014).

Tensions additionally emerged over use of the Island. When people are displaced from locations that they value, there is a risk of cultural erasure. While relocation processes seek to move people out of harm's way, these processes can also increase financial and emotional stress and weaken social structures (Adger et al., 2012). Uprooting communities and their associated livelihoods can endanger individual and social identity, resulting in fragmented social networks that challenge community continuity (Oliver-Smith, 2009). Since physical and social losses compound each other, retaining connections to previously held land is critical for the cultural survival of many Indigenous communities like the IDJC tribe (Burkett et al., 2017; Marino, 2015). The IDJC Tribal Council's goals included the continued ownership of family parcels and protection of the Island even after relocating. However, this desire clashed with federal agencies' floodplain hazard mitigation expectations and regulations of returning residential areas involved in buyouts to open space (Ajibade et al., 2020; Jessee, 2020). Tribal leaders fought to maintain access to the Island for ceremonial, cultural, and recreation purposes, but remained concerned about restrictions to access and potential continued ownership.

The State's approach to administering this relocation process created uncertainty for the IDJC tribe and the future of the Island. As Jessee (2020) aptly questioned for future communities reflecting on this process: "Will communities need to choose between adaptation and identity? Between relocating out of harm's way and maintaining their collective self-determination? Or between safety from coastal flood and justice?" Administrative delays, lack of transparency, and disempowerment of the tribe's right to make informed decisions for their community led to the tribe's loss of trust in the process. The State's top–down approach reinforced existing inequalities and power structures – valuing ethnocentric notions of market value and efficiency over tribal continuity and livelihoods – and it added challenges to the tribe's ability to organize and relocate, even when they were willing to do so voluntarily.

15.4.2 Privilege and influence in Staten Island

It is important not to view Indigenous people as the "romantic other" and recognize that many multigenerational communities are threatened by the impacts of climate change (Barthel-Bouchier, 2013). Staten Island, New York, is home to multiple communities threatened by rising seas and climate-induced flooding. When Hurricane Sandy made landfall in New York City in 2012, Staten Island was hit particularly hard. Although Staten Island is the city's least populous borough, nearly half of the city's 53 deaths from the storm occurred there, and an estimated 300,000 homes in the borough were ruined (Baptiste, 2017). In the waterfront neighborhoods of Staten Island where the biggest storm surges in the city were recorded, many residents were outspoken about their multigenerational ties to the shore, recalling roots to their Italian heritage through their fruit and vegetable gardens (Koslov, 2019).

Residents of Staten Island mobilized after Sandy to demand a buyout from the government rather than rebuild their homes. As many of these neighborhoods were prone to recurrent flooding, groups of residents had previously organized a flood victims' committee to lobby for protections in the neighborhood's wetlands (Koslov, 2016). Though their efforts before Sandy had little success, the committee organized after the storm and, with unanimous support from nearly 200 neighboring households, created a buyout plan (Koslov, 2016, 2019). With federal disaster recovery funds, the State sponsored a buyout program, enabling homeowners to sell their land to the State and restore wetlands in its place to prohibit future development and provide protection from future storms. As petitioned by the community committee, the State named a portion of Oakwood Beach as the buyout program's pilot site. Homeowners who chose to participate would receive the prestorm value of their house and other incentives to encourage widespread uptake of the program.

Organizing neighborhood-wide support for the buyouts, which garnered the attention of the State, required coordinating action across political ideologies, within and between residents and government officials. Part of this process and discourse centered on remaining silent on the topic of climate change – a decision more likely aligned to values-based disputes and partisan ideology than disagreements about the underlying science (Corner et al., 2014). Not discussing the impact of climate change on the storm was more politically enabling and effective, minimizing conflict throughout their community and maintaining social norms and relations of power (Koslov, 2019). Instead, the community identified over-development, which had exacerbated flooding in the area in the years leading up to Sandy, as the scapegoat for the storm's damage.

The buyout process, however, offered protection to only a select few. The majority of households affected by the storm across New York City were low-income renters or immigrants whose status made them ineligible for government aid (Koslov, 2019). Because buyouts only apply to homeowners, the program disadvantaged low-income households that also may have attempted to relocate. The government buyouts gave the most money to the wealthier families who had the highest prestorm market value. The buyout policies (and all buyout programs) prioritized the cost and protection of flooded property instead of supporting the most at-risk people in need of continued protection (Marino, 2018).

In the neighborhoods where organizing for the buyouts occurred, homeowners outnumbered renters. Beyond this circumstance, the homeowners who participated in the buyout program held relative positions of power with greater political access and influence due to their race and social status. Staten Island is New York City's most politically conservative and only predominantly white borough. Many residents work as police officers, firefighters, sanitation workers, or in other unionized blue-collar occupations (Koslov, 2019). For the small number of recent immigrants included in the buyout program, overcoming the language barrier was a challenge, as was some residents' fear of signing government documents in an era of anti-immigrant public policy in the United States. Often, these residents relied on neighbors to help them understand both English and the technical aspects of the buyout process (Moore, 2018). Some struggled to qualify for a new mortgage because of poor credit and having limited funds due to the costs of repairs from previous floods. Guided by principles of capitalism, property ownership, and individualism (Marino, 2018), the State buyout program best served the wealthiest, predominantly white residents included in the program.

The pilot neighborhood committee did not need to include – and in fact moved quickly to avoid competition with – broader or more diverse stakeholders. The State's support of Oakwood Beach encouraged at least seven other Staten Island neighborhoods and hundreds of people in similarly hard-hit Queens neighborhoods to organize their own buyout groups, hoping also that the State would recognize their risk and include them in the buyout program (Barr, 2013; Koslov, 2016). Some of these other residential neighborhoods – as well as environmental justice organizations – were focusing on advancing equity for marginalized populations and getting equal buyout offers for communities of color (Koslov, 2019). But after months of meetings and petitions, the State only extended the buyout program to two additional white middle-class communities, preventing lower-income communities of color from participating. The State program covertly perpetuated systems of white dominance – evading consideration of vulnerability, power disparities, justice, and redistribution of resources to other communities in need. Neither the successful communities nor the State examined their ethics or values in the process.

Even with targeted disaster recovery funding and with community intent to relocate, some people lose. The buyout program privileged property owners, and by extension whiteness (Marino, 2018). The challenge remains to include diverse values in the implementation of relocation programs.

15.5 The role of policy makers and practitioners

Although both cases involved bottom–up planning, the positioning and relationship of the IDJC tribe and the Staten Island communities relative to government agencies and practitioners led to worse outcomes for people with less power. So what role do these outside actors play in decision-making, and what personal responsibility do they have to correct existing injustices? I offer three recommendations below centering the need for equity in relocation processes and practices.

15.5.1 Examine your own privilege and values systems

At the time of writing this chapter, the world is grappling with a global pandemic and racial injustices. But these crises expose systems of oppression that have existed long before 2020. In service of racial equity and justice, many white people have begun to acknowledge that racism takes many forms – including privilege, access to resources, and apathy – and have responded by examining their internal, personal biases and educating themselves on how to be actively antiracist.

We cannot divorce the systems in which we live from those in which we work. Socially constructed vulnerabilities running systematically along lines of class, ethnicity, gender, race, and political status create deleterious and discriminatory outcomes in relocation processes (Marino, 2018). Disparities in such outcomes are linked to disparities in the subjectivities and ideologies of the governments and practitioners managing relocation processes themselves. In the case of Staten Island, the Oakwood Beach community's principles more easily lined-up with the State's, which led to more political opportunity. On the other hand, Indigenous communities in Alaska and Louisiana have actively lobbied for increased public resources with limited results, remaining embedded in a colonial infrastructure that does not value their approach to safeguarding their communities from the impacts of climate change.

As policy makers and practitioners, we must understand the worldviews and privileges within which we operate and the cultural values we bring to our work, as well as the power we hold (access to resources and perceived influence, respect, and expertise) at these decision-making tables. Of course, this issue is not limited to individual bias but includes professional action. "Color blind adaptation planning" has ignored the role of racism in creating climate risk and vulnerability (Koslov, 2019). The refusal of professionals to address historic injustices to frontline communities has undermined efforts to build meaningful partnerships between communities/tribes and government agencies. The failure to explicitly acknowledge local and national histories of racism, the politics of disaster response, and the lack of self-determination can invalidate people's lived experiences (Jessee, 2020). In determining how and when to relocate, partnership-building efforts with communities need to focus not only on supporting community capacity to withstand the impacts of climate change but also to confront the systems that have harmed them. Without meaningful acknowledgment of these cultural contexts we may – unintentionally – silence certain voices and perspectives, perpetuating the oppression of already marginalized people.

15.5.2 Redefine success

We need to move beyond white-centered frameworks in advancing justice and use our power to support frontline communities. This notion includes considering how the process we are using was designed for particular people, which means that not everybody was held in mind when that process was made. By focusing exclusively on economic rationale and relocation funding structures, we are advocating for dollars or houses saved without considering the desires of the communities. And by applying standardized top–down frameworks or processes in different geographies or cultural groups, we perpetuate underlying power dynamics that may fail to distinguish nuanced values, needs, and desires within and between communities (Neef et al., 2018).

Who succeeds in relocation? Governments? The people relocating? The receiving community? Relocation processes, if led by government officials and practitioners, can lead to changes that put particular beliefs, traditions, and cultures at risk – like with the IDJC tribe. Allowing communities to reassert alternative ways to assess value, relationships, and protocols in the relocation process may lead to more possibilities for success. In the case of Staten Island, the State was able to check a box for expending federal funds and razing damaged homes, but the buyout program privileged some communities while ignoring others, further ingraining inequity in the communities that were not supported. To claim success, governments should consider how their policies and practices relieve vulnerabilities in the most at-risk communities. If governments and practitioners can position themselves as supporters rather than leaders of relocation processes, they can prioritize community values.

15.5.3 Support community-driven and community-responsive processes

The ways governments and practitioners develop grant implementation processes and bill hours impedes relationship- and trust-building with communities. In my career, I have inhabited privileged spaces like the White House and New York consulting firms that are too far removed from communities. I have been on teams that set out to advance community goals, but without having time to understand them fully we worked from an abstract vantage point. In acknowledging tensions between government and community goals in relocation, we need to adjust planning conventions and government regulations to lift-up human experience to the same level as technical expertise.

A fundamental principle in justice for frontline communities is the right to self-determination – people have the right to decide when, how, where, and if relocation occurs (Bronen, 2015). Governance institutions in turn need to collaborate and be transparent in decision-making processes, beyond an "illusion of inclusion" (Adger et al., 2012). The greatest barriers to incorporate plural values fully are the perception of loss of control for those with long-standing positions of power (such as governments) and the lack of inclusion for those who are oppressed (Adger et al., 2012). Incorporating cultural values in relocation processes requires naming the diverse perspectives at hand – for example, prioritizing economic efficiency versus maintaining a sense of belonging – and considering collective decision-making at multiple scales.

Government agencies and practitioners can promote "adaptive capacity" of communities to balance power and engage in iterative learning that can generate an exchange of knowledge and flexible solutions (Bronen, 2015). This shift can serve as an opportunity to position communities as leaders and build relationships that will support their influence in future decision-making processes. At the same time, governments and practitioners have access to many resources that communities need. Though the IDJC tribe had been planning for relocation before the NDRC, they needed support in understanding government agencies' policies and procedures. Without leading the process, practitioners can help communities in developing letters of support, applying for grants, and – in the case of some tribes – getting federal recognition to open additional funding opportunities.

Including local knowledge, values, and ways of living in relocation considerations can harness the diverse strengths of community capacities (Thornton & Comberti, 2013). Traditional ecological knowledge is often integral to a community's culture and is a large part of its multigenerational repertoire of habits, skill, and styles from which people construct their livelihoods (Adger et al., 2011). Similar to issues of relocation, the formal concept of TEK has been limited to Indigenous communities; however, TEK also can be useful for identifying culturally appropriate processes in nonindigenous communities.

Known and unknown blind spots can be mitigated when more voices are at the table. Frontline communities directly impacted by climate change often have the best ideas for transformative alternatives. In addition to developing locally appropriate solutions, community-driven processes can increase the local stewardship of plans, limit conflicts due to early involvement, and create greater trust in government (Scott, 2012). We should legitimize and affirm community lived experiences and values and provide the conditions for them to have an active voice and presence.

15.6 Conclusion

Climate change affects not only people and power, ethics and morals, and environmental costs and justice, but also cultural survival (Crate & Nuttall, 2009). The challenge remains to address cultural dimensions of relocation. Decisions about relocation will involve trade-offs regarding economic efficiency, equity, and political legitimacy, as well as pluralistic values. Currently in the United States, economic metrics of value and technocratic approaches to planning take precedence over consideration of cultural losses. Additionally, power dynamics in relocation processes often lead to winners and losers. We must consider who is driving relocation processes and to what extent the process accommodates diverse values and social structures. Relocation processes must directly uphold culture and justice, centering marginalized voices and creating livable communities for all.

15.7 Recommendations

The following recommendations are for policy makers and practitioners to further the movement toward equitable relocation practices.

- *Examine your own privilege and values systems.* As policy makers and practitioners, we must understand the worldviews and privileges within which we operate and the cultural values we bring to our work, as well as the power we hold (access to resources and perceived influence, respect, and expertise) at decision-making tables. In determining how and when to relocate, partnership-building efforts with communities need to focus not only on supporting community capacity to withstand the impacts of climate change but also to confront the systems that have harmed them.
- *Redefine success.* Allowing communities to reassert alternative ways to assess value, relationships, and protocols in the relocation process may lead to more success. By focusing exclusively on economic rationale and relocation funding structures, we are advocating for dollars or houses saved without considering the desires of the communities. If governments and practitioners can position themselves as supporters rather than leaders of relocation processes, they can prioritize community values.
- *Support community-driven and community-responsive processes.* Incorporating cultural values in relocation processes requires naming the diverse perspectives at hand – for example, prioritizing economic efficiency versus maintaining a sense of belonging – and considering collective decision-making at multiple scales. Government agencies and practitioners can promote adaptive capacity of communities to balance power and engage in iterative learning that can generate an exchange of knowledge and flexible solutions.

Note

1 The term "community" is complicated in this case because of the displaced members of the IDJC tribe and other Indigenous nations from the region who have ties to the Island (Jessee, 2020).

References

Adger, W. N., Lorenzoni I., & O'Brien, K. (2009). *Adapting to climate change: Thresholds, values and governance*. Cambridge University Press.

Adger, W. N., Barnett, J., Chapin, F. S., & Ellemor, H. (2011). This must be the place: Underrepresentation of identity and meaning in climate change decision-making. *Global Environmental Politics, 11*(2), 1–25.

Adger, W. N., Barnett, J., Brown, K., Marshall, N., & O'Brien, K. (2012). Cultural dimensions of climate change impacts and adaptation. *Nature Climate Change, 3*(2), 112–117.

Ajibade, I. (2019). Planned retreat in Global South megacities: Disentangling policy, practice, and environmental justice. *Climatic Change, 157*(2), 299–317.

Ajibade, I., Sullivan, M., & Haeffner, M. (2020). Why climate migration is not managed retreat: Six justifications. *Global Environmental Change, 65*, 102–187.

Baptiste, N. (2017, October 30). When a hurricane takes your home. *Slate*, Climate Desk. Retrieved from https://slate.com/technology/2017/10/the-staten-island-community-that-didnt-rebuild-after-sandy.html

Barr, M. (2013, October 28). After Sandy, only a lucky few getting buyouts. *Salon*. Retrieved from https://www.salon.com/2013/10/28/after_sandy_only_a_lucky_few_getting_buyouts_2_ap/

Barthel-Bouchier, D. (2013). *Cultural heritage and the challenge of sustainability*. Left Coast Press, Inc.

Baum, H. (2015). Planning with half a mind: Why planners resist emotion. *Planning Theory & Practice, 16*(4), 498–516.

Bronen, R. (2015). Climate-induced community relocations: Using integrated social-ecological assessments to foster adaptation and resilience. *Ecology and Society, 20*(3), 36.

Burkett, M., Verchick, R. R. M., & Flores, D. (2017). *Reaching higher ground: Avenues to secure and manage new land for communities displaced by climate change*. Center for Progressive Reform, Washington, D.C., USA. Retrieved from http://progressivereform.org/articles/ReachingHigherGround_1703.pdf

Crate, S. A., & Nuttall, M. (Eds.). (2009). *Anthropology and climate change: From encounters to actions*. Left Coast Press, Inc.

Comardelle, C. (2020). Preserving our place: Isle de Jean Charles. *Nonprofit Quarterly*. Retrieved from https://nonprofitquarterly.org/preserving-our-place-isle-de-jean-charles/

Corner, A., Markowitz, E., & Pidgeon, N. (2014). Public engagement with climate change: The role of human values. *Wiley Interdisciplinary Reviews: Climate Change, 5*(3), 411–422.

GAO [U.S. Government Accountability Office]. (2020). Climate change: A climate migration pilot program could enhance the nation's resilience and reduce federal fiscal exposure. *Report to Congressional Requesters GAO-20-488*. Retrieved from https://www.gao.gov/products/GAO-20-488#summary

Grannis, J. (2011). *Adaptation tool kit: Sea-level rise and coastal land use*. Georgetown Climate Center, Washington, D.C., USA. Retrieved from https://www.georgetownclimate.org/files/report/Adaptation_Tool_Kit_SLR.pdf

IPCC [Intergovernmental Panel on Climate Change]. (2012). Summary for policymakers. In C. B. Field, V. Barros, T. F. Stocker, D. Qin, D. J. Dokken, K. L. Ebi, M. D.

Mastrandrea, K. J. Mach, G.-K. Plattner, S. K. Allen, M. Tignor, & P. M. Midgley (Eds.), *Managing the risks of extreme events and disasters to advance climate change adaptation* (pp. 3–21). Cambridge University Press. Retrieved from https://www.ipcc.ch/pdf/special-reports/srex/SREX_Full_Report.pdf

Jessee, N. (2020). Community resettlement in Louisiana: Learning from histories of horror and hope. In S. Laska (Ed.), *Extreme weather and society: Louisiana's response to extreme weather* (pp. 147–184). Springer.

Koslov, L. (2016). The case for retreat. *Public Culture, 28*(2), 359–387.

Koslov, L. (2019). Avoiding climate change: "Agnostic adaptation" and the politics of public silence. *Annals of the American Association of Geographers, 109*(2), 568–580.

Maldonado, J. K. (2014). A multiple knowledge approach for adaptation to environmental change: Lessons learned from coastal Louisiana's tribal communities. *Journal of Political Ecology, 21,* 61–82.

Maldonado, J. K., Shearer, C., Bronen, R., Peterson, K., & Lazrus, H. (2013). The impact of climate change on tribal communities in the US: Displacement, relocation, and human rights. *Climatic Change, 120,* 601–614.

Mandel, K. (2017, October 17). America's climate refugees have been abandoned by Trump. *Mother Jones,* Environment. Retrieved from https://www.motherjones.com/environment/2017/10/climate-refugees-trump-hud/

Marino, E. (2015). *Fierce climate, sacred ground: An ethnography of climate change in Shishmaref, Alaska.* University of Alaska Press.

Marino, E. (2018). Adaptation privilege and voluntary buyouts: Perspectives on ethnocentrism in sea level rise relocation and retreat policies in the U.S. *Global Environmental Change, 49,* 10–13.

Moore, R. (2018, May 16). For Sandy survivors this program made all the difference. *Natural Resource Defense Fund,* Expert Blog. Retrieved from https://www.nrdc.org/experts/rob-moore/title

Neef, A., Benge, L., Boruff, B., Pauli, N., Weber, E., & Varea, R. (2018). Climate adaptation strategies in Fiji: The role of social norms and cultural values. *Elsevier World Development, 107,* 125–137.

O'Brien, K. (2009). Do values subjectively define the limits to climate change adaptation? In W. N. Adger, I. Lorenzoni, & K. L. O'Brien (Eds.), *Adapting to climate change: Thresholds, values and governance* (pp. 164–180). Cambridge University Press.

Oliver-Smith, A. (2009). Climate change and population displacement: Disasters and diasporas in the twenty-first century. In S. A. Crate & M. Nuttall (Eds.), *Anthropology and climate change: From encounters to actions* (pp. 116–136). Left Coast Press, Inc.

Scott, A. (2012). *Measuring wellbeing: Towards sustainability?* Routledge.

State of Louisiana [Office of Community Development, Disaster Recovery Unit]. (2016). *Press Release: LA receives $92 million from U.S. Dept. of Housing and Urban Development for coastal communities, disaster resilience.* Retrieved from https://www.doa.la.gov/OCDDRU/NewsItems/Louisiana%20Receives%20NDRC%20Award.pdf

Thornton, T. F., & Comberti, C. (2013). Synergies and trade-offs between adaptation, mitigation and development. *Climatic Change, 140,* 5–18.

Veerkamp, A. (2009). The impacts of climate change on the Chesapeake Bay. Statement presented at the U.S. House of Representatives Committee on Natural Resources' Subcommittee on National Parks, Forests and Public Lands and the Subcommittee on Insular Affairs, Oceans and Wildlife. Retrieved from https://forum.savingplaces.org/

HigherLogic/System/DownloadDocumentFile.ashx?DocumentFileKey=73dee219-9e4b-ffc8-3665-1a59b9359b59&forceDialog=1

The White House. (2015). *FACT SHEET: President Obama announces new investments to combat climate change and assist remote Alaskan communities*. The White House, Washington, D.C., USA. Retrieved from https://obamawhitehouse.archives.gov/the-press-office/2015/09/02/fact-sheet-president-obama-announces-new-investments-combat-climate

Williams, B. (2002). The concept of community. *Reviews in Anthropology, 31*(4), 339–350.

Interlude 4 Tool Sharpening

Martha Lerski

Both the contours of the piece and the infinity-shaped tool markings on the sharpening-stone base harken to ancient origins.

The sharpening stone reflects a language of process; it is included as a base in the final presentation in recognition of its function as a foundation stone. It evolved, through indentations and markings caused by chisel sharpening, in tandem with the wood carving.

The innate colors/markings of the acacia wood emerge through direct carving[1] – rather than being subordinated or substituted by imposed shapes or foreign pigments.

Among other possible perspectives, the sculpture can be viewed through a geometric lens, as a simplified human figure, or as an exercise in trusting process. What is the role of planning, or of organic process – in art, and in science? Both require content knowledge and training to facilitate creative breakthroughs. Designs or solutions can emerge via an openness to inevitable iterations and imperfections, to new data and narratives, perspectives, and methodologies of multiple disciplines.

Which understandings are innate, long-ago learned and now intuitive – or repressed by history? Some have noted Buddhist elements in this piece, perhaps reflecting my earliest memories of visual art, viewing sculptures in Sri Lanka with my Art Historian mother when I was a toddler. That island's diversity of languages, religions, commercial traditions, and art can be traced back for millennia, well before European colonial rule. Current ethnic and religious tensions there reflect and contribute to ongoing challenges posed by migrations, violence, and dislocations.

Rushing to burnish this piece with a shard of glass ahead of a show, I cut my finger. Crimson mixed with the darker grain of the wood. My immediate concern was with the integrity of the sculpture's colors, damage to my hand of secondary importance. There is a vulnerability to this piece, a solitude and acceptance of markings or irregularities. Pencil drawings that I made on the wood remain from its composition. Its origins endure.

DOI: 10.4324/9781003141457-104

Figure Int 4.1 Tool Sharpening, acacia wood (view (a) of the sculpture by Lerski); photos by Javier Agostinelli.

Tool Sharpening 211

Figure Int 4.2 Tool Sharpening, acacia wood (view (b) of the sculpture by Lerski); photos by Javier Agostinelli.

Note

1 A method of sculpting taking direction from the material, carving directly without external models.

Interlude 5 Untitled, 2013: Storm Series

Martha Lerski

Walking a beach on Long Island Sound weeks after Hurricane Irene, I found a displaced piece of marble strewn by the storm. After lugging the stone to my studio, it sat for more than a year before I shaped it according to its evolving story, as part of an ongoing Storm Series.

In New York City during the following year's Hurricane Sandy, my partner – Bob, and his mother, who was visiting from Pittsburgh, sheltered with me, our dog, my daughter, and her stranded friend from Texas. We huddled through that storm in my fortress-like pre-war midtown Manhattan apartment building. My city block was distant from both the Hudson's and East River's flooding. Meanwhile, Bob's coastal Connecticut home flooded for the second year in a row.

As I composed the sculpture, visits to the shore reminded me of the violence of Irene and Sandy. Many homes were destroyed by violent winds, waves, and tides. The sculpture would reflect strong movements, unexpected twists – and yet there would be a resilient anchor.

Ultimately, what sustains this form is its cantilevered grounding and momentum. In the recesses of my mind was a visit to Frank Lloyd Wright's *Fallingwater* – a recollection of improbable stability while suspended above water. Can engineering and design carry us through all climate changes? During a re-visit to *Fallingwater* this year, decades after my last tour, I experienced an unforeseen dynamic between that architecture and nature: The structure's original placement had not anticipated extreme river flooding more common today.

DOI: 10.4324/9781003141457-105

Untitled, 2013: Storm Series 213

Figure Int 5.1 Untitled, 2013: Storm Series, marble (sculpture by Lerski); photo by Jim Chervenak.

Part IV
Finding hope

16 Voices of Enseada da Baleia: Emotions and feelings in a preventive and self-managed relocation

Giovanna Gini, Erika Pires Ramos, and Comunidade Enseada da Baleia

16.1 Introduction

Enseada da Baleia is a Caiçara community of artisanal fishers located on the south-eastern coast of Brazil. Fishing and subsistence agriculture constitute part of the inhabitants' personality and shape behavior toward the environment in which they live. Cardoso Island has been home for the Enseada community for the last 170 years, as well as to other 26 caiçaras communities and one Guarani Indigenous village. The number of communities decreased to 7 caiçaras communities since 1962 when the island was decreed as a conservation park of the State of São Paulo. The conversion of the island into a national park avoided real estate speculation. However, the administrative decision did not prioritize or protect the rights and well-being of the traditional inhabitants. For instance, the first State Park Cardoso Island (PEIC) 1976 management plan did not mention the presence of local communities on the Island (Diegues, 2005) leading to rights violations that coerced their outmigration. The continued presence of traditional communities is a sign of their commitment to remain in their territory and to preserve their culture, which is intricately linked to place.

The Enseada da Baleia community is located between two worlds – the sea and the estuary – which gives the place peculiar beauty and traditionality. This peculiarity has become socioenvironmental vulnerabilities over the years. In the past decade, environmental stress and climate change have become daily life issues for Enseada members. It is difficult to foresee the more frequent and out of season high tides and cold fronts, which affects their fishing and harvesting rhythm.

As a result of increasing risk and vulnerability, the community started a process of relocation in 2016 that is still ongoing. After conversations with members of the community, it is evident that relocation is a process beyond what is considered to be material and the effort to adapt to a new life. However, many narratives of relocation are limited to investigating the strategies and results, somehow dehumanizing and depoliticizing the experience. For this reason, this chapter intends to explore the emotional spectrum of the process of what we call preventive and self-managed relocation of the Nova Enseada Community (Gini, Cardoso, & Ramos, 2020). Emotions are key for

DOI: 10.4324/9781003141457-16

understanding how social space is created, understood and lived it (Anderson & Smith, 2001). Emotions are key to explore how bodies that are marketed through history and encounters with racialization, gendered, capitalism and imperialism feels about themselves and others (bodies, places, environment, etc.) (Ahmed, 2004). Emotions "stick" to bodies and carry one through life (Ahmed, 2004). For this reason, the relocation is not seen as an event but as the consequence of slow violence (Nixon, 2011) that manifested in the form of increasing precarity, invisibility and climate change over the years. Taking emotions in consideration can help to develop a more compelling notion of Climate Justice.

This chapter was constructed based on intimate and personal letters from the women of the community and is illustrated with drawings inspired by the children's impressions and memories. The chapter is also informed by insights from informal conversations that happened in the span of a year (2019–2020) between the researchers and the inhabitants of Enseada. In this particular case, the relationship with Enseada started as a research project, after taking a role as activism to finally becoming friends. On a cold winter afternoon, while speaking about this chapter and how to convey their emotions, the women proposed to each write one letter to the researchers telling their histories, as a way to create the intimacy to express their emotions. The chapter aims to respect the view of its protagonists and all views have been approved by them (Figures 16.1–16.4).

For several years, the Enseada members resisted social and environmental vulnerability. Three events mark a new moment in the village: The death of their leader Malaquias caused radical changes in the economic and political structure of the community; the accident of the catamaran which destroyed several houses of the village; and a cyclone that destroyed their land. These moments led to the preventive and self-managed relocation of the community inside Cardoso's Island.

16.2 A funeral, a boat accident, and a storm

The Enseada da Baleia members begin to tell stories about the relocation process from September 2010, the year in which the community lost its father. Malaquias was the adopted child of the founders of the community. He was brought to Enseada in the 1940s (Nordi, Cardoso, & Beccato, 2005). After working his whole life for his adopted father, he became the leader of the community when the sons of the founders left for the city in the 1970s. He stayed to carry on with the traditional life with his four biological sons and daughters and two adopted children. For decades, he managed all the economic activities of the community, giving work and dealing with external clients who purchased fish products produced by the community. Upon his death, the commercial side of the community also died and many worried that it was the end of the community. A sad event that radically changed the configuration of the community, *"foi difícil de emfrentar a vida sem ele, nada mais era o mesmo"* (it was difficult to face life without him; nothing else was the same). The late

Voices of Enseada da Baleia 219

Figure 16.1 Coast of Comunidade de Enseada, Brazil. Image by Gabriel Bitar, based on drawings of the children of Enseada; used with permission.

Malaquias constantly repeated that the island was going to disappear and that Enseada would need to relocate (Vazantes, 2010). Malaquias' knowledge came from living on the island all his life, which led him to master the patterns of movements of water and sand (see Figure 16.1). Erosion was slow but evident to those who grew up in Enseada over the years. On the estuary side, Malaquias had built a seawall over the years using sandbags purchased with the profits from the sale of dried fish – a traditional product of the community which has been produced for more than 200 years and is passed from generation to generation by the women of Enseada (see @balaiodeproa Instagram page). However, the members of the community perceive that after Malaquias' death, "*a partir daí a Ilha começou a diminuir, cada maré ela ficava menor, começamos a nos preocupar*" (from then, the island began to shrink, with each high tide, it became smaller,

Figure 16.2 Between the beach and mountains in Comunidade de Enseada, Brazil. Image by Gabriel Bitar, based on drawings of the children of Enseada; used with permission.

and we began to worry). The land was not enough to carry out their daily activities, such as harvesting orchards or playing at the end of the afternoon on their volleyball court where they would gather together as a large family composed of 11 households that share kinship. One household lost hope and left the community days after Malaquias' death.

Malaquias died of complications from chronic diseases, the embodiment of dispossession, abandonment and precarity (Povinelli, 2011). Until then, Enseada had a vertical leadership system; all decisions passed through Malaquias, who oversaw the distribution of work and secured the dried fish clientele. He was the engine and the union of Enseada. As a father, father-in-law and grandfather to all members of the community, he encouraged respect and love among all; as a job provider, he ensured the well-being of the families. Without him, Enseada members were left without their emotional supporter and economic leader. The community went through a crisis of hunger and feared that was the dawning of the end.

A critical moment was when Erci, the wife of Malaquias, in an attempt to evade hunger, asked to move to the city. Her grandchildren say she could not bear to spend one day outside Enseada, especially in the city where "*não há nada para fazer e não há espaço suficiente*" (there is nothing to do, and there is not enough space), so Erci's

Figure 16.3 Trees as connection to family and land. Image by Gabriel Bitar, based on drawings of the children of Enseada; used with permission.

desperate act became a tipping point for urgent intervention. Fortunately, Erci did not need to move out of the community. The women started to organize. Erci's granddaughters took the initiative. Some of them had migrated to the city to study and returned with new knowledge they applied to try to keep the community together. They formed a women's association to make and sell handicrafts. All the women sat together, and because Erci knew how to sew, they started making clothes (see @enseadadabaleia Instagram). Erci took the role of leader in that moment. Step-by-step with a vision of social solidarity-based economy, each woman produced crafts (Yamaoka et al., 2019). After a while, they put their identities into the pieces using materials from their environment, such as fishing nets that were no longer of use, found on the white beaches of Cardoso Island (see Figure 16.2). They began to have

Figure 16.4 Nova Enseada, Brazil. Image by Gabriel Bitar, based on drawings of the children of Enseada; used with permission.

an income again, hunger decreased, and the women learned to work as a team, to provide for their families, to organize themselves, to treat each other as equals, respecting and appreciating different skills. Thus, the women gained the respect of men, and trust regarding the organization and their work. Today they are organized in the Enseada da Baleia Residents Association – AMEB, acronym in Portuguese (Yamaoka et al., 2019). Through AMEB, women organize their work, ensuring that all have access to a minimum monetary income for the subsistence of families (Yamaoka et al., 2019). When possible, they also involve men, valuing the artisanal work of fishing. When they need more workforce, they invite other fishers of the island to participate. This is how these women, in an attempt to save their community, over years of hard work, became the leaders of the community.

In 2015, a tourist catamaran decided to dock in the village without permission or knowledge of the inhabitants. The ship crashed into the island, causing material damage to the Caiçaras, destroying 2 houses and 3 community spaces. It also affected the landscape itself and the esthetic value of the area (Hayama & Cardoso, 2018). The displacement of a 20-meter portion of land worsened the delicate vulnerability of the community.

In one of those houses was Erci's. When she came out urgently from the collapsing house, she saw one of her trees falling into the water (see

Figure 16.3). In an instinctive attempt to save the tree, she tried to jump into the water behind it, but her sons stopped her. In Enseada there was a custom of planting one tree for each new member of the family. Therefore, every corner of the Old Enseada had a special meaning that marked the passage of loved ones through the place. To this day, children miss their swings in these trees and family lunches under their shadows. With pain, they talk about how those trees no longer exist physically, but they are still alive in their memories. Small saplings from the trees were transplanted in the Nova Enseada and are being cared for with commitment and love, which symbolizes continuity in history and tradition.

The catamaran accident tested the new organization AMEB for the first time. The women approached a public defender and made a legal case against the tourism agency (Hayama & Cardoso, 2018). That litigation is still open, but asserting their rights by pursuing the case was a first step toward learning how to defend their territorial rights judicially. This fighting strategy would become essential at the moment of the relocation.

At 2 in the morning of 16 August 2016, furious waves began to beat against the shore; on the estuary side, the water level rose, entering the houses. Some people went to the narrowest point of the strip of sand to see what was happening. It was possible to see how the sand was being eaten by the sea. That night is marked by memories of tasting saltwater, feeling cold and desperate in the dark. No one slept. They stayed awake, trying to communicate with relatives in the city. The next morning, civil defense agents arrived with the intention to evict the inhabitants under a declaration of emergency. The members of Enseada opposed leaving their houses and their land for fear of not being able to return. The civil defense agents made them sign a document, which stated that in the eventuality of another storm, the responsibility of their lives was in their own hands. The community had to *"assumir o risco"* (assume the risk). They agreed that if the situation became worse, they would use their own boats to go to a safe place. Of the 12 meters that had previously separated the estuary from the sea, only two remained, *"foi o momento de olharmos ao nosso redor, com tristeza e ver que tínhamos que mudar o quanto antes"* (it was the moment where we looked around, and sadly we saw that we had to move to a safe place as soon as possible).

16.3 Between *"fear and strength"*: The experience of community relocation

After the cyclone, it was clear that relocation was urgent and necessary; however, this did not make the task of internal agreement easier. Many members resisted the idea of having to leave, particularly the elderly, thinking that they still had years before the sand strip breaks. Many did not have the resources or the physical strength to build a new home elsewhere. The youngest members of the community were those who first insisted on leaving and were therefore the engine of the relocation, suppressing their fears that materialized as "sempre o peito apertado" (always tight chest) and hidden tears; "*várias vezes chorei no banheiro por medo de não dar certo*" (many times I cried in the bathroom for fear of not having it right).

Above all, the most difficult part was "*deixar para trás um sonho de vida onde eu nasci e cresci e tive meus filhos deixar minha casa, minhas árvores e minha história*" (leaving behind a dream of a life where I was born, grew up and had my children). Leaving behind the beauty of "*esse lugar era maravilhoso, sossegado tudo perto, mar, rio, praia, família*" (that place that was wonderful, peaceful, everything was close, the sea, the river, the beach, and the family). The relationship with trees was particularly important because it was like leaving the roots of loved ones, especially those trees which represented people who are no longer with the community. Leaving Old Enseada also felt like putting aside the sacrifice made by their grandparents, Malaquias and Erci, who dedicated their lives to building a family with roots in that place and defending them against everything and everybody. Erci's role as a leader was crucial in convincing members to relocate together and stay united.

Erci also chose the location for the Nova Enseada, a place about five kilometers north of the old one (see Figure 16.4). They called it the place of *Casa Preta* – the Black house. During her youth Erci had lived there and knew that it was a good area to plant. It was full of fruit trees and close to fishing spots.

However, one of the biggest obstacles was obtaining permission from state authorities to relocate within the park, the only place that could provide continuity for the community with respect to its culture and tradition. The island is protected as a park managed by the State of São Paulo under a conservation vision that restricts the freedom of traditional inhabitants guided by the "myth of the untouched nature" (Diegues, 1996). In meeting after meeting the members of Enseada defended their right to relocate inside the Cardoso Island Park. One member of the community remembers a statement that Erci gave in one of the meetings:

"*E, em dezembro, a gente acha o território escolhido pela avó, 'essa terra é boa, vai'. E eu me lembro [...] lá, na reunião que não tinha sido autorizado e ela falou, 'estão pensando o quê?' Ela subiu e ela bateu na mesa assim e falou: 'vocês estão esperando o que para liberar a terra para meus filhos? Eu estou dizendo que a terra é boa, eu estou dizendo. Vocês estão esperando o que? Eu to falando. Que o mar venha, que leve os meus filhos embora? Eu estou falando para vocês que isso aqui é bom'. Então, a gente confiou na vó, a gente confiou no vô que dizia que a barra ia romper, a gente não olhou para as pesquisas, a gente não olhou para isso. A gente tinha ouvido isso a vida toda.*"

(Interview as cited in Yamaoka, 2019, p. 129).

And, in December, we find the territory chosen by the grandmother, "this land is good, go." And I remember [...] there, at the meeting that had not been authorized and she said, "what are you thinking?" She went up and she hit the table like this and said: "You are waiting for what to release the land for my children? I am saying that the land is good, I am saying. What are you waiting for? I'm speaking. May the sea come, take my children away? I'm telling you that this is good." So, we trusted grandma, we trusted grandpa who said the island would break, we didn't look at the research, we didn't look at that. We had heard it all our lives.

From a geographical point of view, the new place seemed ideal because it had all the requirements to be able to maintain traditional activities and generate sustainable income as well as security against floods. These facts were corroborated based on a study carried out by a group of researchers from the NUPAUB-USP (Support Center for Research on Human Populations in Brazilian Wetlands)[1] at the community's request. *"Precisa vir à academia comprovar, o que a sabedoria caiçara dizia e que o governo não quis aceitar"* (the academy needs to prove, what the caiçara wisdom said and that the government did not want to accept) (Mendonça Cardoso, 2017 as cited in Yamaoka, 2019).

Initially, the government resisted. Instead of allowing for relocation on the island, the state and municipal authorities offered two solutions to Enseada members: (a) Integration within another neighboring community already existing on the island; (b) or the migration to the peripherals of the nearest city. The community rejected both solutions since they would alter their internal and external relationships, their lifestyle, tradition, and sociopolitical organization.

Under the women's leadership, and working with the public defender who they had developed a relationship with because of the catamaran accident, the community launched a legal dispute. Finally, the concession was authorized, but the relocation would need to happen practically without financial support from the state or the municipality. The lack of financial aid presented a great challenge for the community, but they overcame this hurdle with creativity, solidarity, and a lot of work. There were fundraisers through the Catholic community; campaigns through crowdfunding platforms; and solidarity parties facilitated by friends and family from outside the community (Gini, -Cardoso, & Ramos, 2020).

A definitive moment was when the Brazilian Institute of the Environment and Renewable Natural Resources (IBAMA) donated tons of reused wood to build the new houses. Of course, transporting tons of wood was a challenge because it was located in the distant city of Guarujá, and the wood had to be transported enormous distances by land and water. To organize this transport, community members had to learn new skills, and their enthusiasm for a new future gave them the strength to do it. They invited *mutirãos*, mutual aid activities that involved participation of tourists, friends and family from other regions and countries (Ajude a Nova Enseada! | Ilha do Cardoso – YouTube, 2017). There were 2 months of mutirões for the transport of the material destined for the reconstruction. The men had initially been skeptical of the relocation, but when they saw that they had the material to work with, they started to get excited.

Slowly, the fear of not making it right, of not achieving the enormous task of relocating an entire community without money, was fading. Each success – the land, the wood, food donations, installments to buy the materials, etc. – gave the community more strength and confidence. The community celebrated each one. It was important to stay uplifted. Especially after endless days of strenuous work, every little victory was crucial. Step-by-step, they were closer to building a place to receive a future that could bring a better quality of life and continue with the solidarity-based economy projects that are giving equality among the members of Enseada. However, sadness continues to intertwine with the joy: "*foi difícil demais deixar nossa antiga morada e ver para nova morada, cada coisa que ia buscar lá, voltava triste demais*" (it was too difficult to left the old house and come to the new house, everything that I when to search there, I came back very sad).

After cleaning the land with several friendly arms, the first house was Erci's, a symbolic construction for the matriarch. Unfortunately, before seeing her house finished, Erci passed away in April 2017, leaving her children and grandchildren to continue the work. Those days intensified the pain and *saudade* (nostalgia) for an ending chapter of Old Enseada's story. Even the physical pain resulting from the work of clearing the land, transporting materials, and building houses increased. "*Com a realocação o que mais me abalou foi não trazer a minha sogra junto para começar nossa história de vida e luta*" (What was most depressing was not bringing my mother-in-law along with us to start our story of life and struggle). Erci's house will belong to the youngest daughter, who is still underage. Meanwhile, the house is empty, but it is used in the activities of communitarian tourism and for the AMEB. Entering Erci's house gives goosebumps to the inhabitants. Every time they enter the house, they remember how they all lived together there while the other houses were in construction: A happy memory. Hands were limited, so they built one house at a time – the order of construction followed a pyramid of priorities with the most vulnerable first. For weeks, the whole community slept on the floor of the grandmother's house and like a hug that house brought them together and gave them courage to continue to exist as a caiçara community.

In August 2018, another high tide finally broke the strip of sand at Old Enseada. In just a few hours it opened 600 meters, separating the island in two and unifying the two worlds, the estuary and the sea. With this storm, the members of the community made peace with their decision to relocate preventively, appeasing the feeling of guilt that appeared in their dreams. "*Nos se consolamos mais quando a Ilha separou, ai gente viu que não tínhamos condições de morar mais lá*" (we consoled ourselves when the island was separated, and it was then that we saw that we had no conditions to continue living there). It was clear then that there was no other option, and the nightmares with the ancestors stopped.

Today the opening is more than 3-kilometers wide, and there is no sign that the Old Enseada ever existed. Each material memory is underwater. It is no longer the same place. However, the old place is still alive in the memories, "stuck" in their emotions, and that helps the community to survive. As one of the Enseada members wrote:

> "É nesse novo cenário que a natureza nos propõe se vão águas dentro de uma infância vivida nesse pedaço de terra que não existe mais, de brincadeiras no 'combro' como assim falávamos, escorregando num barranco alto e forte que tínhamos a certeza que nunca despencaria, passamos hora e horas subindo e descendo, catando 'araçá', 'camarinha', simplesmente vivendo e sendo feliz. Hj quando ali foi conhecer a nova barra me deparei com um sentimento tão forte misturado com tristeza e felicidade, onde as lágrimas sem nenhum reforço começaram a sair...tristeza em saber que esse lugar onde vive momentos que são inexplicável não existe mais e que vai levando a cada hora, dia e provavelmente chegue a uma mês a minha antiga comunidade, onde não estamos mais morando, mas nossa história, nossas árvores ainda estavam lá, e um sentimento de felicidade em saber que estamos num lugar seguro, unidos e refazendo e construindo um novo começo, que não foi fácil mais que tem sabor de vitória, de conquista e fé muita luta. Orgulho imenso em fazer parte dessa comunidade 'família' e de poder mostrar e ensinar pra minha filha o verdadeiro sentido da vida."

In this new scenario that nature puts before us, water goes into childhood that lived in that piece of land that no longer exists, playing in the "combro" as we used to say, sliding down a high and strong ravine that we were sure would never fall, we spent hours and hours going up and down, picking up "araça," "camarinha," simply living and being happy. Today when I went to see the new opening, I had a strong mix of feelings of sadness and happiness, where the tears without any resistance started to come out... sadness in knowing that this place where moments that are inexplicable no longer exist, and that it goes away every hour, day and probably in a month will reach my old community, where we are no longer living, but our history, our trees were still there; and, a feeling of happiness in knowing that we are in a safe place, united, remaking and building a new beginning, which was not easy since it has the flavour of victory, conquest and faith, a lot of struggle. I am immensely proud to be part of this "family" community and to be able to show and teach my daughter the true meaning of life.

Nowadays, the community center and one more house are still missing to imitate the structures that they had in the Old Enseada. However, there is still a long way for the Nova Enseada to become a complete home. It is necessary to fill the place with memories and emotions, and only time will help with that. For the inhabitants, relocation will only be completed when the Nova Enseada appears in their dreams. Meanwhile, "*agradeço a Deus por termos a chance de morar na Ilha novamente, cheios de saudades, mas tranquilos*" (I thank God for giving us the opportunity to continue living on the island, full of nostalgia but calm).

16.4 Conclusion notes

The relocation of the Enseada community, which relied on Erci's drive and knowledge, shows the way in which traditional communities move because of climate change. This chapter highlights the key role of emotions in the process of relocation and adaptation. Mobility as a community strategy for adaptation in the context of disasters and climate change should be institutionally supported by public actors and should involve different perspectives (financial and psychosocial), according to needs, community demands and rights.

It is necessary to understand that relocation involves losses far beyond the material. This is because life stories are abruptly disrupted. The impact of the environment on emotions accompanies mobility; it is essential to recognize the need for a respectful reconstruction of place and history. Respect for community leadership and participation are essential in this process, and that is only possible by listening; by recognizing their self-determination, their ways of being, living and existing; by breaking the myth of vulnerability and immobility that undermines action; and by effectively incorporating their traditional knowledge in the processes of policy formulation and decision-making, through free and informed participation and consent. The consideration of emotions is key in this process and to achieve climate justice, as they show how people create and give sense to their world. The collective reconstruction and relocation of the Nova Enseada Community highlight the need for a change in narratives. It is essential to incorporate the human and sensory experience of those on the frontline, which allows a real and broad understanding of the complexity of mobility dynamics, climate change, and shows the path to create fair and durable solutions.

16.5 Lessons learned from Enseada

The following list was created in a meeting with the sole participation of the members of the community Nova Enseada.

- Local invisibilities, small communities, low access to basic rights during years.

    The State of São Paulo and the Municipality of Cananéia ignored the needs of the community, neglecting participation in the process of decision-making; lack of help to relocate and any other form of assistance to families either food, basic sanitation and access to health. This abandonment exposes the community to a situation of total vulnerability.
- Reversal of responsibilities: Relocation and adaptation.

    We knew that the state wanted to evict the community from Cardoso's Island and we understood that the time invested to look for the state's help would be better concentrated on completing all processes to enforce aid. It would also be frustrating for families and might create many conflicts and risks with the authorities. Therefore, we seek to concentrate, time, strength

and energy on how to achieve the reconstruction and adaptation of traditional activities, creating strategies that lead to hope and more results, where we were only able to achieve it with the help of friends and partners.
- Community leadership, traditional knowledge.

  We seek shared accountability and responsibility for each action, we created several fronts with different representatives across the activities dividing functions during the process of relocation. Working internally with an emotional component, emphasizing the harmony of all processes among the members of the community. Equality was the main aim during the process of relocation; therefore, we were prioritizing the most vulnerable people, using our traditional knowledge to value the different skills of each member of the community.
- During the relocation, central were the culture, traditions, and the maintenance of the means of subsistence and environmental conservation.

  The main strategy that we used for the relocation was the *mutirão*, it is the moment that brings together the shared knowledge, the union of the different skills and celebration with food and traditional music gave us joy and force to continue the process.
- Community participation as a tool for climate justice.

  Each action carried out was built collectively, creating a methodology for community interaction and participation. This makes it possible that everyone feel part of every step in the reconstruction process, seeking a priority for families. We try to put ourselves in the place of the other one as a way to strengthen the participatory process.
- Consultation protocol as an instrument for the challenges ahead.

  As a way to strengthen ourselves and our fight for recognition of our life, our rights and our territory, we created our consultation protocol (to be published soon at http://observatorio.direitosocioambiental.org/). This document gives guiding points to the government and those interested, in how we want to be consulted about any action that may affect our life, our culture, our territory. Following the protocol will ensure that the consultation has our participation and consent.
- Planned and preventive relocation must involve a participative and trusting relationship. The key steps identified by the community to achieve a participatory relocation are as follows:

  i community strengthening work: everyone in the community needs to be together to face this process, no one can be left behind;
  ii analysis of the new territory: what the new territory can bring good and all bad with improvement strategies;
  iii analysis of the current territory: measurements and survey of everything that was built and everything that is necessary to live;
  iv organization of needs: step-by-step of everything that will be done so that a new place brings the ideal conditions for families;
  v budget: raise the general cost of adaptation;

vi define priorities: raising what is most necessary to take the community out of the risk situation;
vii mutirões: define in the network of partners people qualified for work, celebration at each stage of the process, space for articulation and unity;
viii partnerships: raise each potential partner for each stage of the process, take them to the community so they can see the situation, understand the process, be sensitized and apply the search for partners;
ix fundraising: set up a pilot project with all the needs of the community and is being updated at each stage of construction;
x division of work: divide the function between community and partners to avoid overload of work;
xi strategy: to be in constant dialog so that people feel part of the same struggle and seek together to maintain hope;
xii manpower: gather all the community skills in a school job where the older ones are teaching the younger ones and can continue working in their home;
xiii logistics: design each stage so that everything reaches the community and has no losses;
xiv food: the new territory is unknown, there is no known fishing spot and there is no time for it during the reconstruction, so you need to find ways that food is not lacking, this can make people discouraged and return to the old place; and
xv value each achievement and design each strategy.

Note

1 Núcleo de Apoio à Pesquisa sobre Populações Humanas em Áreas Úmidas Brasileiras (NUPAUB-USP), 2016. Technical evaluation report of the resettlement area of the Enseada da Baleia community under the aspects of anthropological, environmental and geological security against the erosion process in Cardoso Island, Cananéia-SP, Administrative Process of Collective Protection No. 07/15/PATC/CDR/DPVR/UR

References

Ahmed, S. (2004). Collective feelings: Or, the impressions left by others. *Theory, Culture & Society*, 21(2), 25–42.
Anderson, K., & Smith, S. J. (2001). Editorial: Emotional geographies. *Transactions of the Institute of British Geographers*, 26(1), 7–10.
Ajude a Nova Enseada! | Ilha do Cardoso - YouTube. (2017). Directed by Xino Xano. Available at: https://www.youtube.com/watch?v=qu1b5AhfWIc&feature=youtu.be [Accessed 11 Nov. 2020].
Diegues, A. C. (1996). *O Mito Moderno da Natureza Intocada*. São Paulo: HUCITEC/NUPAUB.

Diegues, A. C. (2005). Esboço de História Ecológica e Social Caiçara. In *Enciclopédia Caiçara Volume IV. História e Memória Caiçara* (pp. 273–319). São Paulo: HUCITEC-NUPAUB-CEC/USP.

Gini, G., Cardoso, T. M., & Ramos, E. P. (2020). When the two seas met: Preventive and self-managed relocation of the Nova Enseada community in Brazil. *Forced Migration Review*, 64, 35–38.

Hayama, A. T., & Cardoso, T. M. (2018). Comunidades Caiçaras da Ilha do Cardoso, Conflitos Socioambientais e Refugiados da Conservação. In L. LyraJubilut, E. Pires Ramos, C. de Abreu Batista Claro, & F. de Salles Cavedon-Capdeville (Eds.), *Refugiados Ambientais* (pp. 608–638). Boa Vista: UFRR - Editora da Universidade Federal de Roraima.

Nixon, R. (2011). *Slow violence and the environmentalism of the poor*. Harvard University Press.

Nordi, N., Cardoso, T. M., & Beccato, M. A. B. (2005). Histórico da pesca nas comunidades Enseada da Baleia e Vila Rápida, Parque Estadual da Ilha do Cardoso, Cananéia, São Paulo. In *Enciclopédia Caiçara. Volume IV* (pp. 349–356). HUCITEC-NUPAUB-CEC/USP.

Povinelli, E. A. (2011). *Economies of abandonment: Social belonging and endurance in late liberalism*. Duke Press University.

Yamaoka, J. G. (2019). *Resistência pela Permanência no Território: O caso da comunidade Caiçara da Enseada da Baleia, Cananéia - SP*. Universidade Federal do Paraná.

Yamaoka, J. G., Cardoso, T. M., Denardin, V. F. , & Alves, A. R. (2019). A comunidade caiçara da Enseada da Baleia e a sua luta pelo território – Cananéia (SP). *Guaju*, Matinhos, 5(1), 138–165.

Vazantes. (2010). [HDV]Tom Laterza and Directed by D. Sampaio. Available at: https://curtadoc.tv/curta/cultura-popular/vazantes/ [Accessed 11 Nov. 2020].

17 Hope, community, and creating a future in the face of disaster

Claire-Louise Vermandé

As I stood in my parent's driveway on the evening of 4 January 2020, I looked up as a dark charcoal-colored wall of bushfire smoke tumbled down on us in an instant. Earlier in the day we had taken photos of the cloud that had formed in the distance, billowing a mile into the sky, appearing apocalyptic. Moments before the southerly wind hit, we were in late evening light, then suddenly it was pitch Black. Black like the world no longer existed. Had the house lights not been on I would not have been able to find my way back. By the time I walked the 100 yards or so back to the front door, my feet were blackened from walking through the ash that had fallen. My mouth filled with grit, my eyes were stinging, and my throat was burning. After making sure embers were not falling, we retreated indoors to listen to emergency broadcasts and trawl through social media feeds for updates. We had suitcases packed and ready by the front door in case we needed to leave in a hurry. I read a notice by the emergency department stating the wind change would have no impact on residents in our community. So, with that I went to bed.

Unfortunately, this advice was incorrect. When I awoke the next morning, I reached for my phone and saw an emergency text message: "It is now too late to leave, seek shelter as the fire approaches." It was not for my suburb, but for suburbs only about 5 miles away. The Southern Highlands, my hometown where I was born and raised, was now on fire on its southern end as well as to the north. Twenty homes were lost in our region that night, adding to the hundreds lost across the country that spring and summer.

The Southern Highlands is a stretch of countryside consisting of a dozen or so quaint townships and vast areas of farmland and is surrounded by national parks. I was born in the biggest town, Bowral, and lived on the outskirts until I moved to Brisbane, Queensland, where the weather is more temperate, and the buzz of the city offers more opportunity for a corporate professional. Being home at Christmas, however, is something I cherish. That year, I had decided to drive rather than fly. As I set off, the skies above Brisbane were a murky sepia color from bushfires that had raged in the sunshine state since September. I did not see blue sky for the entire 12-hour drive, or did I see it for the 3 weeks that I was home. As I made my way through the Hunter Valley to the north of Sydney, I could see the enormous cloud of smoke stretching up into the sky from the

DOI: 10.4324/9781003141457-17

Gospers Mountain fire. As I drove, I listened to the Radio National emergency broadcast where some residents called in to detail their firsthand experience as fire fronts consumed their properties.

Hundreds of fires raged across Australia. As I made the journey home the Green Wattle Creek fire was the closest to my hometown, and in the days before Christmas, it devastated the entire township of Balmoral. The Currowan fire on the south coast started hundreds of miles away, but by the evening of 4 January, it was threatening homes near Nowra. That night, the fire jumped over the Shoalhaven river, raced up the escarpment and into the township of Bundanoon, covering hundreds of square miles in just a few short hours.

Australian's are no strangers to bushfires. This year, however, I came to understand what it means to be faced with the decision of rebuilding in the same spot or leaving your home after a disaster. To this day, many people continue to struggle to rebuild burnt homes and communities. While I thankfully did not have to make a decision on whether or not to rebuild, for me, the hardest moment came when I had to leave the Southern Highlands and drive back to Brisbane just 3 days after the 4th January fire. I did not stay and help rebuild, I chose to return to my life in the city, and it broke my spirit.

That same spring and summer, I had been helping a former colleague, Jamie Simmonds, as he wrote a memoir about his experience moving a town to higher ground after it was destroyed during the 2011 Queensland floods. Through our countless conversations, I began to see connections between the relocation of Grantham and my own emotional struggles. My experience with the fire that Christmas gave the Strengthening Grantham Project a completely new meaning for me. I began to wonder who in the many towns affected by bushfires would step up to the challenge of rebuilding or relocating these communities. What lessons could they learn from Grantham?

In 2017, I worked at a Brisbane-based real estate investment company, where I was placed in a trial mentoring program with New York native Jamie and we quickly formed a bond over our shared love of sport. Although the official mentoring program never took off (a coincidence rather than a failing on our part, I am sure), the mentoring continued long after I left the company. When Jamie told me that he was writing his memoir, I offered to proofread it, and thus began a new venture for the two of us. I introduced him to the media game, and he gave me a crash course in managed retreat.

During the summer of 2010–2011, flooding devastated Queensland communities. Homes and businesses were besieged by water, inundated for days on end. In early January 2011, the Lockyer Valley, a rural region made up of small farming communities outside of Brisbane, received a month's worth of rain in the space of just 5 days (Queensland Flood Commission, 2011, p. 3). On the fateful day of 10 January 2011, towns in the Lockyer Valley, including Grantham, were all but wiped off the map when a weather system dumped up to 100 mm of rain in a matter of hours on already saturated land (Queensland Flood Commission, 2011, p. 3). Residents described an inland tsunami that tore its way through the valley. The force of the wall of water was so fierce that it tore houses off their foundations and carried them hundreds of yards away, dumping them in fields. Cars, boats, machinery, and shipping containers drifted off in the torrent and became tangled and wedged in railway bridges. People scrambled to get onto the roofs of their homes or sought higher ground by running up to the railway line. Those who could not run or climb perished in the flood. Nineteen people sadly passed that day.

Following that fateful day, the mayor of Grantham, Steve Jones, created a vision for his community so bold that many questioned if it could be done. It was Jamie who would be tasked with the challenge of taking that daring vision and turning it into a reality. Almost a decade after the relocation, Jamie's memoirs put Grantham back in the spotlight by sharing a story of hope with a country who desperately needed it.

Grantham's example of recovery is important not only for the tangible lessons it provides for other disaster recovery efforts but also for the hope it provides. The relocation of Grantham was widely reported throughout the process, but unfortunately what lasts in one's memories are the scenes of chaos and destruction that led to the relocation. Life returning to normal, even thriving, is less memorable to anyone other than those directly affected. Yet, the story of Grantham is one of hope, and it is a story our nation needs – that I needed in the aftermath of the bushfires. I needed to know how people recovered, emotionally as well as physically. As I sat in my parents' home a few days before Christmas, bushfires ablaze around the countryside, blackened leaves raining down upon our property, I took heart from the story of Grantham. Talking to Jamie, to someone who had been there for a vulnerable community in the aftermath of a natural disaster and led them to recovery, offered a sense of comfort. It also gave me insight into what is required to achieve a successful managed retreat. At its core, the story of Grantham is the story of a mayor who wanted to make sure his residents could sleep at night knowing they would never go through another flood event.

Through countless conversations with Jamie about the relocation of Grantham, I have learned that success lies in keeping a community together. The key with relocation is not letting them abandon their livelihoods and move away from their social connections. To keep everyone together, Jamie outlined how a recovery effort needs to create a vision for the future, the importance of communicating with residents, and the need for strong local leadership. These "pillars of recovery," I believe, are essential for rebuilding or relocating efforts, and bushfire affected communities could look to Grantham as a leading example of recovery.

As tempers flared in the aftermath of some of the New South Wales bushfires in the last few days of 2019, I spoke with Jamie about what he experienced when he arrived in Grantham in the days after the flood. He described a scene of disbelief. "You see anger, you see fear, you see frustration, you see people who are very quiet and reserved," he said. "You see all kinds of different emotions bubbling up to the surface because people are going through a traumatic event and they do not know what tomorrow is going to look like. A lot of times the community has not had a chance to grieve yet or absorb what is going on."

He went on to elaborate, "when the clean-up commences the community is all together, meeting and talking. There's mental health support, shoulders to cry on, residents are getting their meals cooked for them. In a lot of ways straight after the disaster is a really positive time. But it gets a lot harder in the weeks and months after. That's when all the help starts to go away and people return to their normal lives," he explained. The bushfire event we experienced last summer stretched on for months and affected numerous communities across the country. As one town went into recovery mode others were yet to face the threat of fires claiming homes and businesses. To date, the task of recovery has been truly enormous.

However, if we look to Grantham we realize they too were just one small town who experienced a natural disaster that affected much of the state of Queensland, including the destruction of thousands of homes in the capital city, Brisbane. This makes Grantham's story of recovery all the more remarkable. Led by Mayor Jones, the people of Grantham stood up and rallied to save their town, even in the most dire of situations when many of them could have given up entirely. They focused all their efforts inwards on their own community and quickly got their recovery underway. It is easy to feel overwhelmed when so much tragedy has taken place over such a vast area, but if each community can rely on their own resources and leadership, and start work immediately as Grantham did, they can greatly improve their chances of a successful recovery, whether it includes a managed retreat or any other adaptation method.

Jamie often uses the phrase, "don't eat the whole elephant at once" to refer to breaking down the task of Grantham's managed retreat into smaller, manageable components. With so many communities devastated during both the floods and fires, and indeed during many other natural disasters that have occurred the world over, it remains important not to become overwhelmed. If you think about how Australia will recover as a whole from the bushfire crisis, the task seems almost impossible, but if you break it down to each township's recovery, you can quickly see actions taking place. Progress, no matter how small, is still progress.

When a disaster rolls through a community, it wipes out tangible infrastructure in an instant. But those structures can be rebuilt. It's the people who suffer physical and mental trauma and are in a state of shock in the days and weeks after. Local leadership, therefore, is critical. Leaders who understand the intricacies of that community, who can stand tall in the aftermath. Leaders who can create a new vision and immediately start the work that needs to be done. For Steve and Jamie, they were only concerned with the recovery of Grantham's residents, not the broader state or nation, and this narrowed vision ensured they could focus to move the town as quickly as possible.

"Straight after a disaster everyone stands up and says, 'We're not going anywhere, we're going to stay and beat this thing, this is my home,'" Jamie said. "But that story changes in the months afterwards when they stop seeing a future. Once everyone leaves and the adrenaline has gone out of the community, you're left questioning 'What am I going to do now? Do I want to stay here?'" For disaster-affected towns such as my hometown, it is important to have local leaders with a clear vision and path forward who can push the recovery all the way through to completion so as to minimize, as much as possible, any thoughts from residents of abandoning their community.

He posed an interesting question: "Do I want to stay here?" Even as he asked, I knew I would have to leave my hometown and return to my job and my life hundreds of miles away in Brisbane. But the desire to stay was consuming. I wanted to be reassured that everyone would recover and rebuild, and life would return to normal – even if it was not my home or my life any longer. I wanted to know that I would be able to return for future Christmases. I wanted to go back to Brisbane knowing that my hometown was still there.

But how would that happen? I had read reports of some towns who had experienced bushfires 18 months prior to last summer's where residents were still living in mobile housing and had not yet rebuilt as bushfires approached their towns yet again. People in these towns had not yet recovered from the previous natural disaster before the next arrived. Sadly, some 12 months on from the 2019–2020 bushfires, it seems this is the case again for many residents. As natural disasters increase in intensity and frequency will this only compound the issues that surround recovery? If these towns had had the leadership and vision that Grantham did, would this still have been their unfortunate fate, or would they have moved into new homes and built a more resilient community within twelve months as Grantham's residents did?

"After a disaster people search for meaning and need hope that their community is going to come back and be like it was before," Jamie explained to me. "If a community doesn't have something to aim for, or a vision to grow towards, or a recovery effort that tries to rebuild the foundations, it ceases to be a community." Jamie also highlighted the importance of residents being present in their devastated surroundings to note what elements should be included in the recovery effort. "When you're cleaning up and you're seeing the devastation, it's human nature to think 'If things looked like this, or if we had done this before, we probably wouldn't have had this happen, or this house would have been saved or this street wouldn't have been damaged.' Remember those things, write them down because they're going to be crucial to your vision moving forward," he said.

Why do we wait until after a disaster to create our vision? Vision is important not just for flood or bushfire recovery but for climate adaptation generally to better prepare against disasters before they occur, potentially lessening their impact. We, as citizens, have an opportunity to safeguard our communities now by calling out potential hazards. Communication between residents and leaders should not be left to the days, weeks, or months after a disaster. We should work together now to identify risks and to create a vision for a better community before the damage occurs.

Once a disaster has occurred, however, communication is key in the recovery effort. In Grantham, "We took every opportunity to explain what we were doing to the residents," Jamie explained. "To do this we had regular community meetings and we used these meetings to introduce the idea of the relocation." He and Steve Jones had to bring the community "along for the ride," so they knew what the vision was and what would happen at each stage, every step proceeded with full transparency. Grantham excelled at community involvement and other towns would do well to implement similar communication strategies for their recoveries. There were larger gatherings for the whole community where high-level detail of the relocation was explained, and there were one-on-one meetings between residents and caseworkers to extract essential personal information including financial and insurance details. The combination of the two was key.

"We had separate master planning workshops with the community, and we introduced the final master plan at the weekly meeting in Grantham with the full community present," Jamie explained. Concise messaging during the community meetings is essential to clearly convey the recovery plan to residents. "If you drill down too much into the details during these types of meetings the message gets lost. The meetings were really about understanding from the community what they wanted to preserve in Grantham, what they liked about Grantham before the flood hit, and why they were still connected to the town and the people," he said. These are all items that can be identified by any town to assist with recovery or relocation efforts.

Speaking to the residents of Grantham as individuals was also critical. "While the workshops and community meetings were great for delivering the overall message, we also needed to trickle information to individual people. The only way you can drill down into the detail and talk about individual needs is on an individual level. Most of the real value we had with the community was at that individual level," he said. The Grantham relocation utilized a combination of local leaders and case workers to liaise with residents.

Another key to a successful recovery is how well the community's needs are understood. To do this, meetings should be set up immediately in the aftermath of the disaster to bring everyone together, and they should continue throughout the entire recovery effort. No two communities are the same and each has its own set of needs to be addressed. While funding can be administered at state and federal levels, at the end of the day these levels of government do not understand the intrinsic and individualized nature of each town that has been affected. For example, two towns in the Southern Highlands were gravely damaged by the bushfires. Even though they logistically reside within the same region, each town, Balmoral and Bundanoon, is unique and possesses different requirements for a successful recovery. On-the-ground workers and local leaders are needed to establish these requirements by obtaining intimate knowledge from each set of residents to determine the best path forward.

Working with residents one-on-one is about getting necessary information for the government and about helping residents to take action. As Jamie said, "We had to assist residents with dealing with organizations like banks and insurance companies to get them into the land swap. Added to this, a lot of people did not necessarily understand their finances, and there were people in Grantham who could not read, so we had to help them understand their contracts and insurance claims. There was so much personal information that we were helping people with that you cannot do in a group setting," he explained. For example, some critical information is incredibly sensitive, such as information about finances, insurance coverage, or how a death in the family might have affected their financial or legal positions. "People won't talk about that in a bigger group," he said, adding "but these were the things we had to understand if we were to help people move into the new estate. ... If you don't know that information, it becomes very difficult to bring people into the process. There'd be a lot of people who with assistance could enter the land swap but without assistance wouldn't even know how to begin to get their affairs in order, so it was about helping them understand that they could do it."

Grantham is an exceptional example of what can be achieved when you have a leader who considers the well-being of his citizens as being of the utmost importance. Steve Jones took the time to speak with his residents and made sure they had their say. After the people have had their say, then leadership needs to step up and provide a direction and a vision for the community that is rooted in their values. "Had Steve not envisioned a relocation and seen that plan through to its completion, we feared that Grantham would have become a ghost town and that people would have just left. People who lived through the event, almost all of them would not have coped with living in the floodplain again. Any time it rained they'd be worried. You can't raise a family in that environment. You can't grow old in that environment. Steve knew that for a small town like Grantham, if half your population moves away, you don't have a community anymore," Jamie said. The same can be said for the two towns in the Southern Highlands, Balmoral and Bundanoon, as well as many others around the country that were destroyed by fires. These are not big towns, if a portion of the residents opt not to rebuild and instead move away, the entire fabric of those towns changes.

Since I began working with Jamie, we have had countless conversations about what he and his team achieved in Grantham. That one phone call, however, on that December night, will forever stand out to me. Contextually, it was bizarre to be sitting in the midst of one of Australia's worst bushfire seasons, which could be attributed to one of our worst droughts on record, while dissecting one of our most devastating flood events. But it was also reassuring to know that there are people out there who will look after these devastated communities, that recovery is possible, and that it can happen even when people relocate – as long as they move together.

After returning to Brisbane, I still felt guilty about having left my hometown. I took an opportunity to drive out to Grantham and finally see the town for myself. I drove along the main road into the town on the low-lying part of the land where the houses used to be and which had now been returned to open fields for farmers. I drove under the railway bridge I had seen in photographs with cars and parts of homes lodged into its timber and steel structure. I went up past the old butter factory and finally arrived at the entrance to the estate.

I drove slowly up the winding road to the peak of the hill, admiring the houses and their gardens, the fabric of a small rural Queensland community. Despite having read about it in considerable detail, and having seen photos, I was nothing short of amazed at what had been achieved when seeing it firsthand. People living there were going about their day as they normally would. It was ordinary. Unassuming. And yet it gave me such a sense of hope and relief. I sat there and thought about those devastated communities I had left behind weeks earlier, and I was overcome with the belief that recovery is possible, that you can once again return to a normal life. The proof of a successful recovery was laid out in front of me.

"Grantham today is just like any other small regional town," said Jamie. "Any person going into Grantham now who doesn't know the history probably wouldn't know about the disaster that unfolded there in 2011. There's nothing obvious about the town that indicates what happened there. If you scratch below the surface you see a memorial to those who died, a cafe displaying photos of the flood. You see houses in a new estate on a hill. In every other way though, it's just a normal regional town in Queensland. It has its shop. There are kids playing and people mowing lawns, and when you talk to the people involved in the relocation they're happy. They're satisfied that rebuilding on the hill made a huge difference."

When I met Jamie, I had never heard of the term "managed retreat." In fact, neither had Jamie: He'd simply moved a town out of harm's way. But Grantham is a successful managed retreat. In fact, it is one of the leading examples of managed retreat anywhere in the world. The Grantham relocation implemented Australia's first land swap initiative, kept the community intact, and prioritized the mental health of its vulnerable residents. It is a beacon of hope that can show a person or a nation that recovery is possible.

Recommendations

- **Including the community in the recovery process** – Giving residents whose homes have been damaged or lost during a natural disaster an opportunity to shape the future of their community may assist with not just the recovery and/or relocation effort, but ultimately help with the emotional recovery of individuals. For Grantham, this was done by asking residents what they wanted to see in the new estate as well as what parts of the "old" Grantham should be kept and what should be improved on. Giving the community something positive to focus on and actively involving them in building a future for their town, rather than leaving them to dwell on the disaster, may lead to more residents opting to stay with their community rather than moving away as they start to see what opportunities can be found. A community will be strengthened both physically and emotionally if residents are afforded an opportunity to help strengthen the resiliency of their hometowns by adding their own voice and ideas to the rebuild, recovery or relocation.

- **Create a vision for the future** – In the aftermath of a disaster, it is imperative that the community immediately start recording how they want their town to look in the future. Residents who experienced the effects of the disaster are best placed to identify areas that require attention, such as infrastructure failures, to help fortify their town against future events. In Grantham, for instance, it was noted that an additional road into the town and a second railway crossing would provide an extra escape as well as allow for additional rescue access should that level of flooding occur again. Even services as simple as including a community parkland or walking tracks should be included. Post disaster provides towns with an opportunity to be able to include wish list items when rebuilding.
- **Leadership should be local and needs to act quickly** – It is crucial to ensure that once a vision has been created local leaders move quickly to implement it in order ensure participation in the relocation or recovery by residents is at a maximum. This person or people can be mayors, business owners, spiritual leaders or anyone else who has an active involvement in the community. They will need to dedicate their time to the task and see it through to completion and aim to do so as quickly as possible.
- **Communication should be multilayered** – Town hall style meetings are excellent at allowing leaders to deliver key, broad, high level messages to residents to update the progress of a relocation, however, detailed discussions need to be had in order to assist residents with personal matters including banking, insurance, medical issues and more. These personal conversations can be done by utilizing case workers who are experienced in dealing with vulnerable individuals who may be suffering the effects of trauma. This allows for a two-way dialogue where detailed information can be recorded and relayed back to leadership to assist with the recovery and relocation effort. This also allows the quieter members of communities to have a voice in a setting more appropriate for them. Group community meetings should be convened by local leadership and to avoid anger and confrontation, which can be expected following a disaster, should limit the amount of interaction received by the community in this setting. It is important, however, to reassure individuals that they can talk to case workers and leadership at any point outside of these meetings.

Acknowledgment

Being given an opportunity to learn directly from one of the key leaders of the Strengthening Grantham Project has been invaluable and I have come to have a deep appreciation and interest in the topic of managed retreat. I sincerely thank Jamie for choosing me to work with him as he wrote his memoir and I am forever grateful for the friendship I gained and the education I received. This year I will formally continue that education as I have been accepted to study a Master's of Climate Change Adaptation at a leading Australian university. I want to keep adding my voice to this important conversation, and I want it to be loud. While my work with Jamie continues to evolve, I hope to keep this momentum going to learn about other leaders and communities around the world undertaking similar pursuits.

Additional reading

A river with a city problem: Brisbane river floods. University of Queensland Press, 2019.
Burning bush: A fire history of Australia. University of Washington Press, 1998.
McKinnon, S., & Cook, M. (Eds.). *Disasters in Australia and New Zealand: Historical approaches to understanding catastrophe.* Palgrave Macmillan, 2020.
Queensland Flood Commission. *Queensland Flood Commission Report.* Queensland Flood Commission, Brisbane, 2011.
Rising from the flood: Moving the town of Grantham. Bad Apple Press, 2020.

Interlude 6 *Gratitude*

Martha Lerski

A form evoking fire, harp music, and bones – *Gratitude* is constructed from a piece of marble the thickness of a tombstone. I extracted the stone from a dumpster outside the recently demolished Huntington Hartford Building on Columbus Circle in Manhattan, now the site of the Museum of Art & Design. The piece became a memorial to my mother, who died in 2008.

After an initially fraught relationship with my parent, which had been shaped by her early departure to escape a damaging marriage, I ultimately grew to understand the complexities of marriage and divorce. Above all, I admired her courageous wartime services in occupied Warsaw (assisting the Head of the Underground Movement, as a courier of a samizdat weekly newspaper and, with her mother, harboring a Polish Jew; the latter action was punishable by death). Later, in the States, she once accurately told an abusive employer, "I survived Hitler, I survived Stalin, and I will survive you."

I also respected that she constructed a meaningful life in a new land, where art and scholarship stood in for lost family and place. What do we have when our milieu, our habitat, our families, language, communal knowledge, and traditions; our local sculpture and paintings; the cuisine we crave, and intangible aspects of our culture, are taken from us? Which aspects of culture carry into different environments to sustain migrants, to enrich host communities, to influence ensuing generations, or to provide novel approaches?

Gratitude 243

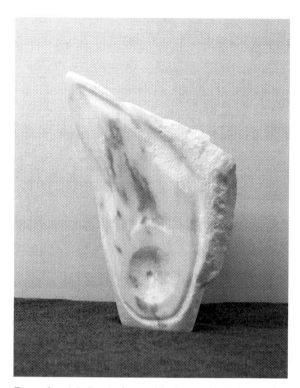

Figure Int 6.1 Gratitude, marble (view (a) of the sculpture by Lerski); photos by Jim Chervenak.

244 *Martha Lerski*

Figure Int 6.2 *Gratitude*, marble (view (b) of the sculpture by Lerski); photos by Jim Chervenak.

Part V
Future directions

18 Retreating from the waves

Orrin H. Pilkey, Sarah Lipuma, and Norma Longo

18.1 Introduction

"So, what are we supposed to do? Throw up our hands and slink away?"

So said a US Army Corps of Engineers (USACE) colonel from the Jacksonville, Florida, District in the 1970s, in response to strong resistance from environmentalists and others to constructing a seawall along a South Florida shoreline. It was symptomatic of the time when the prevailing view, especially among engineers, was that "of course we can hold the shoreline in place and simultaneously maintain a beach."

Perhaps, the colonel's attitude is somewhat understandable given the long global history of shoreline engineering. One of the oldest seawalls in existence is in Batroun, Lebanon, built by the Phoenicians in the first century BCE. Portions of that wall still exist. In the United States, the mightiest seawall is in Galveston, Texas, built around 1910, just 10 years after the great storm of 1900 killed more than 6,000 people there. The wall is 7-miles long, 17-feet high, and 16-feet thick at the base. It has no beach in front of it except where it has been nourished: In other words, where sand has been added.

It must have quickly become evident to the Phoenicians that hardening sandy shorelines with a seawall extracted a price: Complete loss of the beach and increased erosion on adjacent shorelines. It probably also was apparent to early seawall builders that the walls made access to beaches difficult. The problem with seawalls is that they do not address the causes of shoreline erosion. So, when a wall is put in place, the shore continues to erode as before.

Feats of engineering to control flood waters are prevalent inland as well. People have long put floodplains to use for development and agriculture by building levees and dams, draining swamps, channelizing rivers, and building water diversion projects. Floodplains – lands adjacent to water bodies that flood during high-flow events and rainstorms – help to convey, filter, and store water. Over the past century, the US government played a major role in encouraging development of the nation's floodplains through publicly funded engineering projects (Tockner & Stanford, 2003) (Figure 18.1).

DOI: 10.4324/9781003141457-18

Figure 18.1 Tidal flooding impacted the community of Centerport, New York, during Hurricane Irene on 28 August 2011. Buildings like this could be elevated, but the hurricanes of the future will make us rethink the location of residences so close to rivers and wetlands. Photo by Dhaluza. CC-BY-SA-3.0, "Hurricane Irene flooding of Mill Dam Bridge in Centerport, New York." Wikimedia Commons (Creative Commons Attribution ShareAlike 3.0: https://creativecommons.org/licenses/by-sa/3.0/).

The USACE is in one way or another the main driver behind most engineering of the American waterfront. From the 20th century on, opposition to shoreline engineering was considered unacceptable in America because it failed to recognize the many frontiers that engineers had conquered in the past (Pilkey & Dixon, 1996). The idea of *retreat* was simply inconceivable. It has been a long path to reaching the point today when *retreat* from the shoreline is no longer a "dirty" word. And yet, in light of the impacts of climate change, tension remains between those who support protecting development in place and those who support relocation.

Holding the shoreline in place using levees, canals, seawalls, and beach nourishment continues to be central to shoreline management around the world (Pilkey et al., 2016). Two examples of this practice in the United States are described next (Figure 18.2).

The first example is a recently announced project in North Carolina. The USACE has approved $237 million for 10 miles of beach nourishment in two communities on Topsail Island (Office of Senator Richard Burr, 2020). Beach nourishment, also called soft stabilization, has largely replaced seawalls and

Figure 18.2 A 2-to-3-foot high erosion scarp on a Nags Head, North Carolina, beach. Such scarps, some as high as 10 feet or more, are virtually diagnostic of the rapid rate of erosion of nourished beaches relative to natural beaches. In North Carolina, nourished beaches have average life spans of around 3 years. Photo © Margaret Miller Growe. Used with permission.

other forms of hard stabilization, but it is still costly and temporary. Five miles of the nourishment sand will be placed on Surf City beaches in the middle of Topsail Island and another 5 miles will be put on the southern end of North Topsail Beach. This works out to be more than $20 million per mile: The costliest beach nourishment project ever in North Carolina and among the costliest ever on the US Atlantic Coast. The project will be eligible for periodic nourishment over the next 50 years. The average lifespan of nourished beaches in North Carolina is around 3 years, and considering the anticipated intensification of storms and sea level rise, this beach nourishment project is highly unlikely to last 50 years. Adding to the problem is the erosion of the shoreline between these two communities which will certainly affect the erosion rate of the nourished beach.

Funding for this project comes from a hurricane disaster relief package provided by the federal government. Both natural and nourished beaches commonly disappear in storms, but these disaster funds were not originally intended to rebuild eroded beaches (Song & Shaw, 2018). More commonly, the funds are used to aid in repair of homes, businesses, roads, and other infrastructure in coastal and inland communities severely damaged by storms.

If beach nourishment is to be considered a routine part of hurricane relief funding, it could eventually eat up much of the funds, leaving emergency needs unmet. If the federal government provides most of the funding for nourishing a beach, then property owners who imprudently built next to eroding shorelines, and the state and local governments who approved such development, will face no financial repercussions (Gillis & Barringer, 2012; Gaul, 2019). From their standpoint, why should they bother to avoid risky developments or invest in resilience when the federal government (and so all federal taxpayers) will pay to replenish the beach after each storm? Overall, the proposed Topsail Island beach replenishment makes no sense at all.

A second example: The Ike Dike is the largest and most costly shoreline project in American history. Named after Hurricane Ike which damaged Galveston Island, Texas, and surrounding areas in 2008, the projected cost of the Ike Dike is $31 billion (Powell, 2020). The dike consists of a 70-mile-long barrier system of levees and sea gates beginning on high ground north of High Island and running the length of Bolivar Peninsula. It will then cross the entrance of Galveston Bay, run the length of Galveston Island, incorporating the existing Galveston seawall, and end at San Luis Pass. The Ike Dike will protect communities, but it is also heavily premised on the need to protect the petroleum industry facilities along the Texas coast. In theory, the project will prevent storm surge from moving up the Houston Ship Channel and spreading pollution to urban areas (Budds, 2017).

The Ike Dike, despite its cost and massive scale, is not a permanent solution. Alternatively, the $31 billion could be used to clean up pollution around petroleum facilities, to raise, move, or demolish threatened buildings and highways, and to prepare the region for intensified storm flooding and sea level rise. At the very least, planning for the project should include a serious look at spending the money on a truly resilient and long-term approach to the future rather than using walls, tide gates across inlets, and levees, in an attempt to hold the shoreline absolutely still (McNeill et al., 2014).

History tells us that the problem with shoreline stabilization is that it inevitably provides the impetus for increased density of development and an increase in the size of buildings in hazardous areas (Woodruff et al., 2018). A building site protected by a seawall or a newly nourished beach is far more desirable than a site next to a natural but obviously rapidly eroding shoreline (Jin et al., 2015). A 2018 estimate suggests that 41 million Americans live within the 100-year floodplain – a population that will increase by an expected 40% by 2100 (Wing et al., 2018). Although numerous laws exist that manage development in flood hazard areas, the population density has continued to rise, and development intensifies unfettered.

Despite the construction of such major works as those described earlier, climate change and sprawling development are making the devastating impacts of flooding more intense and frequent. Over the past 30 years, flooding has cost on average $8.2 billion per year in the United States, with that cost increasing year over year (Wing et al., 2018). These storms and flood events are all occurring while the background sea level continues to rise. The global average sea level is predicted to rise between 3 and 7 feet by the year 2100 (Bakker, 2017), an estimate that may be conservative given recent observations of accelerating ice sheet melting. All these factors make the steady increase in population density in the floodplain and along shorelines even more problematic.

18.2 Withdrawing from the waves

"We need to act now to save the coastline of tomorrow."

So said the French Minister of the Environment at the beginning of a project to demolish 240 at-risk beachfront homes in response to a 2010 storm that caused tremendous damage. Its flood waters breached old seawalls and flooded several villages (Chadenas et al., 2013). The French government swiftly consulted with communities, reviewed their flood risk, and began taking action: They moved buildings inland, demolished houses, revoked 30 beachfront building permits at other sites, and even prevented property owners from making modifications to vulnerable houses (Griggs & Patsch, 2019).

This same approach to flood risk prevention is happening in many other parts of the world, although implementation strategies differ. The village of Fairbourne in Northern Wales is being relocated inland, in a move that the Welsh government calls "decommissioning" (Buser, 2020).

The most transformative of all managed retreats is that planned for Jakarta, Indonesia. One of the world's largest cities, Jakarta's solution is one that could eventually be repeated in Miami (USA), New Orleans (USA), Honolulu (USA), Dhaka (Bangladesh), Shanghai (China), Guangzhou (China), Manila (Philippines), and many others. Indonesia announced in 2019 that it is moving its capital city to another location because of significant land sinking and the increasing rate and volume of urban flooding (Eliraz, 2020). The increased flooding is largely due to ground water extraction, sea level rise, and compacting sediment from the weight of large buildings (Kimmelman, 2017). In one location, the land sank 10 inches in 4 years. In a working-class district of the city, the land has sunk 13 feet since a building boom in the 1970s (Levine, 2020). These are startlingly fast rates of subsidence (Figure 18.3).

The plan is to move the capital along with 7 million people (one-fifth of the population) from the island of Java to the island of Borneo, 1,250 miles to the northeast. The cost is estimated to be $33 billion (Chappel, 2019). Although this is an unprecedented step for a country to move its entire capital because of flooding, the Indonesian government's previous climate and coastal defense plans favored the privileged elite, neglecting the concerns of the urban poor who are most impacted (Anguelovski et al., 2016). It is expected that the rest of the population will eventually follow the initial move, but evicting and relocating residents of informal settlements will further entrench extreme inequality by breaking up family and social networks and access to work. Only if the plan includes support for poor residents will the relocation be equitable.

In America, there are several cities that are seriously threatened by sea level rise. Thousands of skyscrapers are not going to be moved even if there were a place to move them. Whole populations may eventually have to move to safer areas, leaving abandoned and deconstructed buildings in their wake. This will likely be the case for Miami, Florida. The eventual permanent inundation of the city will come from below through the porous permeable limestone layer

Figure 18.3 Honolulu, Hawaii, is a major city whose shoreline is lined with high rises. Managed retreat may not be a possibility here, as is the case with many important coastal cities. The city may decide to install a massive seawall in the future, depending on the rate of sea level rise, but this will erode the beaches. Some futurist strategic planners envision all buildings with lowermost floors used only for floodable purposes, and transportation within cities by boats or special suspended rail systems. Photo by Edmund Garman, "Waikiki and Honolulu from a doors-off ride with Genesis Aviation Helicopter Tours, 11 December 2014." Licensed under the Creative Commons Attribution 2.0 Generic license. This image was originally posted to Flickr by Edmund Garman at https://www.flickr.com/photos/40315625@N08/16191692896.

underlying the city. The $4.6 billion storm surge protection plan that USACE has proposed with its 13-foot seawalls will not address the existential problem of flooding from below. Ultimately, like Jakarta, the city will be abandoned. Between 4 and 6 million people must move to higher ground or face displacement because of sea level rise by the end of the century (Hauer et al., 2020). Not enough planning is being done in the communities that will need to receive waves of people fleeing from South Florida. In fact, the likelihood of abandonment of a major city is met with the same skepticism that engineers had when it was suggested they could not hold shorelines still forever.

The most common way the United States currently conducts managed retreat is with property acquisition, commonly referred to as a buyout. Accepting a federally funded property buyout is completely voluntary. A buyout is a transaction in which vulnerable structures are removed from areas at high risk of flooding and just compensation is given to the occupants to find new housing, after which the land is "undeveloped" and returned to a natural state. After the structures are demolished and the utilities removed, the land is preserved as open space, often as a park accessible to the public, or for conservation like a restored marsh. Open space acts as a buffer for the surrounding community by providing critical ecosystem benefits, such as runoff filtration and floodwater absorption, for the remaining residents.

The Hazard Mitigation Grant Program (HMGP) is the longest running buyout funding source administered by the Federal Emergency Management Agency (FEMA). Officials from local governments may request money from their state to buy flooded or substantially damaged properties using HMGP funding. Houses that meet the voluntary flood buyout requirements can be purchased by the local government at the pre-flood fair market value of the property. Property owners can also be compensated for relocation of occupants, demolition of buildings, and the value of their land. If a property owner sells a multifamily building, the renters can apply for relocation aid.

18.3 Challenges to retreat

> "They don't understand the anxiety—we've lost all our stuff, we're trying to fix our homes, but we're not getting any information. They want us to live a normal life."

This was the account of a participant of the New York Rising Buyout Program initiated after Hurricane Sandy caused severe flooding in neighborhoods of Staten Island, New York (Binder & Greer, 2016). The participant and many other potential sellers were struggling with the long wait time, uncertainty, and opaque process. There are several pitfalls for community members, program managers, and policy makers in the process of withdrawing from hazardous areas.

18.3.1 Impediments in governance of retreat

Buying out a flood-prone property is a lengthy process. In the United States, it often takes over a year after the initial disaster for funding to become available. Homeowners may have to actively advocate for themselves and their neighbors to the local government if they want to participate in a buyout program, as several neighborhoods in Staten Island, New York, did following Hurricane Sandy in 2012. This requires time, resources, well-organized paperwork, and presence of mind right after a major catastrophe. There is no guarantee that a buyout program will culminate in actual buyouts because there are many points in the multistep and multiyear process where participants can slip through the cracks.

With buyouts often taking over 5 years from disaster to completion, many flood disaster survivors cannot bear living with the uncertainty and instability. After Hurricane Harvey in 2017, for example, thousands of Houston residents said they were interested in a buyout, but the funds did not come until 18 months later (Turner & McFadden, 2019; Moore, 2017). By that time, most property owners had repaired their homes. People are less likely to consider a buyout after all the time, money, and effort they put into making repairs and improvements. Few people can endure the long wait time, least of all people with disabilities or lower incomes. Because of this, retreating is one of the least accessible options for homeowners.

Local governments are not always amenable to the idea that residents may want to retreat, since this not only calls into doubt the "stronger than the storm" narrative but also potentially jeopardizes property tax revenue. Yet, contrary to popular belief, a buyout is not always an undesirable option for homeowners. For example, many people in Staten Island spray-painted signs on their boarded-up homes to get media attention for their desire to be bought out after Hurricane Sandy. In the New Jersey Blue Acres Buyout program, some homeowners were angry about the development practices that had put their house in a vulnerable location in the first place, which had brought heartache and danger to their family (Schwartz, 2018). Many of them felt misled into buying at-risk houses. Often homeowners do not know that they are in a flood zone until they are informed they will have to pay for flood insurance (NRDC, 2018). Less than half of states require that realtors tell a potential buyer the flood history of the property.

To encourage homeowners and builders to move out of risky areas, state and local governments can institute creative policies. For example, the government could disallow reconstruction after a disaster strikes, if a certain percentage of the building is lost or damaged beyond repair. This policy was attempted in South Carolina in the 1970s, but it failed because of intense political pressure from owners of damaged homes. In one case, the local decision-makers decided to allow reconstruction if just the roof of the house could be found!

Another potential policy solution for oceanfront communities is to enact a law creating rolling easements. A rolling easement changes the ownership of beach land from private to public as the sea "rolls" up onto the shoreline (Titus, 2011). The Public Trust Doctrine, which maintains that the government has a duty to conserve and protect the ocean beach for the benefit of the public, is a part of many state constitutions. If added into state legislation, rolling easements would provide advance notice to beachfront property owners that the government would no longer stop the inland migration of the shoreline. Property owners would have an incentive to accept a buyout before or after the next storm if they knew the government prioritized public access to the beach. Despite this, rolling easements have not been widely implemented because of concerns about property-takings lawsuits against the government.

18.3.2 Inequity in retreat

Retreat may be under discussion in many cities and on shorelines around the world, but it is not often conducted in a coordinated manner that considers social equity. In Lagos, Nigeria, the government forced the urban poor to move from flood-prone parts of the city without financial compensation or assistance in finding new housing (Ajibade, 2019). Without job opportunities, some tried to return home only to see new waterfront development projects had been erected in their absence.

In the United States, there are equity issues with the buyout process. Using economic efficiency to decide which properties to protect and which to buy out has created a situation in which higher income residents are given greater assistance to stay in place and rebuild than are socially vulnerable populations. Cost-benefit analyses would seem to be unrelated to the demographics of the communities that are being protected, but incomes and hazards are not evenly dispersed (Hersher & Benincasa, 2019). Because of historic segregation and inequitable land use, low-income people, and people of color are more likely to live in flood-prone areas (Ueland & Warf, 2006).

Government programs that use cost-benefit analysis to distribute recovery funding often end up protecting the assets of affluent homeowners and directly impairing recovery of low-income residents and renters (Muñoz & Tate, 2016). On the other hand, buyout programs that end up displacing lower income people from their communities are inherently flawed. Buyouts are an important and underutilized option in the climate change adaptation toolkit, but should not be proposed as the only option, especially in underserved communities.

The federal government could lessen the impact of this dilemma by giving less weight to cost-benefit analysis and more weight to social vulnerability of the area. Social vulnerability indicators include demographics like socioeconomic status, disability status, race, age, and gender (Cutter et al., 2013). Incorporating these factors can ensure underserved residents are neither underrepresented nor overrepresented in buyout plans.

18.3.2.1 Culture and ties to land in the Arctic

Social justice concerns can also extend to cultural and historic ties to the land. All along high latitude shorelines, including those of the Chukchi Sea, the Bering Sea, and the Arctic Ocean shoreline of Northern Alaska, Canada, and Siberia, the warming atmosphere is creating huge problems. There are 43 Native Alaskan villages threatened by climate change along the ocean or by rivers (U. S. Government Accountability Office, 2009). Managed retreat is being considered for virtually all these villages, most with populations between 200 and 600 individuals (Lester, 2019; Ostrander, 2020). Shishmaref, for example, is located on a barrier island along the Chukchi Sea (Mason, 1996). The community of 560 people is considering moving inland but the cost is prohibitive (Kennedy, 2016). The USACE estimates that the cost of moving inland and reconstructing houses on permafrost terrain would be on the order of $300,000 to $400,000 per village inhabitant (U. S. Army Corps of Engineers, n.d.). An alternative could be moving the inhabitants to nearby towns such as Nome and Kotzebue. But this remedy has unintended consequences because people skilled in fishing and hunting may face serious barriers to acclimating to life in a larger town. Moving Alaska Native Villages is complicated because residents are to some extent dependent on the environment for their subsistence and have strong connections with their fishing, hunting, and cultural grounds (Egge et al., 2020).

18.3.3 Historic buildings and the sentinels of the seas

Lighthouses must be close to the sea to be of use as navigation guides. Even though vessels with modern means of navigation no longer require a guiding light, lighthouses are considered national treasures. In Denmark, the 1,000 ton, 76-foot-tall Rubjerg Knude Lighthouse was moved back 300 feet from an eroding bluff along the northwest coast of the North Sea. The actual move, on wheels and rails in 2019, took place in less than 10 hours and cost slightly over $700,000 (Horgan, 2019).

Not far from the Rubjerg Knude Light, Mårup Church, built in 1250, was perched on a bluff that was succumbing to 5 feet of erosion per year. To plan for the eventuality of its probable demise, the roof and the interior of the church were removed and put in storage. The fate of this famous building differs from that of the lighthouse. For many years, the plan was to rebuild the church, but that plan was given up and the church was demolished in 2015. Only a few items such as the altar chair and some green glazed brick were saved and are in the custody of the National Museum of Denmark. Part of the cemetery is still there, but most of the tombstones have been moved to a safe place "so they don't fall down the cliff to the sea" during a landslide (Overby, 2020).

In the United States, the moving of Cape Hatteras Lighthouse offers many lessons concerning managed retreat and public support for it (National Park Service, 2015). The rapidly eroding shoreline was moving perilously close to the nearly 200-foot, 5,000-ton lighthouse built in 1870 in the Outer Banks of North Carolina (Figure 18.4).

Many North Carolinians were strongly opposed to letting the Cape Hatteras Lighthouse fall into the sea. However, fans and local citizens also were strongly opposed to moving it. To bolster their case, the move opponents found "experts" that claimed it was too large and heavy to move in sand. The National Park Service requested the opinion of the National Academies of Science and Engineering as to the viability of the move. The combined Academy panels declared that the move was essential for the survival of the lighthouse and that it could be done.

This tipped the scales, and in 1999 the lighthouse was moved back 2,900 feet, placing the structure 1,500 feet from the sea, approximately where it had been in 1870. The cost of moving the light and two lighthouse keepers' houses was $12 million (Pilkey et al., 2016). Today more tourists than ever visit the lighthouse, and business around the community is still booming. The lesson to be learned from the Cape Hatteras Lighthouse is that despite initial local resistance to retreating from the battering waves, the move, backed with scientific and engineering expertise, preserved this historically and culturally significant building beloved by all.

However it is accomplished, such important historical treasures, as well as many other buildings and structures, can be moved away from the threat of climate hazards instead of being demolished. Consultation with objective technical experts in the decision-making process is incredibly important to preserve monumental buildings such as lighthouses.

Figure 18.4 The 5,000 ton, 200-foot-high Cape Hatteras lighthouse was moved back from the eroding shoreline 2,900 feet in 1999. Photo courtesy of the US National Park Service.

18.4 Conclusions

"Over the long run, they all will respond through relocation or retreat of some sort, whether managed or unmanaged. Sea level rise will not stop in 2050 or 2100."

This perspective comes from a report on how communities in California can create pathways to adapt to rising seas through long-term planning (Anderson et al., 2020). The previous sections have highlighted the necessity, along with the inherent challenges, of managed retreat. There are still many open questions, as retreat is a complex process. How can a managed retreat process respect peoples' close ties to the land and cultures they cherish? Who makes sure that the sellers are welcomed when they arrive in the receiving communities, and will they have access to commensurate jobs and schooling? How do we make sure there is an adequate stock of housing outside of the floodplains and flood hazard area, so families can move to safer ground? How do we relocate infrastructure, as well as historical and cultural buildings? Will local governments have enough staff and funding to make the relocation process faster, more transparent, and more supportive to willing sellers in their communities? How do we ensure underserved property owners are neither underrepresented nor overrepresented in managed retreat, so that it does not mimic displacement? How long is long-term in a world with a changing climate: Are we planning for the expected sea level rise by mid-century, eight decades from now, or beyond?

The piecemeal buyout process of today does not truly match the vision of a "managed" retreat from the impacts of climate catastrophe. A managed retreat plan would be coordinated between the government and the community members and decision-makers; it would not be dependent on funding after a single major storm; and it would be based on long-term risk to climate threats. A transformative and forward-focused plan for the way we respond to the hazardous impacts of climate events is needed in just about every country.

18.5 Recommendations

Climate change poses a real and existential threat to many communities, especially those threatened by inundation. It will require funding and engagement from all stakeholders, but communities in many vulnerable places will have to start creating plans for climate adaptation that include the possibility of relocation. To do this right, we believe decision-makers must adhere to the following recommendations:

1. Coastal engineering, such as beach nourishment and hard structures, should be viewed as a temporary or short-term response to sea level rise, erosion, and flooding. At some point, by mid-century in many places, beach nourishment will be indefensible, and seawalls will have destroyed most beaches. By that point or much sooner, managed retreat on a broad community-wide front will be required in some highly exposed areas, regardless of massive and expensive engineering works.
2. Work toward managed relocation, not just retreat. People can move, but where they move to and how best to help them relocate are still understudied. Managed relocation would be a holistic approach that assists the seller of a buyout property not only to move to a less risky home but also considers the community they are moving from, and the community they are moving to, along with job opportunities, family, schooling, and other life necessities to make the transition as successful as possible.
3. Reduce inequity issues in buyout policy and planning for managed retreat. Historically underrepresented communities should be given a voice and seat at the table when buyouts are discussed. We need a planning process for equitable relocation that does more than displace those people whose houses are cheapest to buy and demolish.
4. Prohibit reconstruction of structures that have significant storm damage. This will grow increasingly important as seas rise and storms intensify.

Reconstruction of severely damaged homes only covers up the long-term problem and entrenches the belief that we can build our way out of this.
5. Make the case for rolling easements. Rolling easements may be the additional incentive needed to encourage waterfront communities to accept buyouts. Rolling easements will allow for maintenance of some semblance of a beach for the public. Beachfront property owners can recoup their real estate investment by accepting a buyout before the water rises ever nearer to their property.
6. Instead of demolition, move historically significant buildings back to maintain the character of the area. Not every building can be walled off, raised up, or floodproofed, but history still needs to be preserved. It will take a great feat of ingenuity and capital, but many culturally significant monuments, lighthouses, museums, piers, and houses can be moved out of harm's way.

References

Ajibade, I. (2019). Planned retreat in Global South megacities: Disentangling policy, practice, and environmental justice. *Climatic Change, 157*, 299–317. doi: 10.1007/s10584-019-02535-1

Anderson, R., Patsch, K., Lester, C., & Griggs, G. (2020). Adapting to shoreline retreat: Finding a path forward. *Shore & Beach, 88*(4), 1–21.

Anguelovski, I., Shi, L., Chu, E., Gallagher, D., Goh, K., Lamb, Z., Reeve, K., & Teicher, H. (2016). Equity impacts of urban land use planning for climate adaptation: Critical perspectives from the global north and south. *Journal of Planning Education and Research, 36*(3), 333–348. doi: 10.1177/0739456X16645166

Bakker, A. M. R., Wong, T. E., Ruckert, K. L., & Keller, K. (2017). Sea-level projections representing the deeply uncertain contribution of the West Antarctic ice sheet. *Scientific Reports, 7*, Article 3880. doi: 10.1038/s41598-017-04134-5

Binder, S. B., & Greer, A. (2016). The devil is in the details: Linking home buyout policy, practice, and experience after Hurricane Sandy. *Politics and Governance, 4*(4), 97–106.

Budds, D. (2017). Ike Dike: The $15 billion storm surge barrier Houston can't agree on. *Fast Company*. https://www.fastcompany.com/90138474/ike-dike-the-15-billion-storm-surge-barrier-houston-cant-agree-on

Buser, M. (2020). Coastal adaptation planning in Fairbourne, Wales: Lessons for climate change adaptation. *Planning Practice & Research, 35*(2), 127–147. doi: 10.1080/02697459.2019.1696145

Chadenas, C., Creach, A., & Mercier, D. (2013). The impact of storm Xynthia in 2010 on coastal flood prevention policy in France. *Journal of Coastal Conservation, 18*(5), 529–538.

Chappel, B. (2019). Jakarta is crowded and sinking. So Indonesia is moving its capital to Borneo. NPR. https://www.npr.org/2019/08/26/754291131/indonesia-plans-to-move-capital-to-borneo-from-jakarta

Cutter, S., Emrich, C., Morath, D., & Dunning, C. (2013). Social vulnerability and flood risk management planning. *Journal of Flood Risk Management, 6*, 332–344. doi: 10.1111/jfr3.12018

Egge, N., Feurer, A., Jimenez, R., Neumann, H., Shyamakrishnan, K., & Stamson, N. (2020). Alaskan Arctic Coast economic and environmental characterizations and Port narratives. July 16. U. S. Coast Guard, Office of Standards Evaluation and Analysis. CG-REG-1. 155 pp. https://www.dco.uscg.mil/Portals/9/CG-5R/REG/AAC%20Economic%20and%20Environmental%20Characterizations%20and%20Port%20Narratives.pdf?ver=2020-07-16-020052-427

Eliraz, G. (2020). The many reasons to move Indonesia's capital. *The Diplomat*. https://thediplomat.com/2020/03/the-many-reasons-to-move-indonesias-capital/

Gaul, G. (2019). *The geography of risk*. Farrar, Straus and Giroux.

Gillis, J., & Barringer, F. (2012). As coasts rebuild and U.S. pays, repeatedly, the critics ask why. *The New York Times*. https://www.nytimes.com/2012/11/19/science/earth/as-coasts-rebuild-and-us-pays-again-critics-stop-to-ask-why.html

Griggs, G., & Patsch, K. (2019). California's coastal development: Sea-level rise and extreme events — where do we go from here? *Shore & Beach*, 87(2), 15–28.

Hauer, M.E., Fussell, E., Mueller, V., Burkett, M., Call, M., Abel, K., McLeman, R., & Wrathall, D. (2020). Sea-level rise and human migration. *Nature Reviews Earth & Environment*, 1, 28–39. doi: https://www.nature.com/articles/s43017-019-0002-9.

Hersher, R., & Benincasa, R. (2019). How Federal disaster money favors the rich. *All things considered*. NPR. https://www.npr.org/2019/03/05/688786177/how-federal-disaster-money-favors-the-rich

Horgan, R. (2019). Danish Lighthouse moved back from eroding cliff edge. October 24. *New Civil Engineer*. https://www.newcivilengineer.com/latest/danish-lighthouse-moved-back-from-eroding-cliff-edge-24-10-2019/

Jin, D., Hoagland, P., Au, D. K., & Qiu, J. (2015). Shoreline change, seawalls, and coastal property values. *Ocean & Coastal Management*, 114, 185–193. doi: 10.1016/j.ocecoaman.2015.06.025

Kennedy, M. (2016). Threatened by rising seas, Alaska village decides to relocate. August 18. NPR, The Two-Way. https://www.npr.org/sections/thetwo-way/2016/08/18/490519540/threatened-by-rising-seas-an-alaskan-village-decides-to-relocate

Kimmelman, M. (2017). Jakarta is sinking so fast, it could end up underwater. *The New York Times*. https://www.nytimes.com/interactive/2017/12/21/world/asia/jakarta-sinking-climate.html

Lester, M. (2019). A western Alaska village, long threatened by erosion and flooding, begins to relocate. October 18. *Anchorage Daily News*. https://www.adn.com/alaska-news/rural-alaska/2019/10/19/a-western-alaska-village-long-threatened-by-erosion-and-flooding-begins-to-relocate/

Levine, S. (2020). The radical plan to save the fastest sinking city in the world. *Gen*. https://gen.medium.com/the-fastest-sinking-city-in-the-world-has-a-plan-to-save-itself-5f3ce623bd45

Mason, O. K. (1996). *Geological and anthropological considerations in relocating Shishmaref, Alaska*. Alaska Division of Geological & Physical Surveys. https://www.commerce.alaska.gov/web/Portals/4/pub/Geologic_Anthropological_Aspects_Relocating_Kivalina.pdf

McNeill, R., Nelson, D. J., & Wilson, D. (2014). As the seas rise, a slow-motion disaster gnaws at America's shores. September 4. *The crisis of rising sea levels: Water's edge. Reuters investigates*. https://www.reuters.com/investigates/special-report/waters-edge-the-crisis-of-rising-sea-levels/

Moore, R. (2017). Seeking higher ground: How to break the cycle of repeated flooding with climate-smart flood insurance reforms. *Natural Resources Defense Council.* https://www.nrdc.org/sites/default/files/climate-smart-flood-insurance-ib.pdf

Muñoz, C. E., & Tate, E. (2016). Unequal recovery? *International Journal of Environmental Research and Public Health, 13*(5), 507. doi: 10.3390/ijerph13050507

National Park Service. (2015). Moving the Cape Hatteras Lighthouse. Cape Hatteras National Seashore, North Carolina. Last updated April 14. https://www.nps.gov/caha/learn/historyculture/movingthelighthouse.htm

Office of Senator Richard Burr. (2020, January 13). Burr and Tillis announce $321 million in funding to continue disaster relief efforts: Critical projects at Surf City/North Topsail Beach and Carteret County will receive funding [Press release]. https://www.burr.senate.gov/press/releases/burr-and-tillis-announce-321-million-in-funding-to-continue-disaster-relief-efforts

Ostrander, M. (2020). The village at the edge of the Anthropocene. *Sierra Magazine.* https://www.sierraclub.org/sierra/2020-2-march-april/feature/village-edge-anthropocene-newtok-alaska

Overby, H. (2020). Personal communication. November 4. Forest and Landscape Engineer, Danish Nature Agency.

Pilkey, O. H., Pilkey-Jarvis, L., & Pilkey, K. (2016). *Retreat from a rising sea: Hard choices in an age of climate change.* Columbia University Press.

Pilkey, O. H., & Dixon, K. L. (1996). *The corps and the shore.* Island Press.

Powell, N. (2020). As twin hurricanes converge on the Gulf Coast, $31 billion 'Ike Dike' still in planning stages. *Houston Chronicle.* https://www.houstonchronicle.com/news/houston-weather/hurricanes/article/ike-dike-proposal-hurricane-laura-marco-galveston-15510840.php

Scata, J. (2018). Home buyers face stacked deck to learn of past floods. *Natural Resources Defense Council.* https://www.nrdc.org/experts/joel-scata/home-buyers-face-stacked-decks-learn-past-floods

Schwartz, J. (2018). Surrendering to rising seas. *Scientific American, 319*(2), 44–55. https://www.scientificamerican.com/article/surrendering-to-rising-seas/

Song, L., & Shaw, A. (2018). A never-ending commitment: The high cost of preserving vulnerable beaches. *ProPublica.* https://www.propublica.org/article/the-high-cost-of-preserving-vulnerable-beaches

Titus, J. G. (2011). Rolling Easements Primer. *EPA Climate Ready Estuaries Program.* https://www.epa.gov/sites/production/files/documents/rollingeasementsprimer.pdf

Tockner, K., & Stanford, J. A. (2003). Riverine flood plains: Present state and future trends. *Environmental Conservation, 29*(3), 308–330.

Turner, S., & McFadden, M. (2019). Federal aid is reaching storm-damaged communities too late. *The Hill.* https://thehill.com/blogs/congress-blog/politics/464245-federal-aid-is-reaching-storm-damaged-communities-too-late

Ueland, J., & Warf, B. (2006). Racialized topographies: Altitude and race in Southern Cities. *Geographical Review, 96*(1), 50–78.

U. S. Army Corps of Engineers. (n.d.). Alaska Baseline Erosion Assessment: AVETA Report Summary - Shishmaref, Alaska. *Alaska District, Corps of Engineers, Civil Works Branch.* https://www.poa.usace.army.mil/Portals/34/docs/civilworks/BEA/Shishmaref_Final%20Report.pdf

U. S. Government Accountability Office. (June 2009). Alaska native villages: Limited progress has been made on relocating villages threatened by flooding and erosion. GAO-09-551. Report to Congressional Requesters. 49 pp.

Wing, O. E. J., Bates, P. D., Smith, A. M., Sampson, C. C., Johnson, K. A., Fargione, J., & Morefield, P. (2018). Estimates of present and future flood risk in the conterminous United States. *Environmental Research Letters, 13*, 034023.

Woodruff, S., BenDor, T. K., & Strong, A. L. (2018). Fighting the inevitable: Infrastructure investment and coastal community adaptation to sea level rise. *System Dynamics Society*. doi: 10.1002/sdr.1597

19 Climate-induced relocation as a third wave of response to climate change

Patrick Marchman

> Wealthy people like to live, live near the water
> They pass their real estate down to their sons and their daughters.
> Generations of happy just enjoying the view, but
> When the water starts rising you know what they're gonna do
>
> The rich are gonna move to the high ground
> The rich are gonna move to the high ground
> Holy doodle, look at your town
> The rich are gonna move to the high ground…
> Geoff Berner (2008), "High Ground," from the album Klezmer Mongrels

Where will people go? It is a deceptively simple question, but those four words contain a wealth of meaning. From individual households moving a few miles inland to avoid floodplains or encroaching coastlines to millions on millions of people crossing international borders and reshaping geopolitical realities, it all comes back to that simple question.

Climate-induced relocation – here called "migration" for short – represents a third wave of responses to the hyperobject of climate change: That almost-ungraspable event that is pulling the planet out of the relative stability of the Holocene and into a new state potentially very different from that in which every known sedentary agricultural civilization emerged (Baldwin, 2017). Migration is the movement of people and populations, and it is gaining prominence in the climate discourse as a result of the incomplete nature of two previous response waves – the mitigation of carbon emissions and the adaptation of structures to enhance their resilience in place. This essay argues that migration is a necessary third wave but that migration does not replace either mitigation or in situ adaptation: All three are useful and necessary depending on the particular context.

DOI: 10.4324/9781003141457-19

19.1 First wave: Mitigation

The first major wave of mainstream response to climate change was mitigation – in which the focus is on the reduction of CO_2 emissions and other greenhouse gases. The testimony of James Hansen to the United States Congress in 1987 and the release of then-Senator Al Gore's (1992) book *Earth in the Balance* were seminal moments, and for a while in the 1990s, it looked as though humanity just may have gotten the message. The international community appeared to be coming together through conferences such as the Earth Summit in Rio de Janeiro. More importantly, these processes were getting results. In 1992, the Kyoto Protocols to the United Nations Framework Convention on Climate Change committed state parties to reduce greenhouse gas emissions.

Unfortunately, the world did not continue along this trajectory. Emissions continued to steadily climb. Politicians in most Western nations were unwilling to entertain any serious discussion of limiting industrial growth. The faltering Soviet bloc and emerging superpower China continued to place their bets on large-scale industrial investment, even when this caused catastrophes such as the virtual destruction of the Aral Sea in support of doomed schemes to boost cotton production (Bennett, 2008). Organized efforts by governments, industry and neoliberal economists promoted growth as the only path forward in response to the Club of Rome's prescient and still-accurate 1973 report *The Limits to Growth* (Higgs, 2016). Less developed countries quite reasonably wondered why they should forswear the industrial development path that the West had taken and began ramping up their own economies and emissions, especially as the West balked at placing any limits on itself.

As time went on and it became evident that carbon mitigation strategies premised on the reduction of emissions were not meaningfully slowing the increase in CO_2, critics of an excessive reliance on mitigation began to make themselves heard. Military establishments, most notably that of the United States, began to openly assume that carbon emissions would not be reduced and began to plan accordingly to adapt to a new world (Center for Naval Analysis, 2007).

Mitigation's advocates fought back, labeling adaptation and related efforts as, among other things, "defeatism." To encourage adaptation, they argued, would turn important resources and energy away from preventing or reducing climate change. There were powerful calls to arms from this era, such as Al Gore returning with *An Inconvenient Truth* (David et al., 2006). But Oscars and acclaim did not translate to the necessary action.

19.2 Second wave: Adaptation

Around the time of the financial crash of 2008–2009, with the engines of prosperity sputtering and looking like they might not pull civilization up over the top of the next hill with them, the advocates of a theory called "peak oil" had their brief moment, as oil prices shot skyward and caused gas prices in the

United States and elsewhere to increase as well. First proposed by M. King Hubbert (1956) in the 1950s (whose predictions on the peak of US oil production in the 1970s turned out to be correct), "peak oil" was the idea that the production of oil – and sometimes other fossil fuels – was nearing a peak and that its decline would have huge economic, political, and societal impacts. Astutely observing that modern society largely operated on an oil-based infrastructure, adherents to the theory speculated on the impacts that severe disruptions in transportation and food supply would have on society, and often began to prepare accordingly. A common theme was the need to build connections and make change as a community. This focus on interdependence was in direct opposition to the individualistic tendency in the American counterculture on both the left and the right evident since the 1960s in both the hippie and survivalist movements. Peak oil theorists argued that if society collapsed, no one could make it on their own. Retreat to rural communes or fortified bunkers simply would not work. Hoping for technological wizardry to erase industrial civilization's externalities and prevent climate change was recognized as overly optimistic, though techno-optimism remains present today with evocative visions of carbon capture storage in huge underground labyrinths and global-scale geoengineering that will save us all from climate change and prevent us from needing to change our social or political systems to reduce emissions (Alexander & Rutherford, 2019). Nor would individual consumer responses replace collective response. Changing light bulbs was not going to be enough.

As the discourse around adaptation began to come into its own, it began to overlap with other fields, most notably the constellation of ideas coalescing under the term "resilience." And as the concepts and practices of adaptation and resilience developed and grew, so, too, did attention from big institutions, some of them with substantial resources. The Rockefeller Foundation's 100 Resilient Cities initiative took resilience worldwide by providing funding for chief resilience officers in major cities around the globe. New professional groups like the American Society of Adaptation Professionals (ASAP) emerged to serve as networks for a growing number of practitioners whose work fell into the overlapping Venn diagram of climate adaptation, resilience, and hazard mitigation. As of mid-2020, between 500 and 700 practitioners were active members of ASAP. While statistics are hard to come by and hard to generalize, the largest portion of adaptation practitioners come from either environmental, planning, or disaster backgrounds.

For the United States, Hurricane Sandy making landfall in New York City in 2012 was a defining moment. Flooded subways and blocks on blocks of homes with mildewing carpet in their front yards gave notice that adaptation was not an international or future problem: It had to happen here and now. And in halting, tentative ways, it began to. New York City entertained grandiose plans such as a massive set of gates and locks to hold back the ocean. Houses were elevated, moved, or sometimes destroyed. This had happened in the past in the United States – flooding in the Mississippi River valley in the 1990s resulted in numerous home buyouts (Siders, 2019) and even a few entire community relocations, such as Valmeyer, Illinois (Knobloch, 2005) – but these actions had not been explicitly related to future conditions as a result of climate change.

The concept of adaptation at its core is a response to an external reality. It posits a degree of ontological humility: Something is happening, and we must change in response. However, many actual adaptation initiatives are, of necessity, not exactly "humble," and much as mitigation had, many of adaptation's technical solutions became increasingly baroque. The Netherlands developed an industry in exporting adaptation expertise, and for a while, it seemed as though sea level rise would result in floating cities. Water shortages would be dealt with through massive desalination plants. Failing crops in fast-drying parts of the world would be dealt with by genetic engineering, urban skyscraper farms, and lab-grown meat. Nevertheless, the concept of adaptation began as an effort to change human behavior and built environments in response to climate change.

An alternative approach – not adapting ourselves to the new world but adapting the new world to ourselves – can be seen in emerging high-modernist and high-tech adaptation. Yale University professor James C. Scott's concept of "high modernism" (Scott, 1998) is useful to describe this perspective, and the position is exemplified perhaps most clearly in the Ecomodernist Manifesto (Asafu-Adjaye et al., 2015). Drawing inspiration from Stewart Brand's proclamation in the *Whole Earth Catalog* that "we are gods, and we might as well get good at it" (Brand, 1968), ecomodernism set itself against the supposed "romanticizing" of small, local, lower-tech, and ecologically centered responses to the climate crisis in favor of a robust faith in human reason and humanity's capacity to effectively manage the Earth's biosphere – and, in some cases, even improve on how the Earth has operated over the past 4.5 billion years. Humans have always adapted the natural world to themselves. The very idea of a "pristine" wilderness is not only incorrect but also has been used to justify dispossessions and atrocities for centuries. Yet, an important question that is left unanswered by the Ecomodernists is who "we" and "us" actually are, and whether the people who would be making those existentially important decisions, these "gods," would simply be the spiritual descendants of those who made earlier decisions to take land from others or to prioritize industry over environment.

One notable counterpoint to this techno-solutionist logic is the Transition Movement, started in England by Rob Hopkins. Transition embraces the low-tech, local solutions ecomodernism rejected as overly romantic. The movement shared ideas from "peak oil" regarding a transition to life in a world without fossil fuels, and the main objective of Transitions was to guide local communities in anticipating and preparing for this post-oil world and for the shocks of climate change (Hopkins, 2014). Hopkins framed transition not as defensive but as building community and explicitly improving quality of life and human wellbeing. Still very much active, this view of adaptation in the context of collective action deviated from both the crude "lifeboat ethics" that depicted adaptation as a zero-sum game in which finite resources would pit people against each other (see Hardin, 1974; Potts, 2016) and the planetary engineering by unidentified elites championed by ecomodernism.

Yet, even adaptation has limits. Homes can only be elevated so far, walls built so high, social norms stretched so far (Adger et al., 2008; Dow et al., 2013). Recognizing these limits, states in the United Nations Framework Convention on Climate Change have begun negotiating compensation for "loss and damage" – for harms that will not be avoided by mitigation and cannot be adapted (see Mechler et al., 2010).

19.3 Third wave: Migration

Thus, the third wave of response to climate change: Migration. Climate-induced relocation (here called migration for short) has been happening for a long time, of course. Movement in response to environmental change goes back to the dawn of modern humans and even before, but in its modern form climate-induced migration has been largely discussed as a phenomenon of the Global South or something out of the history books, like the mass migration caused by the Dust Bowl in the United States in the 1930s. But in the 2010s, this began to change. Things started to get real for the developed world. Atmospheric rivers dropped immense amounts of rainfall on areas ill-equipped to absorb it. King tides and sunny-day flooding began to encroach on major cities in affluent nations. For millions of people in the Global North, climate change was no longer the subject of scary documentaries about the future or fundraising drives by earnest nonprofits seeking to help people in far-off places. It became real, immediate, and local: A threat that could flood your house or destroy your new car. Coastlines where countless millions had rushed to build their dream homes were suddenly, perhaps, not a great investment.

Almost inevitably, it seems, it is the poor and those who have always been hard done by who are facing the most immediate hardships. Climate gentrification has become more than a hypothesis as neighborhoods such as Little Haiti in Miami, Florida, are invaded by big money and working-class areas near the water are left behind (see Keenan et al., 2018). Seawalls to protect Charleston, South Carolina, do not quite reach historic African-American neighborhoods. Funds for rebuilding and elevating homes are slow to reach the Lower Ninth Ward in New Orleans, Louisiana. And while right now many of the more-documented examples of climate gentrification are in the United States, studies are ongoing of communities in Australia, China, and elsewhere.

Faced with as-yet-unmitigated climate change, and unable to adapt in place, millions of people around the world are being driven from their homes or are relocating as a result of a cascade of individual decisions. Waves of people making individual choices in the beginnings of an unmanaged retreat from the most vulnerable areas. Very few of those moving would say the reason was "climate change," but many would clearly blame rainfall that did not come, or floods that grew worse each year, or heat so intense it challenged the ability of human physiology to survive, or conflict and wars, to which climate change serves as a threat multiplier (IOM, 2008).

At last, the financial superstructure of global capitalism itself seems to be taking notice. Insurers, reinsurers, bond markets and all the rest have begun in small but significant ways to consider how to appropriately price the risk that climate change poses to everything. And it is here that the most significant drivers of migration in the developed world may be emerging. Insurance and mortgage rates could fundamentally change the coasts (see Colman, 2020; Flavelle, 2020). "Unmanaged retreat" is already happening in a growing number of places, and it is likely to continue and to grow.

19.3.1 Why climate-induced relocation?

The nature of climate change makes relocation to some degree inevitable. It is only a question of how many people will have to move and keep moving. Movement is fundamental to life on Earth – the very earth beneath us migrates at a few millimeters or so a year, and the phrase "glacial slowness" still implies movement (see Duckert, 2013; Shah, 2020). For the vast majority of human history, humans moved as a way of life. Even after the establishment of settlements, many of them, too, continued moving in various rhythms.

So how is it that now, at the very moment when billions can be reasonably expected to move in response to post-Holocene conditions, is it seemingly so difficult for many to take the idea of moving seriously, or if it is taken seriously, to classify it as one more of those things that we dare not talk about too deeply out of the belief that the only true way to drive action on climate is to be relentlessly positive and avoid scaring ordinary people at all costs? Relocation is qualitatively different from either mitigation or adaptation, and it contains within it the potential of disorder (as other chapters in this book can attest). It also calls into question what has been taken as a foundational assumption since the dawn of "sustainable development": That the climate crisis could be managed in such a way to avoid the need to make choices that call into question foundational elements of the modern worldview such as growth, progress, and the idea of development itself.

Western modernity, especially, is immensely powerful, but in many ways it has become less resilient and more brittle than other ways of being. We see glimpses of this brittleness during financial crises or when supply chains are suddenly disrupted during a pandemic and our local grocery shelves are barren. Mostly, we ignore this evidence. We put it out of our minds.

But relocation whispers into everyone's ears. It speaks of displaced persons, of receiving communities where the displaced come to make a new life. It reminds those watching to wonder about their own futures, to remember that "you must change your life," to paraphrase Rainer Maria Rilke. Relocation upends the comfortable habits of settled life. It upends the economic system in which one of the greatest sources of wealth resides in real estate and vulnerable homes.

Shaking the foundations of our worldviews is not necessarily a negative thing, however. Done sensitively and contextually and well, climate-induced relocation can offer the chance to rebuild and reorder communities in different and better ways, not only mitigating the damage that our ways of life inflict on the planet and on the web of life, but also potentially mitigating the social damage of inequality, of poverty, of the unhappiness that wide disparities in status produces throughout the social order. Relocation does not have to be managed as some sort of grand chess game to offer these chances – it can be guided and nudged and incentivized in a thousand different ways that just might deliver positive change.

Relocation can be an opportunity, and it is probably best to look at it as such, because not only is it coming, it is already here. Stories of villagers on the coast of Louisiana or refugees in Myanmar are merely the tiniest hint of what is to come. That future will come much sooner than many would like to think – and do not think for a second that it will be merely something that will happen to unfortunates on the other side of the world, safely removed behind smartphone screens and flatscreen TVs. One study widely reported on by the media in 2020 estimates that by 2070 – just 50 years from the time this is being written – between one and three billion people will live in areas that are "left outside the climate conditions that have served humanity well over the past 6,000 years" (Xu et al., 2020). Stop for a second and sit with that sentence. Billions, not millions – almost half of the entire population of the planet today. And 50 years is not far away. For those of us with children, 50 years is well within what we pray will be their normal lifespans. They may be among those billions.

Imagine your grandchildren, or your children, or even you yourself, having to move, whether because a wildfire consumes everything you own, storm surge damages yours and every house you can see, or simply because when the housing market rolled the dice, you ended up losing. Imagine those dear to you trying to find another place to make a life. Imagine shelter generously provided by a government or a charity, queuing up for donated food, sitting, and waiting in a refugee camp for whatever will happen next. Imagine this happening to people who right now live in comfort unparalleled in history, who never thought it could happen to them. Who never thought it could happen to *you*.

Whole countries are confronting the reality that their ancestral land will slip between the waves or become unlivable due to extreme heat. Refugee camps around the world are filled with people who can no longer live in their family homes. Often, relocation does not look like a disaster movie. It looks like chains of migration as one family member after another packs up for a better job and a better life far away. And before they know it, something new has been created and something old has been lost.

Relocation does not necessarily have to be a successor to mitigation or adaptation, but can become a pillar along with them in humanity's overall response to climate change. The combination of these three pillars in new, creative, and even cunning ways potentially opens a path for other opportunities to build and even add to them. What will that look like? The outlines are fuzzy and indistinct. Seeing them requires imagination and creativity. Maybe a sort of bricolage: What looks like disorder, inefficiency, and extravagance from the outside but only in the way that nature itself sometimes appears disordered and inefficient and gratuitous. Less desirable outcomes are certainly possible, maybe even probable, but they are not foreordained – we can choose to meliorate the worst possibilities and possibly even improve the lives and prospects of millions.

John Steinbeck, who wrote *The Grapes of Wrath*, an American classic of climate migration, in his novel *East of Eden* used a Hebrew word, "timshel," as its essential theme – "thou mayest" (Steinbeck, 1952). Choice may be constrained by a thousand obstacles, but the choices that can still be made make a difference. Another world is still possible. It always is.

19.4 Recommendations

- *Build networks:* The American Society of Adaptation Professional's Climate Migration and Managed Retreat member interest group is only one of several groups that have started to form to engage with relocation. The Climigration Network in the United States, the University of Lieges' Hugo Observatory in Belgium, the Arraigo network in Colombia (Chapter 9), and many smaller groups as well as independent researchers and professionals are all doing great work in this area. These are still early days, however, and the pressures that drive relocation are only intensifying. Increased contacts between kinds of networks and a wariness of institutional imperatives to make sharing and collaboration more difficult will be essential to developing the broad knowledge necessary to foster a new generation of researchers and practitioners as migration in all its forms inexorably grows.
- *Search for new precedents and be open to novelty:* Many of the new populist movements and semi-authoritarian figures that have emerged into the open over the past decade are direct responses to the stresses and inequities of an increasingly globalized economy that increasingly is failing to lift all boats, concentrating wealth into the hands of a tiny few while increasing precarity for so many of the previously privileged. And much of their power has come from a reaction against migration in various forms. The analogies used by demagogues (and others) are those of invasion, of occupations, of unwanted change being forced on ordinary folk – and, too often, migration's advocates dismiss this as a moral failing of those who feel threatened. But migration does not have to be viewed merely as the kind of invasion that brings up uncomfortable collective memories of settler societies and the elimination of cultural diversity in the foundation of so many nation-states. Some countries view migration as a source of strength and vitality to support existing structures, but even beyond that, the increased diversity brought by migration in language, in religion, in custom, in worldview, can be looked at as a positive good. Monocultures in nature tend to be much more unstable, outwardly mighty but brittle and unstable. A kind of permaculture of human society could welcome these diversities, those new and differing worldviews providing new lenses through which to address the challenges faced by societies and increasing overall well-being.

- *Rethinking citizenship:* Climate-induced relocation does not have to take the forms that it has taken in living memory, of immigrants, legal or otherwise, queuing up to leave their pasts behind and assimilate to a new society. As the scale of relocation increases, consideration should be given to the maintenance of cultural bonds and traditions, and even of new and different meanings of citizenship. The Sovereign Military Order of Malta, an organization tracing its origins and continuity to the Knights Hospitaller founded in 1099, maintains to this day a kind of transnational citizenship not based on land or borders. This could be a useful model for countries existentially threatened by rising seas or desertification. Other models include the so-called "Nansen passports" that allowed refugees in the aftermath of World War I the ability to relocate to new countries and build new lives (Robinson, 2020). Much of what we all view as common sense – nation-states with a common flag, common language, and defined borders – is only several hundred years old in Europe, and much younger than that in many other parts of the world. This model of organizing global society was not handed down from on high. Others are possible and might be even more conducive to human flourishing.

References

Adger, W. N., Desai, S., Goulden, M., Hulme, M., Lorenzoni, I., Nelson, D., Naess, L. O., Wolf, J., & Wreford, A. (2008). Are there social limits to adaptation to climate change? *Climatic Change*, 93, 335–354.

Alexander, S., & Rutherford, J. (2019). A critique of techno-optimism. In A. Kaifagianni, D. Fuhs, & A. Hayden (Eds.), *Routledge handbook of global sustainability governance*. Routledge.

Asafu-Adjaye, J., Blomqvist, L., Brand, S., Book, B., DeFries, R., Ellis, E., Foreman, C., Keith, D., Lewis, M., Lynas, M., Nordhaus, T., Pielke Jr., R., Pritzer, R., Roy, J., Sagoff, M., Shellenberger, M., Stone, R., & Teague, P. (2015). *An Ecomodernist Manifesto*. Available at: http://www.ecomodernism.org

Baldwin, A. (2017). Climate change, migration, and the crisis of humanism. *WIREs Climate Change*, 8, e460. doi: 10.1002/wcc.460

Bennett, K. (2008, May). Disappearance of the Aral Sea. *World Resources Institute blog*.

Berner, G. (2008). High Ground. On *Klezmer Mongrels* [album]. Jericho Beach.

Brand, S. (Ed.). (1968). Statement of purpose. *Whole Earth Catalog*.

Center for Naval Analysis. (2007). *National security and the threat of climate change*. Center for Naval Analysis.

Colman, Z. (2020, November). How climate change could spark the next home mortgage disaster. *Politico*.

David, L., Bender, L., Burns, S. (Producers), Guggenheim, D. (Director), Gore, A. (Writer). (2006). *An inconvenient truth* [Documentary].

Dow, K., Berkhout, F., Preston, B., Klein, R. J. T., Midgley, G., & Shaw, M. R. (2013). Limits to adaptation. *Nature Climate Change*, 3, 305–307.

Duckert, L. (2013). Glacier.*Postmedieval*, 4(1), 68–79.

Flavelle, C. (2020, June). Rising seas threaten an American institution: The 30-year mortgage. *The New York Times*.

Gore, A. (1992). *Earth in the balance: Ecology and the human spirit*. Houghton Miffler.
Hardin, G. (1974). Living on a lifeboat. *BioScience, 24*(10), 561–568.
Higgs, K. (2016). *Collision course: Endless growth on a finite planet*. MIT Press.
Hopkins, R. (2014). *The transition handbook: From oil dependency to local resilience*. UIT Cambridge Limited.
Hubbert, M. K. (1956). *Nuclear energy and the fossil fuels*. Presented before the Spring Meeting of the Southern District, American Petroleum Institute, Plaza Hotel, San Antonio, Texas, March 7–9, 1956.
International Organization for Migration (IOM). (2008). *Migration and climate change*. IOM.
Keenan, J., Hill, T., & Gumber, A. (2018). Climate gentrification: From theory to empiricism in Miami-Dade County, Florida. *Environmental Research Letters, 13*, 054001
Knobloch, D. (2005). Moving a community in the aftermath of the great 1993 Midwest flood. *Journal of Contemporary Water Research and Education, 130*(2), 41–45.
Mechler, R., Bouwer, L., Schinko, T., Surminski, S., & Linnerooth-Bayer, J. (Eds.). (2010). *Loss and damage form climate change*. Springer.
Potts, M. (2016). A review of *Overdevelopment, overpopulation, overshoot* by Tom Butler. *Philament, 21*, 83–87.
Robinson, K. (2020). *The ministry for the future*. Orbit.
Scott, J. C. (1998). *Seeing like a state: How certain schemes to improve the human condition have failed*. Yale University Press.
Shah, S. (2020). *The next great migration: The beauty and terror of life on the move*. Bloomsbury.
Siders, A. R. (2019). Managed retreat in the United States. *One Earth, 1*(2), 216–225.
Steinbeck, J. (1952). *East of Eden*. Viking Press.
Xu, C., Kohler, T., Lenton, T., Svenning, J., & Scheffer, M. (2020). Future of the human climate niche. *Proceedings of the National Academy of Sciences, 117*(21), 11350–11355.

20 Waves of grief and anger: Communicating through the "end of the world" as we knew it

Susanne C. Moser

20.1 Introduction

"To retreat or not to retreat" is the very heart of what millions of individuals and thousands of communities across the United States and beyond already do or will have to confront. It is a question that must draw on the best available climate change science and contend as well with finance, law, governance approaches, and a host of technical and logistical aspects of planning and implementing adaptive solutions. Advancements in science make it ever clearer that scores of people located in or near coastal regions, wildfire zones, floodplains, and areas exposed to extreme heat will face extraordinary challenges due to the unprecedented scope, scale, and speed of climate change. In a century from now, we may think of this growing crisis as humanity's greatest public works project and humanitarian challenge, experienced first and foremost by those already marginalized racially, socioeconomically, and politically. We must either defend and maintain in place a human-dominated landscape in perpetuity or return once-inhabited places to the forces of nature. And millions of people must find new homes, jobs, schools, a sense of community, and peace of mind.

The fact that "retreat or not" is a question and not a foregone conclusion *requires* that we talk about it: To frame it, explore it, and understand it. It asks us to depoliticize and debate it, envision and examine our options, organize ourselves, come to decisions, and grapple with their implications. It demands that we find solutions, particularly for those at greatest risk and with the least resources. We must locate our collective resolve, build our strength, and support each other through the process. In short, to retreat or not is a question that cannot be answered without communication and deep engagement. And yet, it is one of the hardest issues to talk about.

Many are stymied by the challenge of how to talk about and "message" retreat or relocation – an utterly unpleasant, even unthinkable possibility (particularly, but not only) in the American mind, and particularly the white, privileged mind. And yet, land-use managers, planners, extension/outreach staff, researchers, technical service providers, and other professionals working in communities already confronting relocation need urgent help in finding pragmatic and feasible ways to communicate about it.

DOI: 10.4324/9781003141457-20

The single most-frequently heard communication concern to date has been over "the right words" to use, what alternative there might be to the "R" word (retreat, relocation, resettlement, and realignment) (Koslov, 2016). This narrow framing of the communication challenge reduces the focus to the precarity of guessing the least-triggering, most-persuasive language, which people hope will remove resistance and guarantee "buy-in." This fails to recognize the bigger need that retreat communication has to answer: Namely, to empathetically and respectfully assist people in making a difficult change that severs their ties to place, community, heritage, and livelihoods (Maldonado et al., 2020). Nor does it recognize that human responses to the specter of retreat are rooted in, and complicated by, underlying political dynamics and racial histories, people's socioeconomic realities, personal and communal aspirations, and beliefs. Retreat communication must deal with people's relational needs and emotions in the face of an overwhelming, intractable, and – for hundreds and more years – unstoppable problem. In short, a more comprehensive look at communication and the process of engagement in the face of relocation is needed which addresses the human needs throughout that process.

This chapter aims to take on this broader challenge: How to communicate about relocation in the full and messy complexity in which it unfolds. While the focus will primarily be on "planned" or "managed" retreat, *un*planned retreat also involves communication. Managed retreat, however, entails the ever-present opportunity to take on communication in a deliberate, respectful, empathetic, culturally appropriate, mindful, and reparative fashion. It offers an opportunity to ease a difficult process for all involved. As such it is not only a utilitarian means, but also a humane act amidst the deep structural, policy, and financial reforms necessary to enable the transformative changes necessary to create fundamentally safer, socially just, and environmentally sustainable living conditions for all.

In the next section, drawing on a broad literature survey, I will sketch out the deeply human needs in a transformation process. Thereafter, I will lay out key tasks that transformative communication needs to address (Moser, 2019) and apply them to relocation.

20.2 Insights on communication and engagement from the retreat literature

20.2.1 The retreat experience

In preparing this chapter, I reviewed 63 peer-reviewed articles, two Masters theses, and four reports on coastal[1] relocation/retreat published between 2000 and 2020 for insights on communication and engagement. I examined what keeps people in place or allows them to move; influences on retreat attitudes and behavior; perceptions and emotional experience of retreat; the language of retreat; the geographic, socioeconomic, and cultural contexts; retreat approaches and governance mechanisms used; and any specifics on the communication and engagement methods deployed.

From this review several important insights ensued. First, the initial literature search yielded thousands of articles on relocation. Managed (coastal) retreat constitutes but a small, if growing subset of this much larger multidisciplinary literature (e.g., voluntary or forced relocation due to urban restructuring, war, or infrastructure development; relocation after disaster; and relocation of workplaces). In that broader relocation literature, there is no notable concern with the word "retreat" or the other "R" words. In these other contexts, the word either serves as a neutral descriptor or is part of a larger (often dominant) narrative of progress, improvement, advance, or safety. This linguistic concern emerges thus as a unique feature of the climate-driven retreat literature. Notably, only one coastal retreat-focused study critically questioned the widespread search for the "right word" to name the process (Koslov, 2016).

Second, the broader literature review also reveals extensive attention to people's psychological experience of such relocation experiences (e.g., well-being, life satisfaction, posttraumatic stress, long-term impacts, and social connectivity) as well as to socioeconomic outcomes (e.g., educational attainment of children, housing, income, wealth, and debt). In contrast, the literature on climate-induced retreat is more narrowly focused on conflict, resistance, and conditions of acceptability, not on the lived experience of going through relocation. Explanatory variables often focus on place attachment, sociodemographic factors, design and implementation of retreat policies, lack of adequate financial assistance, and perceived necessity/belief in climate change. These studies do not reveal much about the actual experience of having to give up one's home and belonging to a place, or the experienced retreat outcomes. While they frequently end with pleas for proactive communication and engagement between governments and affected residents to address the observed resistance (and underlying experiences of fairness and justice), there is little depth to trying to understand people's experiences beyond place attachment.

Third, the vast majority of reviewed retreat studies mentions only negative emotional experiences associated with the process, predominantly resistance, refusal, and conflict. Some even acknowledge relocation can be traumatic and note a range of posttraumatic stress symptoms. The majority of these studies relate those negative emotional experiences to people not wishing to let go of home, property, and community (rather than, say, to legacies of injustice). Others make clear, however, that the emotional experience of retreat is just as much driven by the quality of the retreat process (e.g., duration, clarity of communication between authorities and affected communities, ease of process, availability and sufficiency of financial compensation, and interactions with neighbors). This reinforces the crucial importance of effective communication and engagement throughout the retreat process.

Only a very small number of studies look at positive emotional experiences of retreat. Those are typically associated with the process or outcomes at the vacated and new location (Bazart et al., 2020; Gini et al., 2020; Koslov, 2016). One might hypothesize that either there are many more negative experiences, or the negative experiences draw more research (and media) attention than the positive ones, or both. Importantly, however, negative and positive experiences are not separate; rather, what distinguishes them and what can lead from the more painful to the more positive outcomes is conscious psychological processing, social organizing, good governing, and communicative reframing (Gini et al., 2020; Hanna et al., 2020; Koslov, 2016). Drawing on this literature, Figure 20.1 attempts to capture this full-spectrum emotional landscape and the dynamic processing that can lead to

Figure 20.1 The emotional landscape residents may experience in the course of climate-related retreat.

Source: The author.

either the oft-reported negative outcomes or the less frequently reported positive ones.

Many studies report positive emotions associated with the place in question (e.g., love, care, and attachment), in addition to people feeling more or less at risk from climate-related hazards or having experienced hazardous events there (Agyeman et al., 2009; Burley et al., 2007; Carbaugh & Cerulli, 2013; Moser, 2013). Leaving a place (by choice, necessity, or social pressure) frequently involves a wide range of emotions associated with grief, anxiety, and anger. Depending on circumstances, the quality of the engagement process, and government communication, individuals and communities sometimes understandably resort to defensiveness and can end up with unresolved trauma. Some manage to process their emotions, individually and collectively, reframe, retreat, and arrive at what might be described as posttraumatic growth: In an accepting, grateful, positively transformed state of mind, embracing the new location, community, and life. To the extent the retreat trauma is also racialized, the emotional posttraumatic legacies may or may not get resolved. Deeper systemic changes, restorative justice, and reparative measures would be needed.

Regarding the reframing, Koslov (2016) points to the importance of not just finding the right word to name the relocation process, but reshaping the larger narrative that is being enacted in each case. In one case, the narrative was one of righting the wrongs of inappropriate development; in another, it was about realizing the American dream in a safer location.

20.2.2 Limited recommendations for communication in retreat contexts

The reviewed literature does not offer many concrete recommendations of how to implement an effective communication and engagement strategy for retreat. Only one of the reviewed documents holds any (albeit quite generic) suggestions on what that might look like (Plastrik & Cleveland, 2019). Drawing largely on suggestions from the Climigration Network (https://www.climigration.org/engagement), they offer three communication-specific recommendations, namely to:

- design engagement processes for the emotional and social aspects of considering managed retreat;
- expose, rather than hide, climate risks, vulnerabilities, and the implications for retreat; and
- reframe retreat from a loss to a positive redevelopment and improvement story (p. 25).

These recommendations are diametrically opposed to the cognitive-behavioral scientific literature, which emphasizes mostly education about climate change risks and adaptation options. They also diverge from the dominant instrumentalist practice, which tends to focus on eliciting formal input while conveying technical details of climate change and retreat and the logistics of implementing it so as to obtain buy-in. Their recommendations fail to articulate, however, the fundamental shift required toward a relational approach, that is, one that understands retreat as being about changing relationships between residents and their governments, between coastal residents and their neighbors, and people and the places they are attached to (Hanna et al., 2020). They are also silent on how to proactively address the frequently observed "resistance" and "conflict," implying presumably that if the general guidance offered would be followed, such resistance or conflict would not emerge or be sufficiently addressed. This white, privileged stance clearly does not address the lack of safety experienced particularly by communities of color to express themselves or to feel heard in government contexts. It leaves the lived experience of those faced with retreat – often the already socioeconomically and racially marginalized and politically silenced – if not off the table, at least not addressed directly. Other studies emphasize not just "delivering the right message" but also the importance of "meaningful community engagement" (Bazart et al., 2020; Bukvic & Owen, 2017; Piggott-McKellar et al., 2020), but what this means remains largely underdeveloped.

In summary, the insights gained from the existing literature point to the need to take seriously the psychological and relational dimensions, as well as the legacies of injustice, in retreat communication. Relocation – as one manifestation of potentially transformative change in the face of climate disruption – demands that we think much harder about what it means to communicate in the midst of and in support of a societal transformation (Moser, 2019). To date, climate communication research and practice has been rather silent on this question. Taking the psychological and relational needs of transformative change seriously, I develop ten tasks that an effective communication amidst profound change needs to accomplish (Box 20.1).

Box 20.1 Ten tasks of communication amidst a societal transformation

Naming and Framing the Depth, Scale, Nature and Outline of (Necessary) Change
- Fostering the Transformative Imagination
- Mirroring Change Empathetically
- Helping People Resist the Habit of Acquiescing to Going Numb
- Orienting and Course-Correcting Toward the Difficult
- Distinguishing (and Deconstructing) Valuable (Un)Certainties
- Sense- and Meaning-making of Difficult Change Through Story (Not Facts)
- Fostering Authentic and Radical Hope
- Promoting and Actively Living a Public Love
- Fostering Generative Engagement in Building Dignified Futures For All

Next, I explore what each of these tasks means in the context of retreat to make these proposals specific and actionable for individuals and communities involved in relocation discussions.

20.3 Addressing the psychological and relational needs of transformative change

Taking the psychological and relational needs of people facing profound change seriously does not supplant or invalidate the need for a better understanding of climate change-driven risks. The tasks discussed here should be integral to the educational goal already embraced.

20.3.1 Naming and framing the depth, scale, nature, and outline of (necessary) change

As numerous studies have found, when people do not have a good sense of the nature, persistence, and growing magnitude of the risks they are facing where they live, they are far less willing to entertain retreat (Bazart et al., 2020; Dachary-Bernard et al., 2019). Many still have only a limited understanding of the adaptation options available and what their respective requirements, costs, and other pros and cons might be (Jones & Clark, 2014; Kuruppu & Liverman, 2011). Thus, explaining both climate change and related risks for the locality and offering successively more detailed information on various adaptation options are important elements of retreat communication.

However, as with all communication, it would be a (repeated) mistake to assume that simply delivering technical information will help people make sense, fully grasp the implications of, or motivate, action. Delivering such information can, in fact, backfire if not framed and handled carefully. The local context will matter in important ways here: Has the community recently experienced a damaging climate-related event? Was it a "first" in many years or one of several in sequence? Does climate change already produce disruptions or are the impacts still mostly anticipated? Are residents looking at scientific projections on paper or at water coming up through the stormwater pipes? Is there a foreboding sense of dread within the community or first-hand trauma experiences? Was that trauma experienced against a backdrop of ongoing racial, gender, or socioeconomic trauma? What is the history of development and land use in the community and who has control over it? What are the options? How much time is there to choose?

All these questions will relate to how communities (and the experts and officials supporting them) frame the threat they are facing and the choices available to deal with them. But the prospect of ending residence in a particular location should never be solely framed as an ending. To help people come to embrace such profound change, their losses and sacrifices must be placed into a narrative arc toward a new beginning, toward something that is meaningful and good in the eyes of those affected (Berzonsky & Moser, 2017). Such truth telling must lift people up, be part of a fair process, and restore and repair not just physical environments but human communities as well.

20.3.2 Fostering the transformative imagination

Summoning people to a vision of a better future is easier said than done. What framings, what visions for life after retreat can "help […] audiences hold the immensity of what is unfolding without collapsing under its weight" (Moser, 2019, p. 150) *while* simultaneously opening up people's imagination to a desirable if reality-bound future?

As some authors (Paterson et al., 2020; Plastrik & Cleveland, 2019; Shein et al., 2015) suggest, inviting participation in artful expression can be helpful here, but so are visioning exercises (Ames, 2001; Wiek & Iwaniec, 2014), scenario planning (Amer et al., 2013; Tevis, 2010), or raising "futures consciousness" (Adam & Groves, 2011; Kunseler et al., 2015; Sharpe, 2013).

In the retreat context where time pressures, financial constraints, historical legacies, and narrowly focused governance processes all conspire against an expansive and open-ended imagination, facilitators of local relocation conversations must take this task seriously, not because it may foster buy-in and delay action due to unresolved grievances, but because it offers an opportunity to generate novel ideas with which to create something potentially better than what is being left behind.

Whether it is undoing patterns of inappropriate development, restoring natural environments, establishing important services to the community, or ending forms of exclusion of certain community members, imagining how multiple problems may be solved in the course of relocation can be empowering and encouraging for all involved. Clearly, *just* fostering such an expanded imagination is not enough, as many great ideas may exist but fall "on the cutting room floor" of bureaucracy, narrow agency mandates, and lack of funding. It is essential, however, to ensure deep policy and finance reform.

20.3.3 Mirroring change empathetically

Opening up the imagination to develop a vision of a desirable future does not obfuscate the difficulties of facing present realities or future outlooks (much less the relocation process itself). When adaptation professionals (elected officials, government staff, facilitators, or community leaders) muster the courage to have public conversations about retreat, residents themselves may follow in facing the stark realities of irreversible climate change impacts. Insights from psychology tell us that such serious "diagnoses" must be delivered with empathy, time, and sensitivity to the needs and experiences of those affected (Cunsolo & Landman, 2017; Groopman, 2004; Lonsdale & Goldthorpe, 2012). This is made easier if facilitators of such conversations have grown their own comfort with the emotional territory of climate grief (Moser, 2020a, 2020b) and of racial and socioeconomic injustice and antiracist work. It enables them to be with, and hold safe space for, volatile emotional reactions and mirror these responses empathetically back to those involved.

To date most government officials and professionals involved in relocation are not trained to facilitate such emotion-laden processes. While one answer to this challenge is for them to step back and let community members lead, those that remain engaged may not be trained in trauma-aware risk communication, grief work, and the psychology of transformation. This points to the need for critical professional development training for adaptation professionals (Gilford et al., 2019).

20.3.4 Helping people resist the habit of acquiescing to going numb

Many residents will come to the "retreat or not?" question after dikes, sea walls, levees, vegetation breaks, *and* hearts were broken, after homes were shattered, family possessions lost, and life savings burnt or washed away. News outlets will not only repeat their traumatic experiences *ad nauseum*, often adding racist insult to physical injury; stories just like it will be reported with growing frequency from around the country and world. Such is the nature of exponential change curves rising up against densely developed and redeveloped population centers. The temptation will be to "go numb" to growing devastation and loss. The phenomena of compassion and climate fatigue are already being observed (Figley, 1995; Kerr, 2009).

To meet the changing needs of residents throughout the prolonged relocation process as they prepare to leave and try to gain a foothold in a new location (e.g., financial assistance, government bureaucracies, suitable alternative housing, implications for work, schooling, health care, and social connectivity) requires patience, stamina, and emotional resiliency among all involved: From elected officials to support staff on the ground, to the residents themselves. It is critical to attend to both the logistical and the psychosocial needs in the current and new location. And because relocation is not always a wholesale community process, but often involves only some, there are also the needs of those who stay behind. Several reported cases speak to the lack of infrastructure maintenance, services, and neighborliness, and a sense of having been abandoned and forgotten after other neighbors moved away (Purdy, 2019). These are instances of "numbing" and suggest emotional support is an ongoing need throughout the process. Residents choosing to stay or to leave should be made aware of the physical and psychological consequences of their choices and be asked about their support needs.

20.3.5 Orienting and course-correcting toward the difficult

Relocation is a difficult and prolonged process. Even the purely managerial and technical is fraught with challenges. And yet, those managerial aspects are the easier tasks. Oftentimes, especially in the increasingly climate justice-conscious adaptation field, the relocation process will not be constrained to matters of land-use decisions and hazard mitigation. Given structural, economic, and governance interdependencies, it does and should surface questions about systemic racism, housing injustices, gentrification, exclusion from economic opportunity, and deeper questions about community safety in a disruptive climate context (Agyeman et al., 2009; Gould & Lewis, 2017; Keenan et al., 2018).

Rather than assume this is *not* their work, adaptation professionals must expect and embrace these interrelated issues, by doing their own antiracist work and involving skilled facilitators and community members to navigate these difficult conversations. They offer the opportunity to address a pressing environmental concern as well as interrelated socioeconomic ones. While governance processes and funding cycles will have their own timelines, the deeper reckoning and healing will move along at its own pace. This should be openly recognized and commitments sought from those who created and perpetuated those past harms to stay in the process as long as it takes. This task of course-correcting toward the difficult often will fall to career staff and advocates who stay while elected officials – bound by terms of office – may be unwilling to take on such uncomfortable conversations. And yet, career staff require financial and political backing for their work from those elected officials. Voters will need to hold their representatives accountable for putting their political weight behind this crucial reparative work.

20.3.6 Distinguishing (and deconstructing) valuable (un)certainties

Climate change is an observed phenomenon at present and the laws of physics ensure that it is a future certainty as well. However, how much and how fast impacts will manifest in any one location is scientifically uncertain. Nearly everything else about retreat is fraught with lack of information and knowledge as well. In short, "uncertainty is the fundamental condition for [living into our collective futures]" (Moser, 2019, p. 153).

Thus, one task of transformative communication is to foster curiosity, open minds, and ask questions where simplistic answers or searching for premature certainty would be a disservice. Ideally, such difficult-to-answer questions should not be addressed in the wake of crisis when the impulse is to quickly re-establish normalcy and safety, but in the more spacious process of pre-disaster and adaptation planning. Adaptation professionals will need to finely balance this understandable need for safety in reliable knowledge with opening up conversations that avoid quick resolutions. What will help is a willingness to remain in the discomfort of not knowing and together seek meaningful solutions to the retreat problem. If facilitated well and centered around corrective action, such conversations are generative and healing, rather than stymying progress or fostering dissatisfaction, frustration, and hearsay.

20.3.7 Sense- and meaning-making of difficult change through story (not facts)

Facts do *not* speak for themselves. How they are being read depends on the lenses through which they are viewed. By helping communities face difficult facts, generate a vision of a desirable future, and frame the work at hand, the unfolding story of retreat turns from one of loss and ending into one that makes sacrifice meaningful, renders change from a permanent disaster to a time-limited discomfort, and helps to move people to tolerable and maybe even improved prospects (Braamskamp & Penning-Rowsell, 2018; Kozlov, 2016; Purdy, 2019). Through this meaning-making process, people adopt and rewrite the narrative of their lives, from being victims of oppression or fate to being creators of their destinies (Hodgson, 2007).

The more profound a shift is in our lives, the more likely we are to need such meaning-making and the more likely we are to ask deep questions about what is important and meaningful to us. Experience with relocation processes to date show that what is meaningful is highly dependent on the cultural, historical, social, and individual context of those needing to move (Piggott-McKellar et al., 2020). For indigenous communities (e.g., in the Arctic, Louisiana, or Washington), detaching from ancestral land involves questions of tradition, community, and tribal identity (Bronen, 2014; Maldonado et al., 2020; Sakakibara, 2008). For island communities (e.g., in the Pacific), relocation involves questions of national identity, sovereignty, security, belonging, and economic survival (Adger et al., 2011; Weir & Pittock, 2017). For long-time residents of a working-class neighborhood relocation may become meaningful if matters of community, livelihood, justice, and environmental restoration are centered (Koslov, 2016). All communities faced with retreat, however, now stand in the dynamic context of humanity having irrevocably set global changes in motion. As such humans relocating from areas long occupied will write a unique set of chapters in the larger narrative of the Anthropocene (DeSilvey, 2012).

20.3.8 Fostering authentic and radical hope

Ultimately, retreat communication has to engage with how we source and maintain hope – a grounded, authentic hope (Moser & Berzonsky, 2015). Such hope does not negate the seriousness of the present situation nor the difficult outlook, but it embraces the diagnosis, identifies a worthy goal, charts a path between one and the other, and demands active participation in realizing it, deeply anchored in personal and collective meaning, along with steady and strategic support from others.

Importantly, in the case studies of successful relocation experiences (e.g., in Valmeyer, IL; Oakwood Beach, NY; and Louisa County, IA), people not just found hope and empowerment in a positive vision to help them get through the difficult relocation process but from their own sense of agency. This manifested in control over data, information, and decisions, political organizing, shaping their own story, and succeeding not against nature but against authorities, bureaucracies, and retreat opponents (Gini et al., 2020; Koslov, 2016). "Hope," as Tippett once said, "is a function of struggle" (2016, p. 251). For adaptation professionals, the task here is to not allow the conversation to get stuck in unrealistic optimism or diatribe and despair, even when both the future and the way to get there must be radically re-envisioned (Lear, 2006), but to make that struggle productive, honorable, just, and reparative.

20.3.9 Promoting and actively living a public love

Rethinking one's future may be hardest when parties to a relocation process become stuck in interest politics, old power dynamics, and adversity. Historical legacies and persistent social and economic injustices create an uneven playing field for navigating retreat processes. Communication alone cannot undo these inequities. But communication can either worsen and feed into these old dynamics or it can serve a process of reconciliation and healing. That communicative work might be considered a form of "public love." It must consistently foster a sense of solidarity so that community members feel they are going through the challenge together and they are strengthened and restored, rather than diminished, by it. "Living a public love ... would help community members to learn to go through transformation with an open hand, that is, to approach each other from a stance of giving instead of taking; from a place of gifting instead of expecting" (Moser, 2019, p. 158).

There is no illusion about how challenging this will be. Retreat occurs against the backdrop of a widespread litigation culture in the United States, and corruption and human rights violations everywhere. It is for this reason that I emphasize this task. At-risk communities could enter a vicious cycle of costly lawsuits on top of costly adaptation, leading to time-consuming distraction, economic decline, and fewer options, particularly for those who need it most. Thus, this countercultural task requires not just changes in communication among adaptation professionals and affected residents; it requires re-education of legal professionals and others. It also requires involvement of individuals skilled in conflict transformation. Thus, at heart, this task may be the most challenging of all, even if begun simply: By people stopping how they have always interacted with each other, call on their highest selves, and do something different.

20.3.10 Conclusion: Fostering generative engagement in building dignified futures for all

In the end, the task of communicating with people facing the "end of the world as they knew it" and helping them find a way to a new world is about navigating not just inadequate and unfair policies, archaic bureaucracies, and exclusionary processes, but waves of grief and anger. It is about bridging differences, old rifts, and systemic injustices. It is about tapping, in ourselves as communicators and in others, a desire not just to "be right" but to "do right" by each other, and thus maybe – against the overwhelming odds of an inexorably changing climate – not just "do good" but "be good."

Communication cannot accomplish this alone. But a more effective, more deeply engaging, transformative communication should be an integral, crucially important part of deep policy, structural, and financial reforms. This chapter suggests that a transformative communication that effectively assists the retreat process must:

- stop being preoccupied with finding "the right word" and instead assume that the right language is context-specific and found together with the affected communities;
- approach communication and engagement from a relational perspective;
- pay deep attention to the psychosocial needs of individuals and communities;
- work toward a narrative that draws an arc from an ending to a new, better, and reparative beginning; and
- address climate change risks and adaptation, together with the concurrent and often deep-seated legacies of racial, socioeconomic, gender, and other injustices.

Note

1 The literature review drew largely on literature on *coastal* retreat, although there are examples of other floodplain retreat cases, and there is increasing interest in retreat from extreme heat and wildfire-prone locations, but that literature is still sparse and did not yield insights on communication.

References

Adam, B., & Groves, C. (2011). Futures tended: Care and future-oriented responsibility. *Bulletin of Science, Technology & Society, 31*(1), 17–27. doi:10.1177/0270467610391237

Adger, W. N., Barnett, J., Chapin, F. S., & Ellemor, H. (2011). This must be the place: Underrepresentation of identity and meaning in climate change decision-making. *Global Environmental Politics, 11*(2), 1–25. doi:10.1162/GLEP_a_00051

Agyeman, J., Devine-Wright, P., & Prange, J. (2009). Close to the edge, down by the river? Joining up managed retreat and place attachment in a climate changed world. *Environment and Planning A: Economy and Space, 41*(3), 509–513. doi:10.1068/a41301

Amer, M., Daim, T. U., & Jetter, A. (2013). A review of scenario planning. *Futures, 46*(0), 23–40. doi:10.1016/j.futures.2012.10.003

Ames, S. C. (Ed.). (2001). *Guide to community visioning* (revised ed.). APA Planners Press.

Bazart, C., Trouillet, R., Rey-Valette, H., & Lautrédou-Audouy, N. (2020). Improving relocation acceptability by improving information and governance quality: Results from a survey conducted in France. *Climatic Change, 160*(1), 157–177. doi:10.1007/s10584-020-02690-w

Berzonsky, C. L., & Moser, S. C. (2017). Becoming homo sapiens sapiens: Mapping the psycho-cultural transformation in the anthropocene. *Anthropocene, 20*(Supplement C), 15–23. doi:10.1016/j.ancene.2017.11.002

Braamskamp, A., & Penning-Rowsell, E. (2018). Managed retreat: A rare and paradoxical success, but yielding a dismal prognosis. *Environmental Management and Sustainable Development, 7*(2), 108–136, doi:10.5296/emsd.v7i2.12851

Bronen, R. (2014). Choice and necessity: Relocations in the Arctic and South Pacific. *Forced Migration Review, 45*, 17–21. https://www.fmreview.org/crisis/bronen

Bukvic, A., & Owen, G. (2017). Attitudes towards relocation following Hurricane Sandy: Should we stay or should we go? *Disasters, 41*(1), 101–123. doi:10.1111/disa.12186

Burley, D., Jenkins, P., Laska, S., & Davis, T. (2007). Place attachment and environmental change in coastal Louisiana. *Organization & Environment, 20*, 347–366. doi: 10.1177/1086026607305739

Carbaugh, D., & Cerulli, T. (2013). Cultural discourses of dwelling: Investigating environmental communication as a place-based practice. *Environmental Communication: A Journal of Nature and Culture, 7*(1), 4–23. doi:10.1080/17524032.2012.749296

Cunsolo, A., & Landman, K. E. (Eds.). (2017). *Mourning nature: Hope at the heart of ecological loss and grief.* McGill University Press.

Dachary-Bernard, J., Rey-Vlette, H., & Rulleau, e.B. (2019). Preferences among coastal and inland residents relating to managed retreat: Influence of risk perception in acceptability of relocation strategies. *Journal of Environmental Management, 232*, 772–780, 10.1016/j.jenvman.2018.11.104

DeSilvey, C. (2012). Making sense of transience: An anticipatory history. *Cultural Geographies, 19*(1), 31–54. doi:10.1177/1474474010397599

Figley, C. R. (Ed.). (1995). *Compassion fatigue: Coping with secondary traumatic stress disorder in those who treat the traumatized.* Taylor & Francis.

Gilford, D., Moser, S., DePodwin, B., Moulton, R., & Watson, S. (2019). The emotional toll of climate change on science professionals. *Eos, 100*, December 6. doi:10.1029/2019EO137460

Gini, G., Cardoso, T. M., & Ramos, E. P. (2020). When the two seas met: Preventive and self-managed relocation of the Nova Enseada community in Brazil. *Revista Migraciones Forzadas*, (64), 33–36.

Gould, K. A., & Lewis, T. L. (2017). *Green gentrification: Urban sustainability and the struggle for environmental justice* (1st ed.). Routledge.

Groopman, J. (2004). *The anatomy of hope: How people prevail in the face of illness.* Random House.

Hanna, C. J., White, I., & Glavovic, B. (2020). The uncertainty contagion: Revealing the interrelated, cascading uncertainties of managed retreat. *Sustainability, 12*, 736. doi:10.3390/su12020736

Hodgson, R. W. (2007). Emotions and sense making in disturbance: Community adaptation to dangerous environments. *Human Ecology Review, 14*(2), 233–242. https://www.jstor.org/stable/24707709

Jones, N., & Clark, J. R. A. (2014). Social capital and the public acceptability of climate change adaptation policies: A case study in Romney Marsh, UK. *Climatic Change, 123*(2), 133–145. doi:10.1007/s10584-013-1049-0

Keenan, J. M., Hill, T., & Gumber, A. (2018). Climate gentrification: From theory to empiricism in Miami-Dade County, Florida. *Environmental Research Letters, 13*(5), 054001. doi:10.1088/1748-9326/aabb32

Kerr, R. A. (2009). Amid worrisome signs of warming, 'climate fatigue' sets in. *Science, 326*(5955), 926–928. doi:10.1126/science.326.5955.926

Koslov, L. (2016). The case for retreat. *Public Culture*, 28(2), 359–387. doi:10.1215/08992363-3427487

Kunseler, E.-M., Tuinstra, W., Vasileiadou, E., & Petersen, A. C. (2015). The reflective futures practitioner: Balancing salience, credibility and legitimacy in generating foresight knowledge with stakeholders. *Futures*, 66, 1–12. doi:10.1016/j.futures.2014.10.006

Kuruppu, N., & Liverman, D. (2011). Mental preparation for climate adaptation: The role of cognition and culture in enhancing adaptive capacity of water management in Kiribati. *Global Environmental Change*, 21(2), 657–669. doi:10.1016/j.gloenvcha.2010.12.002

Lear, J. (2006). *Radical hope: Ethics in the face of cultural devastation*. Harvard University Press.

Lonsdale, K. G., & Goldthorpe, M. (2012). *Anticipating coastal change: Engaging our senses of loss and hope*. Paper presented at the Adaptation Futures 2012 conference, Tucson, AZ.

Maldonado, J., Marino, E. & Iaukea, L. (2020). Reframing the language of retreat. *Eos*, 101, November 10. doi:10.1029/2020EO150527

Moser, S. C. (2013). Navigating the political and emotional terrain of adaptation: Community engagement when climate change comes home. In S. C. Moser & M. T. Boykoff (Eds.), *Successful adaptation to climate change: Linking science and policy in a rapidly changing world* (pp. 289–305). Routledge.

Moser, S. C. (2019). Not for the faint of heart: Tasks of climate change communication in the context of societal transformation. In G. Feola, H. Geoghegan, & A. Arnall (Eds.), *Climate and culture: Multidisciplinary perspectives of knowing, being and doing in a climate change world*. Cambridge University Press.

Moser, S. C. (2020a). To behold worlds ending. In S. Kaza (Ed.), *A wild love for the world: Joanna Macy and the work of our time* (pp. 79–88). Shambhala.

Moser, S. C. (2020b). The work after "It's too late" (to prevent dangerous climate change). *Wiley Interdisciplinary Reviews: Climate Change*, 11(1), e606. doi:10.1002/wcc.606

Moser, S. C., & Berzonsky, C. (2015). *Hope in the face of climate change: A bridge without railing*. Working Paper. Santa Cruz, CA: Susanne Moser Research & Consulting.

Paterson, S. K., Le Tissier, M., Whyte, H., Robinson, L. B., Thielking, K., Ingram, M., & McCord, J. (2020). Examining the potential of art-science collaborations in the anthropocene: A case study of Catching a Wave. *Frontiers in Marine Science*, 7, 13. doi:10.3389/fmars.2020.00340

Piggott-McKellar, P. J., McNamara, K. E., & Nunn, P. D. (2020). A livelihood analysis of resettlement outcomes: Lessons for climate-induced relocations. *AMBIO - A Journal of the Human Environment*, 49(9), 1474–1489. doi:10.1007/s13280-019-01289-5

Plastrik, P., & Cleveland, J. (2019). *Can it happen here? Improving the prospect for managed retreat by US cities*. Boston, MA: Innovation Network for Communities.

Purdy, B. (2019). *Planning and design scenarios for equitable outcomes in managed retreat* (Master's Thesis). Massachusetts Institute of Technology, Cambridge, MA. Retrieved from https://hdl.handle.net/1721.1/123932

Sakakibara, C. (2008). "Our home is drowning": Inupiat storytelling and climate change in Point Hope, Alaska. *Geographical Review*, 98(4), 456–475. doi:10.1111/j.1931-0846.2008.tb00312.x

Sharpe, B. (2013). *Three horizons: The patterning of hope*. Fife, Scotland: International Futures Forum.

Shein, P. P., Li, Y.-Y., & Huang, T.-C. (2015). The four cultures: Public engagement with science only, art only, neither, or both in museums. *Public Understanding of Science, 24*(8), 943–956. doi:10.1177/0963662515602848

Tevis, R. E. (2010). Creating the future: Goal-oriented scenario planning. *Futures, 42*(4), 337–344.

Tippett, K. (2016). *Becoming wise: An inquiry into the mystery and art of living.* Penguin Press.

Weir, T., & Pittock, J. (2017). Human dimensions of environmental change in small island developing states: Some common themes. *Regional Environmental Change, 17*(4), 949–958. doi:10.1007/s10113-017-1135-3

Wiek, A., & Iwaniec, D. (2014). Quality criteria for visions and visioning in sustainability science. *Sustainability Science, 9*(4), 497–512. doi:10.1007/s11625-013-0208-6

Interlude 7 *Dialogue of the Shattered*

Martha Lerski

Dialogue of the Shattered was created out of discarded marble that one sculptor had thrown out when the discovery of a crack led him to determine that the altered material was no longer workable. My decision to assess the material as it currently existed, respecting its innate qualities but harnessing its new potentials, is typical of the direct carving method which guides me in allowing my materials (stone or wood) to dictate what story or shape will ultimately unfurl.

First, accepting that the stone had a deep fissure that would cause the material to separate, I deliberately broke the piece in two.

At the time that I was carving this, I was undergoing a divorce and was raising a child on my own, a personal displacement. I chose to create a three-dimensional dialogue between two separate pieces rather than dwelling on what was broken – signaling possibilities at times of change and separation. As I write this today, while sheltering in place from a new hazard, I wonder whether common struggles throughout the globe relating to this coronavirus pandemic may unleash imagination, sparking alternate approaches and orientations.

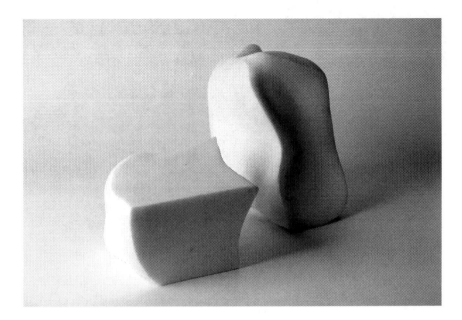

Figure Int 7.1 Dialogue of the Shattered, marble (view "a" of the sculpture by Lerski); photos by Javier Agostinelli; portrayed courtesy of a private collection.

Figure Int 7.2 Dialogue of the Shattered, marble (view "b" of the sculpture by Lerski); photos by Javier Agostinelli; portrayed courtesy of a private collection.

Index

A
Aalbersberg, W. 115, 119
Abel, G. J. 5
Abel, K. 252
Abrar, C. 71
Adam, B. 280
Adams, H. 118
adaptation 264–267
adaptive social protection 182, 183
Adger, W. N. 35, 49, 118, 121, 160, 183, 194–197, 199, 204, 267, 283
Afifi, T. 2, 3
Agenda for the Protection of Cross-Border Displaced Persons in the Context of Disasters and Climate Change 41
Aggarwal, S. P. 68
Aguilar Jr, F. V. 181
Agyeman, J. 277, 282
Ahmad, B. 49
Ahmed, S. 218
Ajibade, I. 2–6, 26, 195, 199, 254
Al-Arian, L. 51
Alaska Native communities 196–197, 255
Albert, J. R. G. 180
Albert, R. 26
Albert, S. 2, 3
Albouy, D. 144
Alexander, S. 265
Ali, J. 5
Ali, S. 37
Alix-Garcia, J. 55
Allen, M. 21
Alvarez, M. K. 3
Alves, A. R. 221, 222
America, climate-induced relocations in *see* United States, climate-induced relocations in
American Society of Adaptation Professionals (ASAP) 265

Amer, M. 280
Ames, S. C. 280
Amit, R. 71
Amoyaw, B. 184
Anderson, K. 218
Anderson, R. 257
Ángel, A. 127
Anguelovski, I. 251
Apte, J. S. 4
Aranda, C. L. 146
Arnall, A. 2, 88, 89
Arnell, N. W. 35, 160, 183
Aronson, E. 23
ARRAIGO platform 128
Aryoso, D. 103n1
Asafu-Adjaye, J. 266
Ash, J. 2, 3
Asis, M. M. B. 181, 184
Assam, migration crisis in 67; breaking the borderscape 74–76; climate migration and borderscapes 71–72; displacement and migration along banks of Brahmaputra 68–71; migration from Bangladesh 67, 71; recommendations 76; surveys and resettlement 72–74
Assam State Disaster Management Authority (ASDMA) 69
Atuyambe, L. M. 51
Au, D. K. 250
Auffray, J. 113
Avelino, J. E. 1
Azad, A. K. 73

B
Bagasao, I. F. 181
Bailey, S. 180
Bainimarama, F. 116
Baja, Kristin 7, 19–30
Bakewell, O. 35

Bakker, A. M. R. 250
Baldwin, A. 263
Banerjee, S. 52
Banivanua Mar, T. 115
Baptiste, N. 200
Barber, P. G. 181
Barbier, B. 36
Bardsley, D. K. 2
Barman, S. 68
Barnett, J. 35, 37, 113, 118, 121, 194–196, 199, 204, 283
Barringer, F. 250
Barr, M. 201
Barry, B. 4
Barthel-Bouchier, D. 200
Bartlett, A. 55
Bassetti, F. 183
Bates, P. D. 250
Battistella, G. 181
Baum, H. 199
Bayliss-Smith, T. P. 119
Bazart, C. 276, 278, 279
Bazeyo, W. 51
beach nourishment 248–249, 258
Beccato, M. A. B. 218
Bedford, R. 115, 119
Beech, H. 52
Begum, A. 5
Bello, W. 181
Bello, W. G. 181
Bender, L. 264
BenDor, T. K. 250
Bengali migrants, in Assam 67
Benge, L. 5, 9
Benincasa, R. 255
Benn, D. K. 9–10, 163–175
Bennett, J. A. 115
Bennett, K. 264
Benn, S. 166, 168
Berchin, I. I. 40
Berkhout, F. 267
Bernard, D. M. 169
Berner, G. 263
Bernzen, A. 35
Bertana, A. 118, 172
Bertocchi, G. 25
Berzonsky, C. L. 279, 283
Bezner-Kerr, R. 5, 6
Bhandari, G. 152
Bharwani, S. 37, 39
Biermann, F. 40
bilateral partnerships 43
Billah, M. 71

Biloxi-Chitimacha-Choctaw tribe, of Isle de Jean Charles (IDJC), Louisiana 25–26
Binder, S. B. 253
Bindi, M. 20
Bird, D. 1, 2, 36
Birk, T. 36, 38
Biswas, A. 152
Black, Indigenous, and people of color (BIPOC) communities 20; *see also* United States, climate-induced relocations in; buyout programs and 21–22; relocation into high-risk areas 25–26
Black, R. 35, 71, 154, 157, 160, 183
Blahůtová, K. 71
Blaikie, P. 58
Blocher, J. 117
Blomqvist, L. 266
Boas, I. 40
Bogardi, J. 113
Bogota, relocation for landslide-affected communities in 127, 138–139; ARRAIGO, concept of 128, 129; informal settlements in risk conditions 127, 128; mitigable and nonmitigable risks 127, 134–138; recommendations 138–139; regulations in Bogota 127–128; risk due to climate change 127; struggles over rights, compensation and housing 129–134; voices of Arraigo 128–138
Böhmelt, T. 5
Bonifacio, G. T. 188
Bonnemaison, J. 116
Book, B. 266
Boon, H. 1, 2, 36
Boruff, B. 5, 9
Boseto, D. 2, 3
Bourne, S. 171
Bouwer, L. 267
Bower, E. 2, 3
Box, P. 1, 2
Boyd, R. 25
Braamskamp, A. 283
Brahmaputra River, periodic displacement along banks of 67, 68; and disaster management policy 69; immediate relief to communities 69; land in Majuli 72; restitution and resettlement assistance 69; shrinking of Majuli due to erosion 68, 70; state cadastral survey 69, 70, 74; unmapped lands 69–70, 72–73

Braithwaite, J. 56
Brand, S. 266
Braun, B. 35
Bravo, M. T. 156
Brazilian Institute of the Environment and Renewable Natural Resources (IBAMA) 225
Bremner, L. 52
Bridle, J. R. 49
Bronen, R. 2, 3, 121, 196, 197, 204, 283
Brookfield, M. 119
Brown, K. 195, 196, 199, 204
Brown, S. 20, 39, 49
Bruun, P. 168
Bryan, C. 10, 180–190
Budds, D. 250
Build It Back (housing recovery program) 144
Bukvic, A. 278
Bullard, R. D. 4, 24
Bumbery, R. 166, 168
Bumpus, A. 113
Buncombe, A. 51
Burkett, M. 37, 39, 199, 252
Burley, D. 277
Burns, S. 264
Burton, I. 7, 34–44
Buser, M. 251
bushfires, in Australia 232–233, 236; see also Grantham, relocation of
Butterbaugh, L. 5
buyout program 21–22, 142–145, 150, 200–201, 252, 253, 255
Bynoe, P. 170

C
Caldwell, E. 36, 37
Call, M. 252
Calzado, J. R. 181
Camailakeba, M. 115
Camilloni, I. 20
Campbell, J. 36, 38, 113, 115
Campbell, L. K. 108
Campos, A. 128
Cancun Agreement 41
Cannon, T. 58
Canziani, O. F. 37
Cape Hatteras Lighthouse 256, 257
Carbaugh, D. 277
carbon mitigation strategies 264
Cardenas, K. 3
Cardoso, T. M. 217, 218, 221–223, 225, 276, 283

Carrizosa, J. 137
Carter, T. 184
Cartier, K. M. 21
Cartwright, A. 138
cash/in-kind transfer 183, 190
Castro August, A. 21
Cerulli, T. 277
Chadenas, C. 251
Chadwick, C. 1
Chapin, F. S. 121, 194, 199, 204, 283
Chapman, R. 2
Chappel, B. 251
Charan, D. 116
Chen, J. 51
Chu, E. 138, 251
Citizenship (Amendment) Act (CAA), India 71
Clark, J. R. A. 279
Clayton, S. 173, 174
Cleveland, J. 277, 280
climate adaptation strategy 1
climate change 113, 156; *see also specific case study* dispossession 41; impact on culture 194; and lives of people on Ghoramara Island 152–155; and migration 35; refugee camps, effects on 49–51
climate change, wave of responses to 263; adaptation (second wave) 264–267; migration (third wave) 267–270; mitigation (first wave) 264; and recommendations 270–271
climate-induced migrants 40
climate-induced migration, reducing maladaptation in 43
climate-induced relocation 263, 267–270; *see also* migration; language for 24–29; and process 20–24
climate niche, effect of rising temperatures on 49
climate refugees 40, 114, 146
climate-related disasters 1
climate-related resettlement, in Malawi *see* Malawi, climate-related resettlement in
Climigration Network 277
coastal engineering 247, 248, 258
Codjoe, S. 36
cognitive dissonance 23
Collins, L. B. 4
Collyer, M. 154, 157, 160
Colman, Z. 268
Comardelle, C. 198

Comberti, C. 196, 204
communication for relocation 273–274; course-correcting toward the difficult 281–282; distinguishing valuable (un)certainties 282; fostering authentic and radical hope 283–284; fostering transformative imagination 280; helping people resist habit of going numb 281; insights from retreat literature 274–279; mirroring change empathetically 280–281; naming and framing depth, scale, nature, and outline of (necessary) change 279–280; promoting and actively living a public love 284; recommendations specific to 277–278; sense- and meaning-making of difficult change through story 282–283; ten tasks to accomplish by 278; transformative 278–285
communication, in recovery effort 236–237, 240
community in recovery process 239
community meetings 237, 240
community needs, in recovery effort 237
1977 Community Reinvestment Act 145
community-supported relocation 29
Conference of the Parties (COPs) 41
Connell, J. 3, 115, 116
Connolly-Boutin, L. 49
Connolly, N. 145
consultation protocol 229
Conway, D. 49
Cooke, B. 89
Cook, I. 41
Corner, A. 199, 200
Correa, E. 88, 89
Costa, C. 128
cost-benefit analysis 255
Cottrell, A. 1, 2, 36
Crate, S. A. 204
Crawford, J. 146
Crawley, H. 180
Creach, A. 251
Crenshaw, K. 6
Crichton, R. 1
Cronin, V. 2
cultural identities, loss of 37–38
cultural values, role in climate-based relocation 194, 255; community-driven and community-responsive processes, support for 203–204; community-driven relocation based on identity and power 199–201; culture and values 194–195; justice and power 196–197; policy makers/practitioners and 202–204; recommendations 205; relocation as multiscalar process 195
Cunningham, M. 146
Cunsolo, A. 280
Cutter, S. L. 182, 255

D
Dabelko, G. D. 35
Dachary-Bernard, J. 279
Dai, A. 51
Daim, T. U. 280
Dalrymple, O. K. 169
Daniel, R. K. 163, 168, 169
Danny, K. 164, 165
Darby, M. 39
Das, D. 70
Das, S. 3, 118
David, L. 264
Davies, M. 183
Davis, I. 58
Davis, J. 4
Davis, P. 37
Davis, T. 277
de Andrade, J. B. S. O. 40
de Campos, R. S. 5, 118, 156, 157
DeFries, R. 266
De Guzman, M. 181
Dela Rosa, Nikki 8, 99–103
Denardin, V. F. 221, 222
DePodwin, B. 281
Dercon, S. 183
Desai, S. 267
Deshpande, T. 138
DeSilvey, C. 283
Des Roches, S. 5
Devereux, S. 182, 183
Devet, R. 189
Devictor, X. 49
Devine-Wright, P. 277, 282
Diaz, C. 128
Dickinson, T. 7, 34–44
Dickson, E. 128
Diedhiou, A. 20
Diegues, A. C. 217, 224
disaster restitution 69
disaster risk creation 39, 42
displacement 34; *see also specific case studies* climate as driver of 35; climate-related disasters and 1

Displacement Solutions 88
distributive justice 4; planned climate relocation and 54–56
Dixon, K. L. 248
Djalante, R. 20
Docena, H. 181
Doevenspeck, M. 2
Doherty, T. J. 173, 174
Dokken, D. J. 37
Dolan, R. 79
Douglas, B. C. 168
Dow, K. 267
Drakes, O. 168, 170, 171
Drolet, J. 182
Dube, T. 5
Duckert, L. 268
Duží, B. 35, 36, 71
Du, J. 1, 2
Dumaru, P. 119
Dunham, J. 184
Dunning, C. 255
Dun, O. 2, 88, 89, 115
Dutta, M. K. 68
Dyson, K. 5

E
Earth in the Balance (Al Gore) 264
Earth Summit in Rio de Janeiro 264
Ebi, K. L. 20
ecological justice 4
Edghill, S. 166, 168
Ediau, M. 51
Edinboro, E. 170
Edwards, J. 116
Egge, N. 255
Eglash, A. 55
Ehrlich, G. 144
El-Hinnawi, E. 40
Eliraz, G. 251
Ellemor, H. 121, 194, 199, 204, 283
Ellis, E. 266
Eltinay, N. 51
emotional landscape, in climate-related retreat 275–277
Emrich, C. 255
Engelbrecht, F. 20
Enseada da Baleia, preventive and self-managed relocation of 217–218, 228–230; catamaran accident 223; cyclone and 223; emotions in 217–218; Malaquias death, impact of 218–220; relationship with trees 224; relocation of community 223–227; women as leaders of community 221–223, 225
Enseada da Baleia Residents Association (AMEB) 222
environmental justice (EJ) 4
environmental refugees 40
Eriksen, S. 37, 39
Esteban, M. 1
Executive Order 9066 (Roosevelt Administration) 27–28

F
Faist, M. 115
Farbotko, C. 2, 40, 114, 115, 117
Fargione, J. 250
Feagan, M. 6
Federal Emergency Management Agency (FEMA) 81, 83, 253
Fernando, N. 3, 121
Ferris, B. 2
Ferris, E. 3, 42, 113, 156
Feurer, A. 255
Field, C. B. 1, 2, 21
Figley, C. R. 281
Fiji, planned relocation in 113, 121–122; climate change adaptation 113–114; controlled resettlements and changes 115; decision-making in rural communities 120; internal relocation in Fiji 113–114, 117; mobility patterns in island societies 114–115; Planned Relocation Guidelines 115–117; and recommendations 122
Filipino migrant workers, in Canada 180, 184
Fischer, R. 19
Flavelle, C. 78, 268
Flick, R. E. 3
Flood Factor 81
flood maps 81
floodplains 247
flood risk prevention, by French government 251
Flores, D. 199
Fonmanu, K. R. 117
forced relocation 21–22, 28
Foreman, C. 266
Forster, P. 88
Forsyth, A. 2
Fountain, H. 51
Fox, M. 113
Fraser, A. 9, 127–139, 138

Fraser, E. 88
Fuchs, S. 2, 3
Fuentes, T. L. 5
Fuller, P. 38
Füssel, H. M. 49
Fussell, E. 252

G
Gallagher, D. 251
Galvez, M. 146
Garcia, J. 40
Gascoigne, C. 36, 37
Gaul, G. 22, 250
GCR *see* Global Compact on Refugees (GCR)
Gebauer, C. 2
Geddes, A. 35, 160, 183
George, S. 51
Geraghty, P. 115
German Development Agency (GIZ) 116
Gharbaoui, D. 117
Ghoramara Island, study on 152; climate change and lives of people 152–155; feasible and just ways for trapped populations 160–161; financial hardships by environmental changes 154; inhabitants trapped 154; loss of landmass due to coastal erosion 152–153; marginalized and vulnerable population 157; plea of people for relocation 155–160; recommendations 161; social attachment to place 158–159; trapped population 154, 157, 160
Ghosh, T. 118, 152
Gibbs, M. 25
Gibson, C. R. 114
Gilford, D. 281
Gillis, J. 250
Gini, G. 10, 217–230, 276, 283
Glaas, E. 35, 121
Glavovic, B. 276, 278
Global Compact for Migration 35
Global Compact for Safe, Orderly and Regular Migration (GCM) 40, 41
Global Compact on Refugees (GCR) 40, 55–56
Goebel, A. 89
Goh, K. 251
Goldsmith, M. 115
Goldthorpe, M. 280
Goodkind, A. L. 4

Good Neighbor Stormwater Park, in North Miami 84
Gore, A. 264
Goswami, P. D. 70–71, 73
Goulden, M. 267
Gould, K. A. 282
government-funded buyouts 83, 201
Govil, R. 3
Grand Coulee Dam project, in Washington 27
Grannis, J. 195
Grantham, relocation of 233–239
The Grapes of Wrath (John Steinbeck) 270
greenhouse gas emissions 264
Greer, A. 253
Greiving, S. 1, 2
Griffiths, A. 3
Griggs, G. 251, 257
Grinham, A. 2, 3
Gromilova, M. 174
Groopman, J. 280
Groves, C. 280
Guggenheim, D. 264
Guiot, J. 20
Gumber, A. 267, 282
Guthrie, P. 2
Guyana's vulnerable coastal communities, resettlement for 163; Almond Beach community 163–165; 2001 Climate Change Action Plan 172; Climate Resilient Strategy and Action Plan 172–173; community losses and uncertainty 167–168; disruptive climate events and few relocation prospects 165–167; little funds for social assistance 168; Mahaica residents in Region V and 171; participatory decision-making in resettlement process 173–174; Plastic City in Region II and 170–171; plight of vulnerable coastal settlements 170–172; priority resettlement needs 173; recommendations 175; sea defense system 168–169; sea level rise and erosion of beach 168; Shell Beach Protected Area 163; strategies for resettlement 173, 174; turtle monitoring activities 164, 165
Gwilliam, M. 119

H
Haeffner, M. 2, 26, 199

Hagelman III, R. R. 3
Hagelman, R. R. 3
Hajra, R. 152
Hall, N. 41
Hammond, L. 3
Hamza, M. 5, 6
Hanna, C. J. 276, 278
Hardin, G. 266
Hardy, R. D. 19
Harris, A. 26, 84
Harris, L. 138
Harris, N. C. 5
Hart, G. 2
Harvey, P. 180
Hassani-Mahmooei, B. 71
Hastrup, K. 38
Hauer, M. E. 252
Hayama, A. T. 222, 223
Haynes, K. 1, 2, 36
Hazard Mitigation Grant Program (HMGP) 253
Hazra, S. 3, 118, 152
Heffron, R. 4
Herrmann, V. 25
Hersher, R. 255
Hewitt, K. 58
Heynen, N. 19
Hickey, S. 89
Higgs, K. 264
"high modernism," concept of 266
Hijioka, Y. 20
Hill, J. D. 4
Hill, T. 267, 282
Hinkel, J. 39
Hino, M. 1, 2, 21, 156
historic buildings, moving of 256, 259
Hita, S. 9, 127–139
Hoagland, P. 250
Hoang, T. 6
Hodgson, R. W. 283
Hoegh-Guldberg, O. 20
Holling, C. S. 80
Holloway, K. 51
Holm-Nielsen, N. 128
Honolulu, Hawaii 252
Hopkins, R. 266
Horgan, R. 256
housing choice voucher 146
Hovelsrud, G. K. 35
Huang, M. C. 182
Huang, S. 181
Huang, S. M. 3
Huang, T.-C 280

Hubbert, M. K. 265
Hulme, M. 156, 267
Hurricane Harvey 22
hurricane relief funding 249
Hurricane Sandy, planned retreat programs after 142–150; buyout programs 142; displaced tenants and challenges 142–143, 145–148; expensive housing outside of floodplain 146–148; housing access, limitations to 145; housing and climate risk 143–145; sociopolitical system and vulnerability 149; Uniform Relocation Act of 1970 and 143, 147, 149

I
Iaukea, L. 274, 283
Ike Dike 250
illusion of inclusion 204
Indian Removal Act, 1830 25, 27
inequity issues in retreat 254–255, 258
Ingram, M. 280
Inks, K. 8, 67–76
institutionalized social protection strategy 182
Intauno, S. 5
international agreements, on migrants 40–43
International Development Assistance 43
Isacoff, R. 10, 194–205
Islam, N. 49
Islam, R. 57
Isle de Jean Charles, disempowerment in 197–200
Issar, S. 160
Iwaniec, D. 280

J
Jacob, D. 20
Jacot Des Combes, H. 117
Jain, S. 174
Jakarta, Indonesia 251
Jakimow, T. 89
Jalata, A. 25
Jamero, Ma. L. 1
Jana, A. 152
Janif, S. Z. 115
Jenkins, J. C. 35
Jenkins, P. 277
Jessee, N. 195, 198, 199, 202, 205n1
Jetter, A. 280
Jimenez-Magdaleno, K. 21
Jimenez, R. 255

Jin, D. 250
Johnson-Bhola, L. P. 169, 170, 172
Johnson, D. 170
Johnson, K. A. 250
Johnson, W. 25
Johnston, E. M. 2, 21
Jones, N. 279
Jordan Compact for Syrian refugees 42
Juhola, S. 35, 121

K
Kahrl, A. W. 21
Kalamadeen, M. 173
Kälin, W. 2
Kandaswamy, S. V. 163
Kanemasu, Y. 119
Kaplan, M. 115
Kapoor, I. 89
Karabanow, J. 182
Karambiri, H. 36
Karanth, K. K. 174
Katonivualiku, M. 2, 6
Kaufman, W. 79
Kaur, M. 116
Keenan, J. M. 21, 267, 282
Keith, D. 266
Keller, K. 250
Kellet, J. 173, 174
Kelman, I. 23, 35, 36, 71, 113
Kennedy, L. 25
Kennedy, M. 255
Keogh, D. 1, 2
Kerr, R. A. 281
Khadka, N. S. 51
Khaleel, Z. 39
Khan, S. 113
Kimmelman, M. 251
King, D. 1, 2, 36
King Hussein Bin Talal Development Area (KHBTDA) 42
Kinoshita, P. J. 184
Kiribati-Australia Nursing Initiative (KANI), Griffith University 42
Kirsch, S. 114, 122n1
Kitara, T. 115
Klein, R. J. T. 267
Klepp, S. 3, 121
Klinenberg, E. 3
Kniveton, D. 36, 71
Knobloch, D. 266
Knudson, K. 37–38
Kochnower, D. 3
Kohler, T. A. 49, 269

Kolbert, E. 62
Koshy, K. 115
Koslov, L. 3, 195, 200–202, 274–277, 283
Kothari, U. 89
Koubi, V. 5
Kraan, C. M. 2, 21
Krehm, E. 53
Kudalkar, S. 174
Kumar, R. 119
Kunseler, E.-M. 280
Kuruppu, N. 279
Kyoto Protocols 264

L
Lahiri-Dutt, K. 70
Lama, P. 5, 6
Lambert, M. R. 5
Lamb, Z. 251
Lampis, A. 127
land loss 78, 84–86; blind spots 82; checkerboard pattern 84–86; as environmental history 81–82; flooded land, benefits of 82; flood risk in North Miami (case study) 82; Hatteras Island, North Carolina (case study) 79–81; and loss of property 78–81; as loss of value 82, 83; matter-in-motion, concept of 85; as physical process 80; as public benefit 83–84; recommendations 86
Landman, K. E. 280
language, for climate relocation 24–29; managed retreat 25–27; relocation 27–28; strategic transition 28–29; vulnerable people 24–25
Laska, S. 277
Lata, S. 119
Latham, M. 119
Lautrédou-Audouy, N. 276, 278, 279
Lawrence, J. 2
Lázár, A. N. 5
Lazrus, H. 40, 114, 196
Lear, J. 284
Leary, N. A. 37
Leatherman, S. P. 168
Leckie, S. 2
Lee, E. S. 35
Lees, D. 169
Lein, H. 172
Lenton, T. M. 49, 269
Leon, J. 2, 3
Lerner, S. 24
Lerski, M. 62–64, 108–109, 209–211, 212–213, 242–244, 289–290

Lester, C. 257
Lester, M. 255
Le Tissier, M. 280
Levine, S. 251
Levy, M. 35
Lewis, J. 58
Lewis, M. 266
Lewis, T. L. 282
lighthouses 256
Lilomaiava-Doktor, S. 116
The Limits to Growth (report) 264
Li Neset, T. S. 35, 121
Linnenluecke, M. K. 3
Linnér, B. O. 35, 121
Linnerooth-Bayer, J. 267
Lins, H. F. 79
Lipuma, S. 10–11, 247–259
Liu, M. 182
Liverman, D. 279
Li, Y.-Y. 280
local leadership 234, 238, 240
Löfgren, H. 156
Logan, J. R. 160
Lonesco, D. 183
Long, K. 35
Longo, N. 10–11, 247–259
Lonsdale, K. G. 280
López, D. 9, 127–139
Lorenzoni, I. 195, 197, 267
Lozano, D. 127
Lujala, P. 172
Lumad people, relocation of 99, 101; challenges faced by displaced Lumad people 101–102; effect of mining operations 101; mining operations in Mindanao 101; poem of Phillipines Lumad 99–101; and recommendations 102
Lustgarten, A. 20
Lutkehaus, N. 3, 116
Lynas, M. 266

M
Mach, K. J. 1, 2, 21
Mafaranga, H. 3
Magnan, A. K. 37, 39
maladaptation: in context of planned relocation 37–39; definition of 35, 37
Malawi, climate-related resettlement in 88; community members voices in process 92–96; flood-prone communities and 90–92; governance framework for disaster management 90; Jombo resettlement site 95–96; Lower Shire Region (study area) 89–90; Nyachikadza residents 92–93; recommendations 96–97; village Mwalija community 94
Maldives, planned relocation in 38–39
Maldonado, J. K. 196, 198, 274, 283
Malig, M. L. 181
managed retreats 251
Mandel, K. 196
Mangubhai, S. 113
Manitoba Provincial Nominee Program (MPNP) 184
Marchman, P. 11, 263–271
Marciano, R. 145
Marino, E. 3, 197, 198, 201, 202, 274, 283
Markowitz, E. 199, 200
Marshall, J. D. 4
Marshall, N. 195, 196, 199, 204
Marter-Kenyon, J. 2
Martin, M. 36, 71
Martin, S. F. 2
Mason, O. K. 255
Matthewman, S. 57
Maxwell, C. 21
Maystadt, J. F. 55
McAdam, J. 2, 115, 117, 156
McBean, G. 5, 6
McCabe, M. V. 39
McCarthy, J. J. 37
McCauley, D. 4
McCord, J. 280
Mcdonald, B. 89
McDonnell, T. 40
McFadden, M. 253
McLeman, R. 2, 36, 113, 252
McMichael, C. 2, 6, 115, 117, 118, 121
McNamara, K. E. 2, 3, 114, 115, 117, 121, 122, 278, 283
McNeill, R. 250
Mechler, R. 267
Mehrotra, S. 20
Melo, L. 25
Mercier, D. 251
Merdjanoff, A. 3
Merone, L. 35
Meusa, S. 171
Mezzadra, S. 71
Midgley, G. 267
migrants and refugees, international agreements on 40–43
migration 34, 267–270; and adaptive social production 187–189; and climate catastrophe 183–184; as climate change

adaptation 36–37; concept of 35; effect of climate change on 35; as failure of climate change adaptation 36; planned 34; push–pull drivers of 35
Milan, A. 160
Milfont, T. 19
Millar, I. 36, 37
Millar, J. 1, 2, 36
Miller, D. S. 5
Miller, F. 2, 89
Milligan, R. 19
Mitchell, T. 183
mitigation 264
Mkwambisi, D. 88
Mohan, G. 89
Mohanty, M.
Moore, R. 201, 253
Morath, D. 255
Morefield, P. 250
Morris, D. 9, 142–150
Morrish, M. 184
Mortreux, C. 37, 118, 156, 157
Moser, S. C. 11, 273–285
Moulton, A. 4
Moulton, R. 281
Moyo, P. 5
Mueller, V. 55, 252
Mukhopadhyay, A. 152, 154
Muller, N. Z. 4
Mullins, K. A. 4
Muñoz, C. E. 21, 255
Musenero, M. 51

N
Nadeau, C. P. 49
Nadin, R. 173, 174
Naess, L. O. 267
Nakache, D. 184
Nakamura, N. 119
Nand, Y. 113
Nansen Initiative on Disaster-Induced Cross-Border Displacement 41
Nansen passports 271
National Adaptation Program of Action (NAPAs) plans 36
National Aeronautics and Space Administration (NASA) 49
National Disaster Resilience Competition (NDRC) 197–198
National Flood Insurance Program (NFIP) 82
National Oceanic and Atmospheric Administration (NOAA) 49

Neef, A. 5, 9, 120, 121, 203
Neilson, B. 71
Nelson, D. 267
Nelson, D. J. 250
Nelson, R. K. 145
Nemani S. 113
Neumann, H. 255
New Zealand Immigration and Protection Tribunal (IPT) 40
Nicholls, R. J. 39, 49
Nichols, A. 117
Nicholson, H. 8, 88–97
Nìmec, D. 35
Niven, R. J. 2
Nixon, R. 218
Nizami, A. 5
Nonko, E. 22, 26
Nordhaus, T. 266
Nordi, N. 218
North Carolina coast, erosion of barrier islands and property rights along 79–81
Norwegian Refugee Council 88
Noy, I. 6
Nunn, P. 114, 119
Nunn, P. D. 2, 115, 122, 278, 283
Nuttall, M. 204

O
Ober, K. 37
O'Brien, K. 156, 195–197, 199, 204
Office of the United Nations High Commissioner for Human Rights (OHCHR) 41
Okada, T. 1, 2
Oliver-Smith, A. 199
Olwig, K. F. 38
Onder, H. 55
O'Neill, S. 121
Oneto, A. 146
Onuki, M. 1
open space 253
Orach, C. G. 51
Ordoñez, M. 127
Ostrander, M. 255
Oswald, K. 183
Oswald Spring, Ú. 35
Overby, H. 256
Owen, G. 278

P
palliative adaptation 37, 39
Pantawid Pamilyang Pilipino Program 180
Paolella, D. A. 4

Parke, A. 122n2
Parks, B. C. 4
Parris, B. 71
participation, in resettlement process 88–89; *see also* Malawi, climate-related resettlement in
Participation: The New Tyranny? (Cooke and Kothari) 89
Pasquini, L. 138
Paterson, S. K. 280
Patsch, K. 251, 257
Pauli, N. 5, 9
Paul, R. 52
Payne, A. 20
peak oil theory 264–265
Pearson, J. 2
Peiser, R. 2
Pelling, M. 138, 169–171
Penning-Rowsell, E. 283
Perera, S. 71
Petersen, A. C. 280
Peters, K. 51
Peters, Laura E. R. 7–8, 48–58
Peterson, K. 196
Philippine Mining Act of 1995 101
Phiri, K. 5
Pidgeon, N. 199, 200
Pielke Jr., R. 266
Piggott-McKellar, A. E. 2, 122
Piggott-McKellar, P. J. 278, 283
Pilkey-Jarvis, L. 248, 256
Pilkey, K. 248, 256
Pilkey, O. H. 10–11, 79, 247–259
Pinter, N. 2
Pitingolo, R. 146
Pittock, J. 283
planned relocation 156; as form of migration 35; international guidance/governance on, lack of 35; as last resort 41–43; maladaptation in context of 37–39; in Maldives 38–39; policy and supportive governance structures for 35; reducing maladaptation in 43; of refugee camps 52–54
planned resettlement, challenges associated with acceptance/resistance to *see* Malawi, climate-related resettlement in
planned retreat 1; feminist decolonial approach in 6; gender and 5–6; importance of 2–3; and justice challenge 4–5; loss and gain from 3
Plastrik, P. 277, 280

Platform on Disaster Displacement (PDD) 41
Polanco, G. 188
Polasky, S. 4
Potts, M. 266
Povinelli, E. A. 220
Powell, J. A. 21, 23
Powell, N. 250
Powell, T. 6
Prange, J. 277, 282
Preston, B. 267
Preto, B. C. 24
preventive relocation 229–230
Pritchard, P. C. H. 163, 164, 168
Pritzer, R. 266
procedural justice 4; planned climate relocation and 54–56
Procházka, D. 35, 71
protracted refugee scenario (PRS) 49
Public Trust Doctrine 254
Pulhin, J. M. 35
Pulido, L. 4
Pulwarty, R. S. 169
Puntub, W. 1, 2
Purdy, B. 281, 283

Q
Qiu, J. 250

R
racial discrimination 146
Ragobeer, V. 170
Ramirez, F. 88, 89, 127, 128, 134
Ramos, E. P. 10, 217–230, 276, 283
Ramsay, R. 171
Ransan-Cooper, H. 115
Ravuvu, A. 115
Rawls, J. 4, 54
recognition justice 4
Reddy, S. M. W. W. 3
Rees, J. C. 2
Reeve, K. 251
refugee camps 49, 58; building of 58; effects of climate change on 49–51; justice-based approach to relocation for 55; planned climate relocation of 52–54; protracted refugee scenario (PRS) 49; purpose of 51–52; Rohingya refugees in 52; warehousing, practice of 49
1951 Refugee Convention 35, 40
refugee crisis 39
refugee movement, restrictions on 51
refugees 48–49; in camps managed by

UNHCR 49; international agreements on 40–43; number of 48; participation in relocation planning 56; relocation guided by social justice 53–57; vulnerable to climate change 51; warehoused 49, 51
Refugees International 40
refugee status, identification of 40
Reisinger, A. 2
relocation 195, 268–270
repetitive loss properties (RLPs) 82–83
resettlement 29, 88, 127; as climate change adaptation 88; guidelines on 88, 89; importance of participation in 88–89
resilience 265
Resilient Edgemere 146
restorative justice 4; planned climate relocation and 54–57
Resurreccion, B. P. 5
retreat trauma 277
Reyes, C. C. M. 180
Reyes, C. M. 180
Rey-Valette, H. 276, 278, 279
Richter, I. 19
right to self-determination 204
RLPs *see* repetitive loss properties (RLPs)
Roberts, J. T. 4
Robinson, K. 271
Robinson, L. B. 280
The Rockefeller Foundation's 100 Resilient Cities initiative 265
Rød, J. K. 172
Roelen, K. 182
Rogers, S. 89
Rohingya refugees, in Bangladesh 52, 53
rolling easements 254, 259
Rothstein, R. 143
Roy, J. 266
Ruano, S. 160
Rubiano, D. 127, 128, 134
Rubjerg Knude Lighthouse 256
Ruckert, K. L. 250
Ruggieri, Beatrice 8–9, 113–123
Rulleau, e.B. 279
Rutherford, J. 265

S
Sabates-Wheeler, R. 183, 189
sacrifice zones 24
Sagoff, M. 266
Said, M. 49
Sakakibara, C. 283

Sakdapolrak, P. 3
Samanta, G. 70
Sampson, C. C. 250
Sanahuja, H. 88, 89
Sanghi, A. 55
Santos, R. 146
Sarma, J. N. 68
Schade, J. 115
Schafer, H. 39
Schaffer, L. 5
Scheffer, M. 49, 269
Schell, C. J. 5
Schinko, T. 267
Schipper, E. L. F. 37, 39
Schlosberg, D. 4
Schmidhuber, J. 49
Schwartz, J. 254
Schwartz, M. 168
Scott, A. 204
Scott, J. C. 266
Scudder's resettlement model 39
Scudder, T. 39
sea level rise 168, 194, 251
Seebauer, S. 3
See, J. 2
Seneviratne, S. I. 20
Seng, D. C. 113
Sengupta, S. 51
Shahan, A. 53
Shah, S. 268
Shareef, A. 39
Sharpe, B. 280
Shaw, A. 249
Shaw, M. R. 267
Shearer, C. 196
Sheena, S. 152
Shein, P. P. 280
Shellenberger, M. 266
Sherlock, R. 51
Shi, L. 251
Shreve, C. M. 23
Shryock, H. S. 35
Shyamakrishnan, K. 255
Sibley, C. 19
Siddiqui, T. 71
Siders, A. R. 2, 3, 21, 265
Siegel, J. S. 35
Simpson, M. C. 169
Sing, P. 116
Sipe, N. 2
Skleparis, D. 180
Smit, B. 2, 36, 49
Smith, A. M. 250

Smith, B. 113
Smith, G. 180
Smith, S. J. 218
Smyth, I. 5
social justice 4
social protection model 182–183
soft stabilization *see* beach nourishment
Solomon, A. 166
Solomon, D. 21
Somé, B. 36
Song, L. 249
The Sovereign Military Order of Malta 271
Spilker, G. 5
Spitzer, R. 165
Stal, M. 88
Stamson, N. 255
Stanford, J. A. 247
Stapleton, S. O. 173, 174
Staten Island, privilege and influence in 200–201
Stathakis, A. 3
Steckley, J. 6
Steckley, M. 6
Stefancu, O. 9, 152–161
Steinbeck, J. 270
Steinberg, T. 85
Stein, I. 8, 78–86
Stern, N. 182
Stojanov, R. 35, 36, 71
Stone, R. 266
Strong, A. L. 250
Suckall, N. 88
Sulakshana, E. 3
Sullivan, M. 2, 26, 199
Sultana, F. 5
Surminski, S. 267
Svendsen, E. S. 108
Svenning, J. C. 49, 269
Sward, J. 36
Sweetman, C. 5

T
Tabari, H. 51
Tait, P. 35
Takahashi, A. 113
Tamir, O. 37
Tan, N. 1
Tanner, T. 183
Tate, E. 255
Tavris, C. 23
Taylor, A. 51
Taylor, M. 20

Teague, P. 266
Tebboth, M. G. 49
Teicher, H. 251
TEK *see* traditional ecological knowledge (TEK)
Temporary Foreign Worker Program (TFWP), Canada 184, 188
Tenzing, J. D. 182
Tessum, C. W. 4
Tevis, R. E. 280
Thakrar, S. K. 4
Thaler, T. 2, 3
Tharoor, I. 53
Thibaut, J. 54
Thielking, K. 280
Thomas, A. 20
Thomas, A. S. 113
Thomas, D. 35, 160
Thomas, D. S. G. 183
Thomas, F. R. 115
Thomas, M. 1, 2, 36
Thornton, F. 2, 115
Thornton, T. F. 196, 204
Ting, L. 117
Tippett, K. 283
Titus, J. G. 254
Tockner, K. 247
Tong, A. 42
Tooler, N. 2, 3
Topsail Island beach replenishment 249–250
Torres, A. 127
traditional ecological knowledge (TEK) 197, 204
Transition Movement 266
trapped population 154, 157, 160
Tronquet, C. 118, 123n3
Trouillet, R. 276, 278, 279
Tsang, M. 8, 78–86
Tuan, Y. F. 48, 57
Tubiello, F. N. 49
Tuinstra, W. 280
Turner, S. 253
Tyner, J. A. 181
Typhoon Haiyan in Philippines, mobile livelihoods and 180; labor export, strategy of 181; migrant worker responses to Typhoon Haiyan 184; migration and adaptive social production 187–189; migration between Canada and Philippines, case studies on 184–187; Pantawid Pamilyang Pilipino Program 180; social protection in

disaster efforts 182–183; state-driven migration-as-development paradigm 181

U
Uekusa, S. 57
Ueland, J. 255
Ullah, A. 71
Ulrichs, M. 182
UNHCR *see* United Nations High Commissioner for Refugees (UNHCR)
Uniform Relocation Act of 1970 (URA) 143, 147, 149, 150
United Nations Framework Convention on Climate Change (UNFCCC) 41, 43, 264, 267
United Nations High Commissioner for Refugees (UNHCR) 48, 49, 88
United States, climate-induced relocations in: Black, Indigenous, and people of color (BIPOC) and 20–26, 29; culturally targeted universalism approach for 23–24; federal aid for second homeowners 22; loss of control in 23; modified and co-developed language for 24–29; policies lacking cultural sensitivity 28; process change in, need of 20–24; recommendations for 30; white supremacy in 19–20, 25, 26, 28
United States, flood experiences in 104–107, 250
unmanaged retreat 268
Urban, M. C. 49
US Army Corps of Engineers (USACE) 247, 248, 255

V
Valduga, I. B. 40
Valenzuela, V. P. 1
values 195
Van Den Hoek, J. 7–8, 48–58
Van der Geest, K. 113
Vandervord, C. 113
Van Hear, N. 35
Van Sant, L. 4
van Weezel, S. 55
Varea, R. 5, 9
Vargas, E. 26
Vasileiadou, E. 280
Vaz-Jones, L. 5, 6
Veerkamp, A. 194
Velasco, M. A. 171
Vella, K. 2
Verchick, R. R. M. 199

Vermandé, Clairé-Louise 10, 232–241
Vikhrov, D. 36
Vincent, K. 5
vision for community 234, 236, 238, 240
Vitullo-Martin, J. 147
Vogel, C. H. 35
voluntary relocation 27, 37
Vries, D. D. 21

W
Wadey, M. P. 39
Waite, M. 189
Walkerden, G. 57
Walker, L. 54
Walker, S. 55
Walshe, R. A. 113
Waraich, O. 51
warehousing, practice of 49
Warf, B. 255
Warner, K. 2, 3, 36, 113
Warren, R. 20
Watson, C. 173, 174
Watson, J. E. M. 122
Watson, S. 281
Webber, M. 89, 113
Weber, E. 5, 9, 113
Weerasinghe, S. 2, 3
Weir, T. 283
Wernerus, F. M. 169
Wester, M. 5, 6
White, I. 276, 278
White, K. S. 37
White, L. 27
Whyte, H. 280
Wiek, A. 280
Wijsman, K. 6
Williams, B. 4, 195
Williamson Ian, P. 117
Williams, P. E. 169, 170, 172
Williams-Rajee, D. 19, 21
Wilmsen, B. 2, 89
Wilson, D. 250
Wilson, M. 19
Wing, O. E. J. 250
Winkel, J. 49
Winkler, C. 3
Winling, L. 145
Wisner, B. 58
Wissoker, D. 146
Woelfle-Erskine, C. A. 5
Wolf, J. 156, 267
women, impacts of planned retreat on 5–6
Wong, T. E. 250

Woodruff, S. 250
workshops 237
World Bank 88
World Bank Operational Directive on Involuntary Resettlement 37–39
Wrathall, D. 2, 252
Wreford, A. 267
Wright, B. 4
Wu, H. 10, 180–190

X
Xu, C. 49, 269
Xu, Z. 160

Y
Yacouba, H. 36
Yamaoka, J. G. 221, 222, 225
Yee, D. 2, 3
Yee, D. K. P. 180
Yeoh, B. S. 181
Yila, J. O. 5
Yuefang, D. 89

Z
Zavar, E. 3, 22
Zehr, H. 55
Zeiderman, A. 127, 128
zero tolerance areas 170
Zhang, K. 168
Zhao, T. 51
Zhou, G. 20
Ziervogel, G. 37, 39, 138
Zoomers, A. 36
Zoromé, M. 36
Zulfiqar, M. 5

Printed in the United States
by Baker & Taylor Publisher Services